W9-BFQ-443

NEITHER ATHENS NOR SPARTA?

NEITHER ATHENS NOR SPARTA?

The American Service Academies in Transition

JOHN P. LOVELL

Indiana University Press Bloomington and London

Manufactured in the United States of America

Library of Congress Cataloging in Publication Data

Lovell, John P 1932-
Neither Athens nor Sparta?

Includes bibliographical references and index.
1. Military education—United States—History.
2. United States. Military Academy, West Point—History. 3. United States. Naval Academy, Annapolis—History. 4. United States. Air Force Academy—History. 5. United States. Coast Guard Academy, New London, Conn.—History.
I. Title.
U408.L67 355'.007'1173 78-9509
ISBN 0-253-12955-9 1 2 3 4 5 83 82 81 80 79

To Joanne, Sara, and David

Contents

PART II. THE CHANGE PROCESS: CASE STUDIES IN THE EVOLUTION OF THE SERVICE ACADEMIES

PART III. ANALYSIS: THE DETERMINANTS AND CONSEQUENCES OF ORGANIZATIONAL CHANGE

Tables

Figures

PREFACE

Personal experience serves the social scientist as a two-edged sword. On the one hand, often it is the source of inspiration for his research which sustains his interest. On the other hand, it can be the source of bias and distortion unless he is careful to treat his personal reactions and impressions as mere hypotheses to be tested against systematically gathered evidence. The notes reveal the care that I have tried to exercise to ensure that this study does not rest unduly on participant observation. On the other hand, there is no subject about which I might have written that has been inspired so directly by involvement over a period of many years.

The service academies were more image than reality for me when I applied for admission to the oldest of them nearly thirty years ago. The Air Force Academy was still no more than a glint in the eye of the nation's aviators. The Coast Guard Academy was unknown to me. I knew West Point and Annapolis from newsreels, but knew no one who had graduated from either school. I now realize that even after four years as a cadet I knew less about West Point than I thought I did. But from the first long walk up the hill from the railroad station on the banks of the Hudson River, I was captivated by this world of contradictions. West Point was romantic, absurd, challenging, tedious, purposeful, superficial, dynamic, mired in the past, grimly serious, hysterically funny. I assumed that the other academies shared these characteristics.

Six years after graduation, having resigned my commission as an Army officer in order to pursue graduate study in political science at the University of Wisconsin, I found myself drawn back to my alma mater for dissertation research. The focus of my Ph.D. study was attitude change among cadets during the four-year socialization process. (Much less change occurs, I found, than either the most severe critics or the most zealous supporters of West Point education are prone to suggest.) Interviews, conversations, and documents which I reviewed provided me with new insight also into the process by which policy changes were being implemented or thwarted.

In the spring of 1963, I made a ten-day visit to the Air Force Academy, having decided to follow up my dissertation research with a brief comparison of the organizational-change processes at the newest and the oldest of the academies. The resultant brief draft manuscript provided an analysis that satisfied neither me nor colleagues who critiqued it for me. So I allowed the manuscript to gather dust for several years while directing my research and teaching energies elsewhere.

Visits to West Point during 1968-69, when I was teaching at Vassar College, revived my interest in organizational changes that were being discussed, implemented, or—with the war in Vietnam escalating—set aside. Upon completion of a foreign policy textbook, I began exploring the feasibility of renewing my academy research on a broader comparative basis. I was determined to include Annapolis in the research and became convinced that the Coast Guard Academy, although smaller and different from the Defense Department academies in other ways, ought to be included to provide a frame of reference from which to identify similarities and differences among the academies.

Fortuitously, early conversations at the Naval Academy to explore interviewing possibilities there led to an invitation to join their faculty as a visiting professor, which I did during the 1971-72 academic year. During the year I also was able to make visits to West Point and New London; and in the following years, I made return visits to each of the four academies. The more extensive my involvement with the service academies, the more fascinated I became, not only with the subject itself, but with its seemingly inexhaustible nuances. Nearly every interview or piece of correspondence led me to new insights, and in some cases to considerable alteration of my views. My appreciation of the intricacies of this learning process makes me aware that even some of those observations which I now have put in print must be regarded as tentative. Yet I also have recognized my responsibility not to shy away from interpretations and conclusions, however unorthodox or controversial, simply because evidence not yet unearthed might lead to further refinements in my thinking.

If this study has value, it reflects the generous investment of time, thought, and energy of the dozens of people who have helped and encouraged my research over the years. Especially vital to the success of the research were the interviewees, who are identified in Appendix 1, and persons with whom I corresponded. The excerpts that are included in the body of the book only reflect a small portion of the knowledge and insight which I derived from the interviews and letters.

Many of my correspondents were deeply involved in the events described in the book; references to their correspondence are provided in the notes. In addition, I benefitted from information from or an exchange of ideas with the following individuals: Ward T. Abbott, Harold R. Aaron, William C. Carr, Stephen Clawson, James C. Clow, Douglas Conte, Arthur B. Engel, Joseph H. Hughes, Jr., Joseph J. Kruzel, Joseph T. Maras, Douglas Marshall, John S. Pustay, Michael T. Rose, Forrest V. Schwengels, Edward Sherman, Michael J. Smith, Judith Stiehm, Rick Strong, Lucian K. Truscott IV.

Comments by Maurice Garnier, Richard F. Rosser, and William R. Stroud were especially helpful to me in the design stages of my research. I am grateful to the number of people who read and offered comments on one or more draft chapters. These include: G. Pope Atkins, Edward E. Bozik, Horace M. Brown, Jr., John Sloan Brown, John S. B. Dick, James R. Golden, Archie Higdon, Thomas E. Kelly III, Philip S. Kronenberg, William M. Lovell, Thomas E. C. Margrave, Allan Millett, Peter R. Moody, Roger H. Nye, Lee Donne Olvey, Rocco Paone, Richard Parkhurst, Freeman Pollard, Wesley W. Posvar, John R. Probert, Paul R. Schratz, John W. Shy, Lewis S. Sorley III, Frederick C. Thayer, Malham M. Wakin, Ronald A. Wells, Anthony L. Wermuth, Roderick M. White, John D. Yarbro. The manuscript in entire draft form was critiqued by John Gallman, Morris Janowitz, David W. Moore, Robert H. Moore, and William P. Snyder. Their comments have been enormously helpful.

The number of librarians, archivists, institutional research directors, and information officers who assisted me over the years is beyond a full recounting. A partial list includes Marie T. Capps, Thomas P. Garigan, Gerald W. Medsger, Joseph E. Marron, and Stanley P. Tozeski at West Point; William W. Jeffries and the library staff at Annapolis, and W. S. Busik and his staff at the Naval Academy Alumni Association; Barbara A. Gilmore at the Naval Historical Center; Elizabeth A. Segedi at Coast Guard Headquarters; Paul H. Johnson and Malcolm J. Williams at the Coast Guard Academy and W. K. Earle of the Coast Guard Academy Alumni Association; George V. Fagan, Claude J. Johns, Alta A. Thompson, and Risdon J. Westen at the Air Force Academy; Robert Goehlert, Thomas Michalak, and Michael Parrish at Indiana University.

The year which I spent at Annapolis was invaluable. My departmental colleagues during the year helped to make it not only productive but thoroughly enjoyable: G. Pope Atkins, Robert A. Bender, Thomas Boyajy, Charles L. Cochran, Ray E. Davis, John A. Fitzgerald, Jr., Joseph A. Jockel, Elmer J. Mahoney, Philip A. Mangano, Rocco M. Paone, John R. Probert, Howard F. Randall, Jr., Robert

L. Rau, William R. Westlake. Elmer E. Inman paved the way for my first visit and offered wise counsel and friendship during the year and subsequently. I am grateful to the Naval Academy for hiring me to teach during the first semester and for permitting me to remain during the second semester to do my research. My sabbatical leave salary during the second semester came from Indiana University, which also funded some of the travel and the typing for the book. There have been no other sources of funding.

Michael Hayes, Steve Majeski, Dorothea Schoenfeldt Miller, and Wayne Stevens provided research assistance as graduate students at Indiana University. I was assisted in the typing chores by Wilma Chambers, Dee Cowden, Gretchen Deal, Juedi Kleindienst, Fran Nasso, Leslie Potter, Sandy Roberts, and Patricia Shirley. The typing proved particularly onerous, because it involved interview transcripts and extensive correspondence in addition to several drafts of the manuscript. Georganna Priest typed the final manuscript. Naturally, none of the dozens of persons who assisted me must be held accountable for any errors or flaws of interpretation that remain; I accept such responsibility fully.

In portions of chapter eight I have drawn freely from my article "Modernization and Growth at the Service Academies: Some Organizational Consequences," in *The Changing American Military Profession,* edited by Franklin D. Margiotta (Boulder: Westview, 1978). Discussion in chapter ten draws upon my essay "The Service Academies in Transition: Continuity and Change," in *The System for Educating Military Officers in the U.S.*, edited by Lawrence J. Korb (Pittsburgh: International Studies Association, 1976). These works are utilized with permission of the original publishers.

Finally I want to acknowledge the contribution of my wife, Joanne Granger Lovell, whose patience rarely faltered, and of my children, Sara and David, who were rarely patient. How could they be? Neither was born at the time of my first trip to the Air Force Academy in 1963, and now both are teenagers. To this treasured trio the book is dedicated.

NEITHER ATHENS NOR SPARTA?

PART I

Introduction

1

The Paradox of Change

The epic of the American service academies is carved into the landscape of American history with the boldness and durability of the sculptures on Mount Rushmore. The academy saga and the American saga are inseparable, linked at critical junctures to the same myths and the same realities. Carefully forged, the linkages to the past are maintained tenaciously at each academy—the more so in this modern era of "future shock," with its societal yearning for roots. Even at the Air Force Academy, a post–Korean War institution with a futuristic image, a sense of continuity with the past is cultivated, although traditions have had to be invented or borrowed. West Point customs and rituals have been adapted to the Colorado setting, and pioneers and early champions of aviation—the Wright brothers, Charles Lindbergh, Billy Mitchell, and others—have become part of the organization genealogy with which cadets are urged to identify.

The older academies come by tradition more naturally, but they are no less ardent about its preservation and embellishment. At Annapolis, for instance, a 96-foot-diameter crypt, prominently displayed in the Academy chapel and containing the remains of John Paul Jones, makes it difficult to forget the father of the American Navy. The demand that each successive generation of Naval Academy freshmen ("plebes") memorize slogans that Jones and other early naval heroes made famous also fosters a pervasive sense of continuity with the past. American nautical lore is prevalent also at New London, where Coast Guard Academy cadets are required to observe a number of rituals and customs that were practiced by their nineteenth-century forebears. As the oldest of the four academies,[1] West Point provides the most conspicuous array of reminders of the role which the Military Academy and its graduates have played in our national heritage—Revolutionary War fortifications, statues of Patton and MacArthur, monuments, paintings of Ulysses S. Grant and Robert E. Lee, battle flags, and other memorabilia.

Tourists are charmed by the aura of tradition that is so ubiquitous at the academies. More critical observers, concerned about the present-day performance of the academies in producing effective officers for the nation's armed forces, may wonder if the academies are *too* firmly rooted in the customs and practices of the past. Students at the academies sometimes wonder themselves. "We have maintained 176 years of tradition unmarred by progress," cadets at West Point tell one another, in terms that seem to confirm the worst fears of the American taxpayer. The idiom varies from one service academy to another only in historical specifics, not in tone.

In recent decades, however, the academies have made a number of important departures from traditional customs, and have abandoned some practices that seemed immutable as recently as the 1950s. Pressure for change and resistance to change have provided the dialectic in the evolution of the American service academies, and it is this dialectic which we shall examine in detail in our study of the dynamics—and the agonies—of organizational change.

The striking contrast between the academies of today and those of an earlier era, and the continuity between present and past as well, are evident in an event that occurred at the Naval Academy in 1972. Officials at the academies in the early 1970s remained resistant to the tide of congressional sentiment toward the admission of women to the academies. Nevertheless, the Naval Academy extended an invitation to Gloria Steinem, editor of *Ms.*, and to Dorothy Pittman, a black activist involved in the organization of urban communities, to address the brigade of midshipmen. Each of the service academies had had supplemental programs and guest lectures by prominent civilians as well as by distinguished military men for many years. However, even a few years earlier, a guest appearance by a noted feminist or by a black activist would have been unimaginable.

It was a warm spring evening when the two women spoke in the Naval Academy field house. The crowd of four thousand midshipmen was augmented by academy staff and faculty members and their families, by townspeople from Annapolis, and by a scattering of visitors from Washington. The two women, each in a pullover sweater and slacks, the very picture of mod casualness, appeared on stage with Academy Superintendent Vice Admiral James Calvert, tall, trim, and dashingly military in his dress-white summer uniform with the gold braid of flag rank.

That the two guest speakers should denounce the military for perpetuating male chauvinism and racism only added to the sense that the door was being slammed, not merely eased shut, on the institutional morés of the past. A faint but ironic echo of tradition was provided earlier in the week by the announcement that attendance at the Steinem-Pittman lecture was compulsory for all midshipmen.[2]

Equally symbolic of change was the appearance at West Point during the same month of Congressman Ronald Dellums, black representative from California whose sharp criticisms and probing investigations often have been a thorn in the side of the American military establishment. Attendance at the Dellums lecture was voluntary, and since other diversions were available on the balmy spring night, less than a hundred cadets, officers, and guests turned out for the speech.[3] However, the facts that Representative Dellums was there at all and that there were black cadets at West Point in numbers large enough to provide the stimulus for the invitation were evidence of significant change.

Both the Steinem-Pittman and the Dellums lectures came through student initiative. Similar initiatives had been stifled at the four academies in recent years, however, and success in these two instances was an indication not only of the growing willingness of cadets and midshipmen to make their views known forcefully, but also of the growing recognition by academy officials that to consistently suppress or ignore student views was detrimental to academy purposes and well-being.

Academy officials in the 1970s have become increasingly attentive to the attitudes and beliefs of cadets and midshipmen as well as to the values and aspirations of the young men and young women whom they seek to enroll. As Senator J. William Fulbright and others in Congress have noted and as the media illustrated in documentaries such as the CBS special "The Selling of the Pentagon," the American military establishment had become "p.r." conscious in a big way.[4]

At the three larger academies, recruitment has changed from an off-hand service function, making information available on request, to a full-scale operation that is awesome in scope and variety and sometimes dazzling in its Madison Avenue sophistication. The array of recruiting techniques includes commercially produced films, TV commercials, press releases, red-carpet academy tours for high school counsellors, speaking engagements in local high schools by cadets and midshipmen, and vast networks of alumni and reserve officers to identify and encourage talented prospective cadets and midshipmen.

The themes that the academies are utilizing in recruitment brochures, films, and speeches reveal the shifts that have been occurring in the images that academy officials themselves maintain of their respective institutions, as well as their changing perceptions of the appeal which various facets of the academy experience or character will have to potential recruits. The focus on patriotism, the heroic exploits of famous academy graduates, and the customs and traditions of the institution still appears. But increasing emphasis is placed on other themes: the advantages of a four-year, tuition-free education with pay; the excellence of the academic facilities and the high quality of the faculty and program; the diversity of options available for intellectual development; the variety not only of athletic but of other extracurricular activities available; the tangible benefits associated with a military career—and in the case of the Coast Guard, the humanitarian as opposed to combat-related characteristics of such a career.

More importantly, the shifting recruitment theme corresponds to substantive reforms that have been effected at each academy. The new Air Force Academy, for example, became both an architectural testimony to modernity and a curricular one as well. Spurred and guided during its first decade by an energetic Dean, Brigadier General Robert McDermott, the Academy introduced an enrichment program that supplemented a core curriculum with a large variety of electives. Moreover, cadets were permitted to select from among twenty-seven academic majors. The Air Force Academy pioneered among educational institutions in 1957 in the establishment of a Department of Astronautics, and became unique among the service academies by offering courses in fine arts and music appreciation. Its program in engineering science was sufficiently solid to win accreditation in 1962 from the Engineers Council for Professional Development. Its overall program was so attractive that only eleven years after its founding the Air Force Academy ranked sixth among 1,629 institutions of higher education in preferences expressed by 140,000 "high-ability males" who had taken the National Merit Scholarship examination.[5]

The Coast Guard Academy had languished near the point of total collapse in the early 1950s, but has experienced a dramatic renaissance since that time. A succession of forward-looking superintendents in the late 1950s and early 1960s—Rear Admirals Frank Leamy, S. Hadley Evans, and Willard Smith — injected a spirit of ambitious reform. The introduction of electives provided the basis for academic "concentration groups," which in turn paved the way for academic majors. Thus, by the 1970s, cadets could choose from among majors that included ocean science, economics and management, history and government, chemistry, physics, mathematics, computer science, ocean engineering, electrical engineering, nuclear engineering, marine engineering, and civil engineering.

Contrary to the general trend, the Coast Guard maintains its unofficial bureaucratic credo, "in obscurity lies security." The Coast Guard Academy operates its recruiting effort on a shoestring budget, although even within these limits increasing sensitivity is being shown in seeking out talented recruits.

Annapolis also has had its renaissance. The academic reform movement began in the late 1950s when Rear Admiral Charles Melson was Superintendent. A full decade of slow, halting evolution followed before major changes were made. A substantial reorganization not only of the curriculum but of the academic departmental structure occurred during the superintendency (1968-1972) of Vice Admiral James Calvert. The extent to which the Academy departed from its long-standing rationale for a totally prescribed curriculum was indicated in a discussion in the Academy catalog of the twenty-six majors programs that were available by the early 1970s.

> The day is long past when every line officer could be expected to embody all the qualifications and specialties desired or needed in a naval career. Today's Naval Academy, therefore, does not seek to give the same all-inclusive educational package to every graduate. Rather, it undertakes to produce in every graduating class a group of individual line officers—all well trained in basic professional subjects—who collectively possess the wide range of knowledge and capabilities demanded of the officers of our modern Navy.
>
> To attain this breadth in his education, each midshipman must satisfy certain minimum course requirements in social sciences and humanities, mathematics and science. To ensure depth, he completes a major sequence from a variety of fields designed to provide him with the academic background necessary for effective leadership in today's Navy.[6]

Even West Point, which in many respects remains the most tradition-bound of the service academies, has made impressive improvements in the quality of its academic program. Indeed, West Point led the way among the academies in shifting the emphasis away from exclusive concern with engineering subjects to a broader general education. Under the headship first of Colonel Herman Beukema and then Colonel George Lincoln, by the 1950s the Department of Social Sciences had developed an enviable national reputation, pioneering in the exploration of such topics as comparative government and the economics of national security. Beginning with a period of creative ferment in the late 1950s under the superintendency of Lieutenant General Garrison Davidson, the Military Academy introduced electives into its program and has expanded the number gradually in the years since then. However, the Military Academy has moved somewhat more cautiously than have the other academies. For instance, the Academy periodically has considered the introduction of academic majors, but it has consistently rejected the idea, permitting only interdisciplinary "concentrations" which avoid the aura of undue specialization.

By the 1970s, West Point as well as Annapolis and New London had completed major expansion programs that gave them physical facilities which rivalled in quality the modernistic ones at the Air Force Academy. Computer centers, closed-circuit television systems, well-equipped laboratories in the natural sciences and in foreign language departments, and spacious new libraries are now among the educational components found at each of the service academies. For example, the Air Force Academy has a planetarium and an observatory that are used in instruction in astronomy. The Coast Guard Academy has a new (1973) ocean science laboratory. A science research laboratory at the Military Academy supports original research on such topics as infrared spectroscopy, planetary physics, and low temperature and high pressure physics. The Naval Academy has a laboratory, used in the study of ocean engineering, that includes an 85-foot towing tank, equipped with a pneumatic wave-maker and sophisticated sensors to facilitate analysis.

Better credentials were demanded of faculty members at each academy concurrently with the move to electives (and thus more highly specialized coursework) and the modernization of academic facilities. A Ph.D. had been a rarity among faculty members at the service academies as recently as the 1950s. By the 1970s the Ph.D. was the rule rather than the exception among the core of permanent members of the faculty at each academy, and virtually all temporary faculty members (those assigned for a short tour of duty) came to their assignments after civilian graduate schooling.

Nor have the service academies neglected the modernization of the military-professional training that makes them distinctive. On the contrary, over the years the academies have kept closely abreast of changes in military technology and weapons systems. Cadets and midshipmen have been afforded intensive exposure, especially during the summer months, to the latest in military techniques, weapons, and equipment. Coast Guard Academy cadets, for example, receive training in helicopter rescue operations, electronic long-range aids to navigation (LORAN), ice patrolling, and the latest antisubmarine warfare devices, in addition to the standard training in seamanship, navigation, and marine safety. West Point cadets receive "hands-on" training utilizing the latest small arms and artillery equipment. Training for Air Force Academy cadets includes an orientation in piloting of jet aircraft, instruction in the latest aerospace weapons systems, and options that range from parachuting to undersea diving. Midshipmen at the Naval Academy not only receive extensive instruction at sea on board modern naval combat vessels, they also receive orientation instruction in nuclear submarines, jet aircraft, and the latest Marine Corps equipment and weaponry.

Finally, since 1976 the service academies have been admitting women as well as men. Despite some initial grumbling, especially on the part of male cadets and midshipmen, the permanent cadre at each academy has demonstrated a determination to make coeducation at its institution succeed.

In short, the changes that have been effected at the service academies since World War II constitute a dramatic success story in several respects. It is by no means a story of unmitigated triumph, however.

Academy officials are quick to insist that neither the modern emphasis on academics nor the expanded freedoms and extracurricular opportunities for students represents a neglect of the traditional Spartan concern for cultivating "the martial virtues." "Duty, honor, [and] service to the country and its people," West Point Superintendent Sidney Berry wrote in mid-1977, have been the core values of the Military Academy since its founding. And yet, he observed, "there has been systematic evolutionary change in response to the needs of the Nation and the imperatives of the times—responsible change within a framework of constancy and tradition."[7]

Implicit in these and similar remarks from officials at the other service academies is an image of the modern American service academy as a creative synthesis of Sparta and Athens—tough, traditional, and yet enlightened and adaptive. But it is not an image that withstands careful scrutiny. Superintendent Berry's statement was made in the wake of the most massive cheating scandal in academy history. The "martial virtues" have been under severe strain not only at West Point but in varying degrees at all the academies. The academies have been plagued also by other problems, such as a student attrition rate of between 40 and 50 percent of entering classes, rising operational costs, and widespread student malaise. The detailed report of the Borman Commission (which was created by the Secretary of the Army to investigate the 1976 West Point honor scandal) and the voluminous follow-up study in 1977 by the West Point study committees, as well as the several volume analysis of the academies by the General Accounting Office, are among the recent high-level reviews that have centered attention on such problems. As the reports make clear, problems such as widespread cheating and high rates of attrition are attributable neither to contamination of the student bodies by "a few bad apples in the barrel" (as some academy officials have claimed) nor to temporary—and therefore easily remedied—waves of student discontent. Rather, it has become evident that to a significant degree the problems are deeply rooted in institutional structures and practices.

The service academies have become modernized, but they have not fully reconciled their traditional and modern components. They have moved well beyond their old "trade school" format in seeking to provide a solid academic education as well as military discipline and training. But in doing so, the contradictions between Athenian and Spartan goals have become magnified. Indeed, the paradox generated the transition which the academies have undergone since World War II is that modernization and reform have served, in part, to exacerbate internal tensions and to stimulate new problems. This study is primarily concerned with the unravelling of that paradox.

The starting point for this analysis is an examination (in chapter two) of the evolution of the service academies in the nineteenth century, when they were essentially military seminaries. These were the formative years of the academies, when traditions were established that have had lingering effects upon contemporary attitudes and behavior. Even at the eve of World War II, the organizational structures and practices that had been established by the turn of the century remained fundamentally intact. However, the currents of change were set in motion by World War II and by the exigencies of the transition to the postwar environment, as described in chapter three.

The changes in the early postwar years were incremental rather than comprehensive. Each academy retained a curriculum that was totally prescribed except for a choice among foreign languages. The approach to learning placed a heavy emphasis on rote, and especially at Annapolis and New London the curriculum had a trade school orientation. Life was severe and regimented, with the intense indoctrination of "plebe year" or "swab year" virtually unchanged from the prewar pattern. Faculty members often prided themselves on being disciplinarians, and few of them had academic qualifications beyond the baccalaureate. Sunday chapel attendance was compulsory. So also, practically speaking, was attendance at football games, where organizational loyalty in its most vocally intense form was displayed.

Yet discussion of some fundamental departures from the seminary-academy tradition had begun. Such departures would occur over a period of approximately a decade, beginning in the late 1950s in a new academy that had been opened for the Air Force. The case studies of Part 2 are designed to provide insight into the process by which major changes were effected at each of the four academies: at the Air Force Academy during its first decade (chapter four); at West Point during the superintendency of Garrison Davidson (chapter five); at New London under a succession of innovative superintendents beginning with Frank Leamy (chapter six); and at Annapolis, from the early reforms of Charles Melson to the climax of reform under James Calvert (chapter seven).

Changes that are examined in the case studies not only include curricular reforms but run virtually the whole gamut of organizational change: modification of formal goals; in the recruitment of students and in the recruitment of staff and faculty; in public relations programs; in systems of discipline and socialization, such as the plebe systems; in the honor systems; in athletic programs; in structures of governance. Each case study reveals change as a process that has elements of drama and struggle. The key actors occupy similar positions in each case: Superintendent, Commandant of Cadets or Midshipmen, Academic Dean. In one or more cases, important roles are played by others, such as academic department heads, other faculty members, the football coach, the chaplain, the director of admissions. External participants include officials at service headquarters, members of Congress, academic consultants, boards of visitors, alumni, and the mass media.

Each case study is designed to assess the relative contribution of the various participants to the change process. Who are the key proponents of various changes, and who are the principal opponents? Is the disposition toward reform (or toward maintaining the status quo) explicable in terms of one's organizational role, one's previous career experience, one's links to academy tradition, or do other factors influence one's point of view? To what extent are officials at one academy cognizant of the practices and policies of one or more of the other academies, and to what extent does such cognizance affect their actions? What strategies do academy leaders employ in attempting to enlist internal and external support for changes which they favor? What kinds of strategies tend to be successful, and what kinds unsuccessful? Under what circumstances do various external participants serve to encourage change, and under what circumstances to bolster internal support for the maintenance of traditional structures and practices? What have been the internal consequences of particular changes, especially in terms of the efforts of the academies to reconcile the Spartan and Athenian components of their mission?

Findings from the four studies are summarized, supplemented, and analyzed in Part 3 of the book. The analysis of organizational change is especially an attempt to determine how, why, and under what circumstances organizations change their goals and the strategies and technologies which they employ to attain their goals. The goals of the academies are describable most meaningfully not in terms of the desired end product (well qualified lieutenants and ensigns) but rather in terms of the array of subsidiary goals which must be accomplished if the overall mission is to be performed. These include the Athenian (academic) and Spartan (military-professional and athletic) goals so prominent in academy statements of "mission," as well as other goals common to all modern complex organizations, such as recruitment of personnel, management, and public relations.

The academies draw upon the society at large for the "raw products" that enter their institutions, and upon external bodies such as the Congress and the Pentagon for essential resources such as funding and authorization for added staffing and facilities. They apply a variety of technologies to the task of transforming the raw products into finished ones, qualified to join the officer corps. Their performance in effecting this transformation is formally appraised by the parent organizations on a continuing basis, and, on a less regular basis, by the Congress, boards of visitors, accrediting agencies, and bodies such as the General Accounting Office, as well as informally by the mass media, alumni groups, and others in the general public. These appraisals have an impact on funding decisions, on future assignments of personnel to key positions in the organization, and on the strategies and technologies that are employed in pursuit of organizational goals (see Figure 1).

The case studies of Part 2 provide many clues regarding the extent to which these patterns are explicable in terms of external influence—mandates from the parent organizations, congressional investigations and legislation, demands for action (or inaction) from alumni, and so forth. Clues are also provided on the relative importance of internal initiatives in shaping the course of change at the academies—the leadership vision of superintendents, the capacity of academic deans or their faculties for innovation, the determination of commandants of cadets or midshipmen to maintain tradition or to make new departures.

Part 3 represents an effort to pull together the various clues from Part 2 in a systematic fashion, supplementing evidence when necessary with additional data from recent decades in order to develop some useful analytical generalizations. The effort to generalize about how and why various changes have occurred or have been thwarted is organized around three working hypotheses.

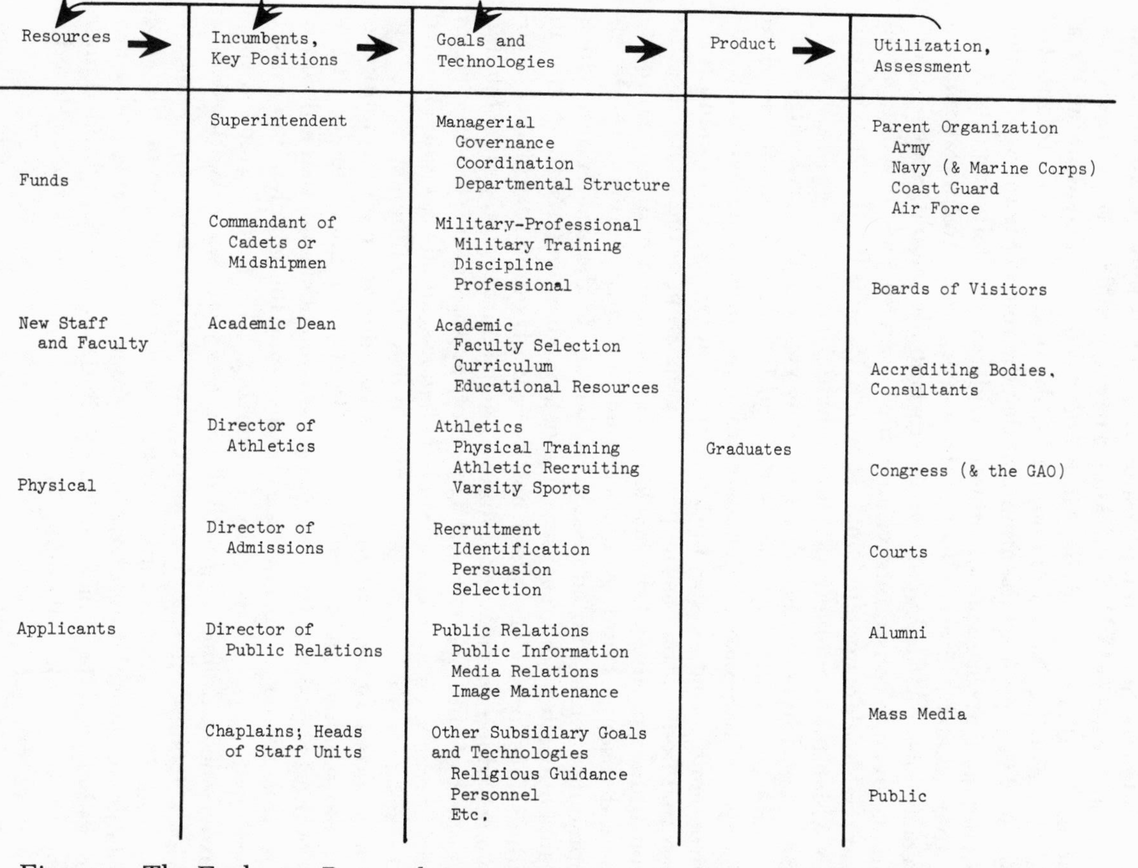

Figure 1. The Exchange Process between a Service Academy and Its Task Environment

The first of these is that goals and technologies will vary according to changes in personnel holding key leadership positions. Insiders at the academies have tended to subscribe to this hypothesis. Historical examples, such as that of Sylvanus Thayer, and modern ones, such as those of McDermott at the Air Force Academy, Davidson at West Point, Leamy at New London, and Calvert at Annapolis, frequently are cited in support of the argument that the role of the internal leader is crucial. The case studies provide evidence of vitally important initiatives taken by internal leaders. However, upon careful examination in chapter eight, the hypothesis is substantially refined. Special attention is given to constraints within which even the most creative organizational leaders operate. Notable among these is the duration of time for which the leadership position is held.

A second hypothesis, which persons viewing the academies from outside frequently espouse, is that goals and technologies tend to change in response to demands from the organizational "task environment." That is, the initiative for change is said to come from external bodies, such as the parent organization or the Congress, that constitute a portion of the academy's environment which controls essential resources and oversees organizational performance.[8] Developments such as the admission of women to the academies, which came only after Congress passed legislation banning discrimination in admissions for reasons of sex, and the end of compulsory chapel attendance, which was made possible through court action, can be cited in support of this hypothesis.[9] However, as the discussion in chapter nine reveals, circumstances such as honor scandals have determined the susceptibility of the academies to external influence. Moreover, the response of the academies to the task environment has tended to be indirect and gradual rather than direct and immediate.

A third hypothesis focuses on a dimension of change that is obvious; but it is one with consequences for academy life which neither insiders nor outsiders have fully recognized. The four academies have experienced dramatic growth during the years since World War II. Goals, technologies, and structures have become more complex. The codification of once informal customs and rites, and the proliferation of rules and regulations, have accompanied growth. The hypothesized links between these facets of bureaucratization and heightened internal tensions especially along Spartan-Athenian lines, are explored in chapter ten.

The book as a whole highlights the distinctiveness of each academy, as well as similarities among them. The analysis makes clear, however, that in varying degrees all four are at critical junctures in their evolution. The reconciliation of Spartan and Athenian goals is by no means the totality of the problem which the academies now face, but it is at the heart of a dilemma which has become progressively acute as the academies have become "modernized."

At various times over the years there have been critics who have argued that the American service academies ought to be radically transformed, if not abolished. That the academies have weathered many storms and have survived severe criticism in the past is no cause for present complacency. For reasons that will be apparent as the study unfolds, proposals for radical change at the academies must be taken more seriously now than in any period since the unification controversy at the end of World War II. In the concluding chapter, a continuum of change scenarios is discussed and assessed.

It will become obvious that some of these scenarios are more plausible than others. It is far from obvious, however, how academy officials and those in Washington who help to control the destinies of the academies will respond to current exigencies. The analysis in this book provides a basis for suggesting which among the various scenarios is most probable for the foreseeable future. Whether this is also the most desirable scenario the reader will decide.

2

Origins of the Seminary-Academy

As tempting as it might be to read into various facets of the development and orientation of the American service academies a direct emulation of classical Spartan and Athenian models, this is not the argument here. It is true that Western civilization in general and Western education in particular have been profoundly influenced by the classical Greek cultures. Some scholars would argue that "everything of importance in our civilization derives from them [the Graeco-Latins]".[1] References to both Sparta and Athens were common in colonial America, and in the formative years of West Point, the earliest service academy. The distinctive service academy terminology (for example, the term *plebe*, as used at West Point and Annapolis for the lowest class of cadets) and symbolism (such as the Military Academy crest, which uses Greek symbols of both valor and wisdom) draw explicitly upon the classical Greeks. The traditional emphasis at the service academies upon educating and training "the whole man," with mathematics the central element of intellectual discipline, is entirely consistent with classical notions of education such as that set forth in Plato's *Republic*.

The terms *Spartan* and *Athenian* are used here metaphorically rather than literally. The Spartan ideals are those of the noble warrior: austerity, discipline, the comradeship of arms, devotion to the state, and, above all, a commitment to heroic deeds and a love of glory. Athenian ideals, in contrast, are especially those of culture and learning. It is not necessary to argue that service academy officials consciously sought to emulate their classical forefears in Sparta to recognize that these ideals have been important elements of the academy subcultures. Nor in identifying an Athenian dimension to the service academies is it necessary to argue that Sylvanus Thayer, the father of the Military Academy, or his successors at any of the service academies drew explicitly upon the writings of Plato, Aristotle, or other celebrated Athenians.

What we do contend is that the concept of what a service academy should be, incorporating Spartan but also some Athenian values, evolved early and greatly influenced the evolution of the academies. The concept, which we shall describe as the "seminary-academy," was first articulated fully in the Thayer years at West Point (1817-1833). To a considerable extent it served as the model not only for the other two academies (Naval, Revenue-Marine) that were founded in the nineteenth-century, but even for the Air Force Academy at its inception in 1954.

Thayer's seminary-academy vision had drawn upon French approaches to technical education which represented the most modern equivalent of the Athenian ideal of knowledge as virtue. However, equally important to the Thayer system was the maintenance of a Spartan living environment. The synthesis of Spartan and Athenian ideals, for the time, was effective. However, as the seminary concept evolved in the post-Thayer decades, its Spartan component became embellished with the legends of the heroic exploits of academy graduates in war, while its Athenian component became dulled from inattention. It was this half-embellished, half-neglected version of the seminary-academy model that continued to mesmerize officials of the academies a hundred or more years after the Thayer era.

A discussion of the historical evolution of the seminary-academy model, with its real and mythical dimensions, and of the growing discordance between the Spartan and Athenian components, is a necessary foundation for the analysis of changes that have occurred in recent decades. In this chapter we shall consider: (1) Sylvanus Thayer's vision of a seminary-academy; (2) the roots of the Thayer system; (3) the Thayer heritage at West Point; (4) the founding and evolution of the Naval Academy, which reflected the influence of the West Point model; and (5) the distinctive early experience of the Coast Guard Academy, which nevertheless had begun to acquire some of the seminary-academy characteristics by the time the new permanent installations were opened in New London in the early 1930s.

Thayer's Vision of a Seminary-Academy

Sylvanus Thayer was Superintendent of the Military Academy at West Point from July 1817 to July 1833. He was in the tradition of the great college presidents of nineteenth-century America, at a time when paternalism was the norm and successful colleges typically reflected the stern fatherly guidance of their presidents.[2] A bachelor, Thayer immersed himself completely in his duties. Thoroughly conversant with the background and performance of every cadet and every member of the faculty, as well as with every detail of academy operation, Thayer was busy sixteen hours a day, seven days a week. For Thayer, as for most educators of the period, the purposes of the educational experience were not merely intellectual. The disciplining and development of the body, the soul, and the mind were at stake; to accomplish this task required creation of a total environment—a mold within which character could be shaped. As the Board of Visitors to West Point in 1820 observed approvingly of the environment that Thayer was creating: "In all ages, military seminaries have been nurseries from which have issued the highest elements of character, and some of the most conspicuous agents in the operation of society."[3]

The practices that Thayer implemented in fulfillment of his seminary-academy vision have been described in admirable detail elsewhere but bear summarizing here.[4] Chiefly, "the Thayer system" refers to the modes of discipline and character building that were employed, the pedagogical techniques that were utilized, and the academic curriculum that was prescribed.

Discipline and Character Building

The Thayer system followed a strict hierarchy. The Commandant of Cadets had primary responsibility under the Superintendent for the supervision of discipline and military training. He in turn was assisted not only by other military officers but also by a cadet chain of command. The ordinal designation of classes meant that seniors were "first-classmen," and it was they who held the key positions.

Insubordination was an adequate ground for dismissal. For example, Cadet Edgar Allan Poe was dismissed from the Military Academy in 1831 upon being found guilty of "gross neglect of duty" and "disobedience of orders." Specifically, Poe had been charged with missing thirteen parades, seven reveille formations, a guard mount, chapel, and all academic duties for a two-week period.[5]

Although the Thayer system emphasized obedience to rules and authority, Thayer tried to ensure equality of treatment of cadets irrespective of their social backgrounds. To do so was a substantial feat, because in Thayer's day a sizable proportion of cadets came from relatively prominent families. (Subsequently, the Military Academy became increasingly an avenue of upward mobility for sons of lower-middle-class families.) Equality of treatment of cadets, once admitted, was promoted by the requirement that all exist on the modest pay they received rather than on support from home.

Cadet life was regimented and isolated. Access to and from West Point was principally by boat, except for trips by foot to the nearby town of Buttermilk Falls (which is now Highland Falls). But the tightened regulations under Thayer prohibited visits even to Buttermilk Falls except with special permission. Over-the-wall sojourns by cadets to Benny Havens's tavern, for instance, could be made only at considerable risk of punishment. Academy records reveal that Cadet Jefferson Davis, who later graduated to become U.S. Secretary of War and then President of the Confederacy, was almost dismissed from the academy in 1825 because he had gone "to a public house or place where spiritous liquors are sold kept by one Benjamin Havens at or near Buttermilk Falls and distant about two miles from West Point." In this case, the court imposed a lesser sentence than dismissal, citing Davis's prior good conduct.[6] As the Board of Visitors to the Academy in 1820 had noted: "The situation of West Point is so favorable that there exists [*sic*] but few of the usual temptations to vice and dissipation. All the inhabitants, being tenants at will, the Superintendent has the power of instantly removing any improper character, whether male or female."[7]

The building of the cadet's strength of character was to be fostered not only through insistence upon adherence to rules and regulations, but also through formal training in ethics and through the requirement of chapel attendance. Chapel services were nominally nonsectarian, but were Episcopalian in format, following the reasoning of Army Chief Engineer General Joseph Swift that "the service of that church was deemed to be the most appropriate to the discipline of a military academy."[8]

Pedagogy and Prescribed Curriculum

Heavy emphasis was placed upon the practical application of education and training. However, in contrast to those who argued that military officers need only an apprenticeship to prepare them for their responsibilities, Thayer and his supporters believed that an extended period of formal instruction (four years) was an essential supplement to field experience. (The apprenticeship view was still prevalent in this era in the Navy and in the Revenue-Marine, as the Coast Guard was then known.)

The cadet was given grades daily for his recitations; in addition, he was graded on general review sessions twice a year. These grades, plus grades for military training and for conduct, contributed to his order of merit according to the weight attached to each (the grade for mathematics, for instance, was weighted more heavily than that for instruction in French). To have each cadet recite daily in each class was possible only if the classes were kept small. The system could be maintained best, it was felt, by those who had experienced it. Thus, Thayer turned increasingly to Academy graduates for instructors.

Governance of the academic program was entrusted to an academic board of senior faculty members from each field, although the Superintendent retained final authority. The curriculum was totally prescribed, with first- and second-year introductory courses serving as the foundation for advanced work later. The emphasis on engineering subjects included drawing, with mathematics at the core of the curriculum. French language instruction was stressed also, primarily because a large number of the textbooks that were used were written in French. There was also instruction in chemistry, natural philosophy (which we would now call physics), geography, history, and moral philosophy.

The Roots of the Thayer System

It should not detract from the significance of Thayer's contributions to note that a good part of the system's basic design which he implemented had been established by rules and regulations promulgated before Thayer was appointed, and much of the rest of the design was modelled closely after the educational system of the École Polytechnique in Paris.

Thayer's Predecessors

Thayer's system of discipline, for example, differed little in formal detail from that of his predecessor as Superintendent, Captain Alden Partridge. Indeed, what is often forgotten by present-day admirers of Thayer is that he succeeded where Partridge had failed in enlisting the support of his faculty largely because they saw that Thayer regarded discipline as but one means of realizing the broader educational goals of the seminary-academy that he hoped to create.

The importance of the first Superintendent of the Military Academy, Jonathan Williams, in laying the groundwork for a forward-looking scientific-engineering program of the sort Thayer would institute also should be noted. Williams's great uncle, Benjamin Franklin, had organized the American Philosophical Society in 1743 to promote intellectual discourse and the ideas of the Enlightenment. Similarly, Williams organized a military philosophical society after being appointed by President Jefferson to head the newly founded Military Academy. Participants in the society's biweekly meetings included officers and cadets at West Point as well as distinguished persons elsewhere, including Eli Whitney, Robert Fulton, Thomas Jefferson,

James Madison, John Quincy Adams, Edward Preble, and Stephen Decatur. Alden Partridge was a participant while an academic staff member at West Point (presenting a series of papers on scientific experiments which he had conducted). Sylvanus Thayer, who graduated as valedictorian at Dartmouth College before his appointment to West Point as a cadet, also was a participant.[9]

Although the War of 1812 precipitated the dissolution of the Military Philosophical Society, the war served to impress large numbers of American leaders with the risks of relying on poorly trained militiamen and officers who were political appointees with little or no military training in combat with a professional army. One consequence of this growing concern for military preparedness was legislation to expand and improve the program at West Point.[10] On July 2, 1816, Secretary of War George W. Crawford initiated further guidance upon which Thayer would soon draw, with a directive providing for (1) a board of visitors to the academy; (2) the subjects in which instruction was to be given; (3) general examinations administered twice a year, with graduation according to order or merit; and (4) a four-year program of instruction for all cadets.[11] (It is of some interest that the embittered Alden Partridge, after leaving West Point and opening his own American Literary, Scientific, and Military Academy at Norwich, Vermont in 1820, rejected both the prescribed curriculum that characterized West Point education and the requirement that all students spend a uniform amount of time completing their studies.)[12]

French Influences

Probably most important in shaping the ideas that Thayer later translated into West Point's program was the French example. For a time, French ideas had been immensely popular throughout American higher education.[13] The French support of the American cause in the Revolutionary War against England, including the personal contribution of such Frenchmen as General Lafayette, aroused popular sympathies for the French among Americans. The French Revolution, perceived as a continuation and perhaps further refinement of the ideals of the American Revolution, fired the imagination of many Americans.

However, especially in New England, the excesses of the French Revolution in its later stages led to a considerable dampening of enthusiasm for French ideas. In many colleges, the influence of French ideas and educational practices were supplanted by German influence. For example, George Ticknor, who had been Sylvanus Thayer's classmate at Dartmouth and remained his friend, traveled to

Göttingen in 1818 in the company of Edward Everett; they were joined by George Bancroft (who later founded the Naval Academy). The three returned to the Harvard faculty to champion curriculum reform, based upon the model of German higher education. By 1850, more than two hundred American students had made pilgrimages to Göttingen, Leipzig, Bonn, Berlin, and the other great German universities.[14]

If, as Samuel Eliot Morison observes, Ticknor's cultural leadership, inspired by the German example, "gave Boston the name of 'the Athens of America,'" West Point, as the closest American approximation of Sparta, remained under the influence of French ideas not only during the Thayer years but until the defeat of the French by the Germans in 1870-71.[15] (Even with the discrediting of French military ideas in the post-Franco-Prussian War period, the French language was retained as a required part of the Military Academy curriculum, and a proposal that German be added was rejected.)[16]

The French influence at West Point, well established by the time Thayer arrived as Superintendent, continued under his leadership. One of the non-French faculty members (who had a reputation as a complainer) was led to grumble, "All the Frenchmen keep a kind of separate community here, they will not conform to our manners and modes of living; whence they are not liked by our people."[17] French influence also was introduced as a result of Thayer's own two years in France immediately before assuming the head post at West Point, and the four-year visit by Dennis Hart Mahan. Thayer brought a storehouse of ideas about education that he had gathered at the École Polytechnique and approximately a thousand volumes of technical studies, textbooks, and other works on mathematics, military art, and engineering, along with a collection of maps of the Napoleonic campaigns.[18]

A few years later, Dennis Hart Mahan, who had become an assistant professor of mathematics at West Point after graduating in 1824 at the top of his class, was sent to Europe for study. Upon his return four years later, Mahan had become a francophile. His enthusiasm for French ideas was expressed in his lectures and writings on fortifications and civil and military engineering, and in his discourses on the principles of war. In addition to introducing the systematic study of war into the classroom, Mahan was the leading figure in the Napoleon Club that was organized among Academy officers near the end of the war with Mexico, a group which continued to meet weekly until the Civil War.[19] Although Thayer resigned as Superintendent in 1833 to protest what he saw as interference by the Jackson administration in

Academy affairs, Mahan stayed at West Point until his death in 1871. A dedicated disciple of Thayer's, Mahan in turn influenced several generations of West Pointers; and through his son, Alfred Thayer Mahan, who was raised at West Point, the influence extended to Annapolis as well.

The Thayer Heritage: Success and Complacency

In his insightful study of Antioch, Reed, and Swarthmore colleges, Burton Clark has observed that "The ultimate risk of distinctive character is that of success in one era breeding rigidity and stagnation in a later one."[20] The observation applies poignantly to the fate of the Thayer system by the late nineteenth or early twentieth century.

While it is not necessary to describe in detail what happened at West Point from the end of the Thayer superintendency to 1940, it is useful to discuss briefly the key characteristics of the transformation that occurred. West Point, sustained by the momentum that had developed under Thayer's leadership, moved into what one historian has described aptly as its "Golden Age."[21] This was followed in the post-Civil War era, however, by a period during which prior accomplishments served as the foundation for self-satisfaction and complacency.

The Curriculum: From the Thayer Era to the Turn of the Century

The curriculum that Thayer instituted at West Point was avant-garde. In all nine colleges founded in the American colonies prior to the Revolution, the deductive and didactical spirit of scholasticism was yielding to the inroads of the experimental and experiential spirit of Newton, Locke, Bacon, and Descartes. However, it would be several decades after the Thayer superintendency before the attention that he and his colleagues devoted to such subjects as mathematics and natural philosophy would be regarded as conventional throughout American higher education. The Thayer system also departed from tradition by requiring study of a modern language rather than the classical Latin and Greek. French had not been taught anywhere in American higher education until 1779.[22]

Moreover, the Military Academy was America's first institution of higher education in engineering—and, after the École Polytechnique, one of the first in the world. With increasing American industrialization and growing transportation needs associated with westward expansion, the demand for engineering skills grew rapidly in the nineteenth century. But throughout the Thayer years, West Point remained one of the few schools that could help meet the demand. Roughly one of every four graduates during the Thayer period (1817-1833) went on to a career in civil engineering—exploration, surveying, road and bridge building, canal design and construction, dam building, railroad engineering.

Engineering remained a prestigious career choice throughout the nineteenth century. From 1891 through 1900, every top Academy graduate chose the Corps of Engineers as his branch of service; and in six of the ten years, the top five men in the class selected the Corps of Engineers. However, by 1900 the number of engineers that West Point was producing was but a tiny fraction of the national output. The enrollment had scarcely increased since Thayer's day (there were only 609 graduates from 1891 through 1900); and although the top cadets were still choosing engineering, the majority made other choices. Assignment to the artillery was more characteristic than to engineering, for instance, as was assignment to the cavalry. The substance as well as the image of West Point were becoming more Spartan, and its reputation as a national pacesetter in science and engineering had declined.

The turn of the century was a time of considerable intellectual ferment and change in American colleges. Many schools were abandoning the totally prescribed curriculum in order to introduce electives. Many were providing students with more freedom in other realms of academic life as well. The intellectual development of the individual was more consonant with modern goals of higher education than the traditional emphasis on strict discipline, moral training, and "character building." In this context, the military Academy, which under Thayer had been a pioneering educational institution, now became a bastion of the status quo.[23]

Thayer's Faculty and the Rise of a Gerontocracy

During the first half of the nineteenth century, there were few colleges that could boast of faculties engaged in scholarly research and publication. Colleges had been staffed principally by clergymen, intent on imparting moral wisdom and fervor, but seldom concerned with the task of creating new knowledge.[24] But at West Point in the Thayer years and in the early post-Thayer period before the Civil War, a number of faculty members had established wide reputations as scientists and scholars.

Virtually all West Point professors in those days were relatively young. Thayer himself was only thirty-two when he became Superintendent. Charles Davies, Albert Church, William Bartlett, Jacob

Bailey, and Dennis Mahan each took charge of an academic department less than ten years after his own graduation from West Point. After the Civil War, however, the average age of West Point faculty members rose sharply and their scholarly productivity declined. Church, Bartlett, and Mahan remained on the faculty until the 1870s—more entrenched than ever on the board that set academic policies, but long past their most productive years.

A change in Army policy in 1839 had made the position of Superintendent at the Military Academy one to be filled for a relatively brief tour of duty rather than for a period that extended until retirement, death, or resignation, as the policy had been for Thayer. With this change, the Academic Board acquired greater importance. Professors had tenured appointments, which meant that they remained for entire careers or until death while superintendents came and went.

It may well be that, on the whole, the generations of faculty members at West Point in the decades after the Civil War were no less gifted intellectually than those in the Thayer years and in the "Golden Age" that followed immediately thereafter. However, in the several decades before the Civil War, faculty members had directed much of their energy to pioneering ideas in engineering education and to the furthering of science; contacts with civilian scientists and educators were numerous (given limitations of transporation and communication at the time). In contrast, after the Civil War, faculty members increasingly withdrew from concern with the civilian sector. Their academic writings typically took the form of textbooks for use by West Point cadets. In addition, they spent much of their energy defending the Academy against criticism and nurturing a growing West Point saga.

Development of the West Point Saga

"No other period," Samuel Huntington writes of the post–Civil War decades, "has had such a decisive influence in shaping the course of American military professionalism and the nature of the American military mind." He notes that this was a period in which "The military . . . were divorced from the prevailing tides of intellectual opinion." He makes the provocative argument that "the very isolation and rejection which reduced the size of the services and hampered technological advance made these same years the most fertile, creative, and formative in the history of the American armed forces.[25]

However, as Huntington acknowledges, this was also a period in which "West Point . . . gradually lost contact with the rest of American education to which it had made such significant contributions, and went its own way."[26] Moreover, it was a period of the romanticization of the military profession, the veneration of war heroes, and the canonization of institutional practices of the Academy.

In the period just before the Civil War, the glorification of the military art embodied in Dennis Hart Mahan's teachings as well as in the sessions of the Napoleon Club in which he participated, had helped to instill in West Pointers a romantic image of a military career. Furthermore, the dozen or more military academies that had sprung up in the South in the 1830s, 1840s, and 1850s as carbon copies of West Point "both reflected the martial spirit and contributed to its growth." As John Hope Franklin has observed, the "West Points of the South . . . used, insofar as possible, the West Point curriculum, West Point texts, the West Point system of grading, and West Point discipline."[27]

The war itself provided a major stimulus to the romanticization of military ways in general and of West Point in particular. One important effect of the war was the elevation to prominence of dozens of West Point graduates for their exploits in battle. Before 1861, none of the men who had attained the rank of general officer in the American Army was a West Pointer.[28] However, during the Civil War, virtually all of the important commands in the Confederate Army as well as in the Union Army were held by graduates of the Military Academy. In the decades after the war, the wartime exploits of illustrious graduates came to dominate the lore of cadets. Because nearly all the noted graduates made return visits to their alma mater, the presence of the celebrated Civil War heroes was felt physically and spiritually.

Not everyone looked upon the rise to prominence and influence of West Pointers during the Civil War as a favorable trend. Some persons, such as John A. Logan, who gained command of the Army of Tennessee as a citizen-soldier major general during the war only to lose the command to a West Pointer, saw the trend as an unhealthy movement away from Jacksonian egalitarianism and toward elitism. Logan, who became U.S. senator in the postwar period and founder of the Grand Army of the Republic, a veterans organization, deplored the growing influence of West Pointers in a widely circulated 700-page diatribe.[29]

Ironically, the net effect of these and other criticisms of the Military Academy was to accentuate the tendency of graduates to depict in glowing tones the Academy's contribution to society. The initiatives of George Washington Cullum in gathering biographical data on all West Point graduates and in publishing accounts of their accomplishments illustrate the point. Cullum had graduated in the final year of the Thayer superintendency. He returned to West Point in the

1850s to serve as adjutant under Superintendent Robert E. Lee, and then served as Superintendent himself from 1864-1866. In 1868, the first volume of his *Register of Graduates* was published. Cullum not only dedicated himself to ensuring that history would record the important deeds of Military Academy graduates, but also became one of the initiators of an association of graduates. During the late nineteenth and early twentieth centuries, faculty members and other graduates followed Cullum's example by writing glowing biographies of distinguished West Pointers, extolling the virtues of Academy education in national magazines, and publicly rebutting critics of the Academy.

The Twentieth-Century Thayer System

The Thayer system which West Pointers identified with and practiced in the 1920s and 1930s was different in many respects from the system of Sylvanus Thayer a hundred years earlier. But the sense of continuity was maintained to a far greater degree than is typical of institutions of higher education—most notably through the accumulation of Academy lore and traditions and through the communication from one generation to the next of a spiritual sense of community (the "Long Gray Line") in which all West Pointers participated. Moreover, the Thayer system legacy to some extent became the legacy of all the service academies. The Naval Academy, more than the Coast Guard Academy, drew explicitly upon the West Point example. Annapolis experienced much the same evolution, from an institution that within a few decades of its founding had established a reputation as a center of experimentation and learning to one in which, in time, the Spartan emphasis and tradition became dominant.

Early Evolution of the Naval Academy

For many years prior to its founding in 1845, a naval academy ashore for training prospective officers had been sought by leading naval officers and interested civilians. The War of 1812 had provided the stimulus not only for reforming the War Department and the program at the Military Academy but also for establishing a seagoing navy on a firmer basis. Beginning in 1814, a succession of Navy secretaries proposed, with the concurrence of ranking naval officers, the creation of an academy ashore.[30] Congressional support was lacking, however, especially among representatives of the inland states. Naval appropriations from the first to the second Monroe administration were cut, as was the number of naval officers and enlisted men that were authorized. With the presidency of John Quincy Adams, a New Englander, pro-Navy sentiment revived. A bill providing for creation of a naval academy passed the Senate by a 28 to 18 vote, but was defeated in the House. The Tyler administration managed in 1841 to get support for a dramatic increase in appropriations and in the authorized personnel strength of the Navy. A bill to expand the Navy still further, including creation of a training school, was defeated in 1842 on a vote that split along seaboard (pro-naval-expansion) versus inland (opposition) lines.[31]

The Founding

A four-year naval academy was established in 1845, by a virtual sleight of hand. The celebrated historian George Bancroft (Ticknor's long-time colleague at Harvard) had become Secretary of the Navy in March 1845. Bypassing Congress completely, Bancroft arranged with Secretary of War William L. Marcy for Fort Severn at Annapolis, to be used as a naval school. President Polk approved, and in October the Naval School (as it was called until 1850, when it was renamed the United States Naval Academy) was opened.

Two circumstances made it possible to enlist the necessary support of Congress to legitimize and fund the "coup" that had been effected. The first was the sudden increase in public concern for the conditions under which midshipmen were trained. In 1842, acting midshipman Philip Spencer, son of the Secretary of War, was hanged for mutiny aboard the U.S.S. *Somers*, along with two enlisted seamen. The investigation which followed was widely publicized and led to pressures on Congress to rectify conditions that had made such a tragic incident possible. Secondly, and more fundamentally, the technological ferment of the Industrial Revolution was making sailing ships obsolescent as the basis for a major naval force. In 1837, the *Fulton*, the Navy's first steam-driven warship, was launched, and two years later Congress appropriated funds for three additional steam vessels. Clearly, new kinds of expertise for future naval officers were needed. Thus, in 1845 the argument for the kind of school Bancroft had created by decree was persuasive.[32]

Influence of West Point and the Thayer System

Major credit for the intellectual direction and impetus that were provided for the new academy must go to its most influential faculty member, William Chauvenet. An energetic and imaginative man, Chauvenet would gain a reputation as "the ablest mathematician and astronomer which Yale has produced."[33] Moreover, only a few years out of Yale, his reform efforts as president of a school designed to prepare prospective midshipmen to pass their precommissioning examinations brought him to the attention of the Secretary of the Navy. Chauvenet's school, located in a Philadelphia home for aging seamen known as the Asylum, had succeeded where similar schools elsewhere had failed.

Still, Chauvenet's appointment to the new academy came through the intercession of Alexander Dallas Bache, a mutual friend of Bancroft and Chauvenet. Bache was a West Pointer (and grandson of Benjamin Franklin) who had graduated during the Thayer years at the top of his 1825 class. The combination of his Military Academy affiliation and his ties to Bancroft and Chauvenet is illustrative of the numerous linkages that developed between the new academy at Annapolis and the older one on the Hudson.

The design of the new academy reflected such influence. Shortly after Bancroft took over as Secretary of the Navy, he sent Passed Midshipman[34] Samuel Marcy, son of the Secretary of War (Spencer's successor), to West Point to study its program and to report back conclusions relevant to the launching of a new naval school. Passed Midshipman Marcy accompanied his father on a visit to West Point in May 1845, and upon his return submitted a detailed description of the Military Academy program to Secretary Bancroft. In his report, Marcy observed that:

> The Military Academy at West Point undoubtedly furnishes the best model in the country. How near an approach to its character the present means of the Navy will admit is the first inquiry to be made, and if they are found sufficient to warrant the taking of the first step towards the erection of a similar institution it is to be hoped that it may be taken and followed by such additions and improvements as will secure to the Navy relief from one of its most pressing wants.[35]

Although Marcy's description of the Military Academy program and his proposal for emulating it contributed greatly to the program initially adopted at the Naval Academy (when Marcy joined the faculty), a firm curriculum was not established until a reorganization in 1850. The war with Mexico (1846-1848) had disrupted the newly established school. For example, a number of midshipmen left for active combat service. The Annapolis atmosphere has been described as one in which midshipmen "engaged in brawls with the townspeople, indulged in a few duels, and performed all manner of high-jinks."[36] Enforcement of discipline was made difficult by the lack of familiarity of staff and midshipmen with the new rules and procedures, and also by the great disparity in maturity between the older midshipmen who were in their middle or late twenties, and the younger midshipmen who were in their mid teens. In 1849, a board of distinguished naval officers was asked to study the Academy and make recommendations. Army Captain Henry Brewerton, who had graduated from West Point during the second year of Sylvanus Thayer's superintendency and was now Superintendent of the Military Academy himself, was a consultant to the board. Their recommendations, which included increasing the authority of the Naval Academy Superintendent, creating the position of Commandant to oversee military training and discipline, moving to a four-year prescribed curriculum, and restricting admission to boys ages fourteen to sixteen, were implemented.[37]

As at the Military Academy, more weight was assigned to mathematics than to any other academic subject in determining a final order of merit among midshipmen. (However, as a means of enforcing strict discipline, conduct was given a weight equal to mathematics.) William Chauvenet was the first head of the Department of Mathematics at Annapolis, but he brought with him as assistant professors two members of his faculty at the Asylum School: Lieutenant James Ward, a graduate of the Vermont Military Academy, and Henry Lockwood, an 1836 graduate of West Point. In 1853, when Chauvenet became head of the newly created Department of Astronomy and Navigation, Lockwood took over the mathematics department and sat on the Academic Board (a body patterned after that at West Point). Lockwood's presence at Annapolis was so prevalent that a century later midshipmen were still making snide remarks about the West Point influence. At his instigation, infantry drill was introduced into midshipman training, with Lockwood eagerly serving as the drillmaster. It was a ritual that endured.

French language instruction was included in the curriculum, as it was at the Military Academy. Chauvenet was able to arrange to have his French-born father, who as a boy had served as secretary to the chief commissary in Napoleon's army, hired as an assistant professor of French.[38] Another similarity to the Military Academy curriculum was in the inclusion of natural and experimental philosophy in the Naval Academy program of instruction. William P. Hopkins, the first instructor of the subject, had graduated from West Point in 1825, where he had taught chemistry, minerology, and geology from 1827 to 1835.[39]

From 1845, the officer primarily responsible for military training and discipline was known as the Executive Officer. Commander Sydney Lee had held that title upon reporting to Annapolis in 1848 and until he assumed the new title of Commandant in 1850. By 1852, Sydney Lee had left for an assignment as captain of Commodore Perry's flagship, which would initiate the opening of Japan to the West. Concurrently, Sydney's younger brother, Captain Robert E. Lee, returned to West Point for a three-year tour as Superintendent.[40]

Post-Civil War Stabilization and Reform

Although Chauvenet and his contemporaries laid the basic foundation of the Naval Academy, the stabilization and reform that were provided in the early post-Civil War period were of equal importance to that institution's successful development. Because of its exposed location, the Academy had been moved to Newport, Rhode Island, at the outbreak of the Civil War, with a large number of midshipmen and officers resigning to join the Confederacy. In 1864, a board of visitors recommended abolition of the Academy in favor of the establishment of seven separate schools; but in 1864 Congress approved the return of the Academy to Annapolis.[41]

Postwar renovation began under Superintendent David D. Porter. Admiral Porter created a Department of Steam Engineering and made curricular changes which, according to later accounts, transformed the Naval Academy from "a high school to a college."[42] His Commandant, Commander Stephen B. Luce, is described in present-day writings at Annapolis as "unquestionably the most able seaman of his day, and perhaps the greatest seaman of all time."[43] Luce had served as head of the Academy's Department of Seamanship for a year during the Civil War. Finding existing instructional materials inadequate, he wrote his own text, one that was used for nearly two decades and became adopted at the Coast Guard Academy as well. His tour as Commandant reflected his continuing concern for the development of naval professionalism, to which he made an enduring contribution with the founding of the Naval War College in 1884.

Christopher Rodgers, the son of Commodore George Washington Rodgers and the nephew of Commodore Oliver Hazard Perry, was well known from the beginning of his career and had already established a formidable reputation by 1874 when he returned to Annapolis as Superintendent.[44] He had been a predecessor of Luce as Commandant, but his credentials had been established at sea, where he was, "next to Farragut, probably the most highly regarded naval officer of that age."[45] He had served off the Florida coast in the Seminole War, at Vera Cruz in the war with Mexico, up and down the coast of the Confederacy during the Civil War, and had held half a dozen ship commands. His foreign tours ranged from the Mediterranean to the Pacific. He returned to the Naval Academy a firm believer in education, and convinced that the midshipmen's program of instruction needed to be more rigorous. Under Rodgers, mathematics instruction was expanded to include differential and integral calculus, and Spanish was added to language instruction (as it had been at West Point). In a notable departure from the Thayer system, Rodgers instituted electives for students who were doing especially well in the prescribed courses. In 1882, however, congressional abolition of the distinction that had been maintained between "cadet midshipmen" (the regular student body at the Naval Academy) and "cadet engineers" (who had pursued a different course of study) led the Academy to abandon electives.[46] The prescribed curriculum would then survive for over seventy-five years.

A number of Naval Academy students during these years later established reputations as inventors or educators. The year following his graduation, a member of the class of 1878 developed the electric motor that came to be used to run trolley cars. A classmate established the Engineering School at Harvard and became its first dean. Another led in the development of engineering education at the University of Pennsylvania.[47]

Intellectual ferment also was evident in the activities of Annapolis faculty members. Alfred Thayer Mahan (whose middle name reflected the esteem of his father, Dennis, for the West Point Superintendent with whom he served) taught under Rodgers. As an instructor at the Naval Academy, Mahan wrote the first of the essays on naval education and seapower that within two decades would make him world-famous.[48] The year after Mahan's first essay was published, a fellow instructor, Ensign Albert A. Michelson (USNA 1873), designed and conducted experiments to determine the velocity of light and thus became the first American to win the Nobel Prize in physics.[49]

The head of the Department of Physics and Chemistry under whom Michelson served was William T. Sampson, who had graduated first in the Naval Academy Class of 1861 and became commander of U.S. forces in the Caribbean during the Spanish-American War. The assignment as department head was Sampson's third academy tour. In September 1886, Sampson returned again, this time as Superintendent, to serve in that capacity until June 1890. The period of naval expansionism, and of the "new navy" outlook, had begun. Although these developments were especially evident at the Naval War College and in the writings of Mahan, Sampson imparted to the Naval

Academy increased concern for professionalism and for scientific studies.[50]

Formative Years of the "New Navalism"

The basic organizational structure and mode of operation that the Naval Academy maintained into the post–World War II period had been established by the time of the Sampson superintendency. But in the early years of the twentieth century, the character and goal orientation of the institution were especially shaped. These were heady years for American naval professionals, as the "new navalism" caught hold of the popular imagination. In the Spanish-American War the role of the Navy had been paramount. The United States emerged from the war not only as a major world power, but as one with far-flung colonial possessions. Hawaii had been annexed. The Philippines, Guam, and Puerto Rico were acquired through the defeat of the Spanish, with special privileges acquired in Cuba (including a 99-year lease of a naval base at Guantanamo). Mahan's dictum regarding the necessity for control of the seas became gospel, and the means for achieving such control were being attained.

Theodore Roosevelt, the key evangelist on behalf of American naval power, occupied the White House. Roosevelt, whose early interest in nautical affairs was reflected in his book *The Naval War of 1812*, was a long-time friend of Mahan and a devotee of Mahan's writings and theories. As Assistant Secretary of the Navy, Roosevelt had been instrumental in getting the Navy into a state of readiness for war with Spain; his secret orders to Commodore George Dewey anticipating the Battle of Manila Bay are the best-known example of his efforts. As President, Roosevelt continued to promote the naval cause, in rhetoric and in policy. The Navy League, founded in 1902 as an interest group designed to arouse continuing public support for the Navy, received Roosevelt's enthusiastic endorsement.[51] Under Roosevelt, American intrigue and the American Navy enabled Panamanian conspirators to achieve independence from Colombia. Roosevelt was able thereby to facilitate the movement of a two-ocean navy through construction of a canal (directed by George Washington Goethals, who had graduated from West Point in 1880) across the Panamanian isthmus. Most significantly, Roosevelt pushed a series of bills through Congress providing for naval expansion. Within four years, annual appropriations for the Navy had increased by nearly 50 percent, to record peacetime levels. Authorization was provided for ten battleships, four cruisers, and seventeen other vessels of various classes. Whereas the United

States had been negligible as a naval power as recently as 1890, by the end of Roosevelt's administration, America had one of the top three navies in the world. The world tour of the Great White Fleet from 1907 to 1909 was a colorful attempt by Roosevelt to demonstrate American capability to implement the Mahanian vision.

A turn-of-the-century expansion in the numbers of midshipmen at the Naval Academy was accompanied by a major building program, with architectural renovation and expansion directed by Ernest Flagg. New buildings, named after famous graduates—Mahan, Sampson, Maury—were symbolic of the romantic aura of the period, in which midshipmen were infused with the glories of past accomplishments of the American Navy and its future importance in world affairs. Admiral George Dewey came to Annapolis in 1904 to lay the cornerstone of the new chapel, which was opened in 1908. In 1913, a twenty-one ton sculptured sarcophagus, said to contain the body of John Paul Jones, was placed beneath the chancel of the chapel.[52]

Origins of the Coast Guard Academy

The heyday of popular enthusiasm and congressional support for the Navy at the turn of the century, which was so important in shaping the character of the Naval Academy, had come only after a period of neglect extending back to the end of the Civil War. Expansion of the Navy during the war had been followed by a sharp contraction. Attrition among midshipmen at Annapolis was high (often over 50 percent) during the post–Civil War decades, and those young men who remained until graduation faced bleak career prospects. Promotions, which had been rapid for many officers during the Civil War, became a rarity. Whereas George Dewey and Alfred Thayer Mahan had become lieutenant commanders within a few years of their graduation from Annapolis, the top twelve graduates of the class of 1868 remained in grade as lieutenants for twenty-one years. Naval officers were encouraged to resign, even after only a few years of service. Indeed, in 1882 Congress reduced the number of officers authorized for each rank, with the result that for several years only the top 25 percent of each graduating class from Annapolis received commissions as naval officers.[53]

It was in this context that the Secretary of the Navy, William E. Chandler, proposed that the Navy take over the functions of the Revenue Service (as the Coast Guard was known then). Doing so would provide additional billets for naval officers; moreover, the new school that had been created to train Revenue Service cadets could be eliminated because the Naval Academy alone would suffice. Chandler's suggestions evoked an angry reply from Ezra W. Clark, Chief of the Revenue-Marine Division of the Treasury Department, who described the Navy as "a floating mass of incompetency." Moreover, he noted caustically, the handful of officer appointments in the Revenue Service each year certainly would not provide much relief for the "nearly nine hundred clamorous idlers" among Annapolis graduates.[54]

The episode brought to a head long-smoldering frictions between the Navy and the Revenue-Marine. The Revenue-Marine retained its distinct identity. By and large, Chandler's proposals for a merger of Navy and Revenue-Marine were defeated, although a number of Naval Academy graduates *were* granted commissions in the Revenue-Marine. The need for the Revenue-Marine to argue for an existence separate from Navy control would recur. Moreover, for the School of Instruction of the Revenue-Marine (the forerunner of the Coast Guard Academy), the problem of establishing a distinctive program and *raison d'être* had hardly begun.

The Revenue-Marine had been given great initial impetus by the first Secretary of the Treasury, Alexander Hamilton, who was convinced that the new nation was in need not only of arms of service to fight on land or at sea, but also of a service to police the customs and to enforce navigation laws. Throughout the nineteenth century the Revenue Service was plagued by the absence of an effective merit system to govern the recruitment and promotion of officers. Under the prodding of President Grant's Secretary of the Treasury, George S. Boutwell, and Sumner I. Kimball, appointed to the newly created post of Chief of the Revenue-Marine Division, legislation was enacted to reform the Revenue-Marine. As part of this reform effort, in 1876 a school of instruction was created to provide a two-year period of instruction for Revenue-Marine cadets.[55]

Founding of the Coast Guard Academy

Although 1876 is the date still cited in Coast Guard publications as that of the founding of their academy, the fact is that for several decades thereafter, the School of Instruction of the Revenue-Marine was little more than a two-year apprenticeship supplemented by tutoring in technical subjects. Instruction was conducted primarily at sea, first aboard the topsail schooner *J. C. Dobbin*, later (from 1878) aboard the sailing vessel *Chase*. It was not until 1900 that permanent quarters ashore were established (at Arundel Cove, near Baltimore, until the move to Fort Trumbull, at New London, Connecticut, in 1910). In 1903, a third year of instruction was added for cadets who were to become line officers. The appointment of prospective engineering officers was authorized in 1906. Whereas previously they had been commissioned directly from civilian life, they were now required only to take a six-month program at the school. The total number of cadets at the turn of the century remained small—five to ten per class. Nevertheless, further development of the program came with congressional authorization in 1906 of the employment of two civilian instructors.

In 1915, the Revenue Service and the Life Saving Service were merged in an act of Congress proclaiming the creation of the Coast Guard. The School of Instruction, which in 1914 had become the Revenue Cutter Academy, thereby became the U.S. Coast Guard Academy. The goal of the Academy, the Board of Instruction advised in 1917, should be the development of a program "in every way comparable in its completeness of courses, instruction, and educational facilities with either West Point or Annapolis."[56] Congress encouraged the trend toward emulation of the other academies in 1926 when it authorized a four-year course of instruction at the Coast Guard Academy (implemented beginning in 1931). In 1929 funds were appropriated for building a completely new academy on a recently acquired site in New London. Because the line and engineering corps of the Coast Guard were now merged, the four-year program was required of all cadets. As at the other academies, it was a prescribed curriculum, with heavy emphasis on technical and engineering subjects.

Distinctive Subculture

Although the Coast Guard Academy modelled many of its practices after those of the older academies, it nevertheless developed its own distinctive character.[57] Part of this distinctiveness stemmed from minuscule size. Even on the eve of World War II it was still the intimate seminary-academy that West Point and Annapolis had been in the pre–Civil War period. There were just over two hundred cadets at New London in 1940, and thirty staff and faculty members.[58] One of the handful of civilians on the faculty, Chester E. Dimick, had been there since the first two regular civilian appointments to the Academy had been made in 1906. Life for cadets and staff and faculty alike continued to be governed far more by custom and informal mores than by formal rules and regulations.

The unique responsibilities of the Coast Guard also contributed to the distinctive character of its academy. Although the Coast Guard became a subsidiary arm of the Navy in time of war, most of the time in the pre–World War II period it operated under the Treasury Department, with responsibilities for the enforcement of maritime law (including efforts to prevent smuggling and rum-running), for the installation and maintenance of navigational aids such as lighthouses, for plotting the course of icebergs, and, perhaps more notably, for search and rescue operations at sea.

The organizational lore that evolved in the Coast Guard, therefore, was filled not with battles won or enemy vessels destroyed, but with daring rescue operations and with the skillful performance of duties under hazardous conditions of weather and sea. The heroes with whose exploits Coast Guard Academy cadets were inculcated were seldom those of internationally celebrated troop commanders leading their men into the "teeth" of entrenched enemy positions, as much-decorated Brigadier General Douglas MacArthur had done on several occasions with the famous Rainbow Division in World War I.[59] Nor was the model likely to be that of the ship captain in battle, outwitting adversaries and bringing fame and glory to his nation, like Commodore George Dewey at Manila Bay. Rather, the heroes whom Coast Guard cadets typically were asked to emulate were men like Lieutenant D. H. Jarvis, who went from Coast Guard cutter to deersled and dogsled to lead an expedition for 1500 miles across the frozen Arctic in 1897-98 to rescue marooned whalers near Barrow.[60] Or they were men like Commander Earl Rose, captain of the cutter *Tampa,* who led an operation to rescue 137 people from the burning passenger liner *Morro Castle,* off Asbury Park in September 1934.

Yet despite this distinctive emphasis in the Coast Guard and Coast Guard Academy subcultures, there was never any doubt that the school at New London, like those at West Point and Annapolis, was a *military* academy. Reveille, morning inspections, drill, and the general regimentation were constant reminders of this. And with the outbreak of World War II, the Spartan ethos became paramount.

3

World War II and the Postwar Transition

The years from the turn of the century to the eve of World War II were formative for the Coast Guard Academy. It is reasonable for our purposes to omit a detailed discussion of developments at West Point and Annapolis in these decades. Some changes were occurring during this period, to be sure. World War I led to an acceleration of programs. The interwar years were ones in which military training was revised in an effort to accommodate the lessons of the First World War. Instruction in the physical sciences and in engineering was updated, and there was a modest expansion of offerings in history, government, economics, and English. The hazing of plebes was curtailed, under pressure from Congress; moreover, West Point took the lead among the academies in codifying its plebe system and its honor code.

Still, especially as compared to the pace of change in civilian higher education generally, the evolution of West Point and Annapolis during the first three decades of the twentieth century clearly was one of cautious incrementalism rather than of any sharp departures from the organizational format that already had been developed. All three academies remained essentially military seminaries on the eve of World War II. The effects on the academies of World War II, however, were profound. It is essential to describe the impact of the war in some detail as a basis for describing and analyzing the dynamics of change at the academies in the decades since the war.

The Impact of War

By the mid-1930s, events had begun to move relentlessly toward the outbreak of another world war. The German remilitarization of the Rhineland in 1936, the formation of the Rome-Berlin Axis of that year, the Anti-Comintern Pact between Germany and Japan, Japanese abrogation at the end of 1936 of her commitment to the Washington and London naval arms limitation agreements, and the full-scale offensive of Japan against China in 1937 all foretold impending global conflagration. President Roosevelt followed his "quarantine the aggressors" speech of October 1937 with a plea to Congress in January 1938 for increased American defenses. The emphasis of the plea was on strengthening America's "two-ocean navy," no doubt reflecting Roosevelt's long-standing naval bias, his experience in the Wilson administration as Assistant Secretary of the Navy, and perhaps his recognition of the kind of appeal to which a gun-shy Congress might respond.[1] Congress, despite its continuing commitment to the Neutrality Acts of 1935 and 1937, appropriated funds for a 20 percent increase in the Navy budget. The size of the entering class at the Naval Academy in the summer of 1938 increased by nearly a third over that of the previous year (Table 1).

Table 1 Admissions to the Service Academies, 1937–1943

Academy	Year (most admissions occurred in July of the year)						
	1937	1938	1939	1940	1941	1942	1943
USMA[a]	424	352	373	563	446	863	856
USNA[b]	560	741	741	908	1035	1201	937
USCGA[c]	48	65	124	148	146	158	151

[a]Figures compiled from USMA Alumni Assn., *Register of Graduates* (1973).

[b]Figures compiled from USNA Alumni Assn., *Register of Alumni* (1971). Whereas the USMA *Register* identifies the dates of admission for all graduates and nongraduates, the USNA *Register* lists only the totals of graduates and nongraduates by graduating class. The USMA *Register* allows one to identify "turnbacks" (those who, because of academic failure, were "turned back" to graduate with a later class), whereas the USNA *Register* does not. Thus, the figures for the Naval Academy in this table may be slightly inaccurate.

[c]Figures reported in "U.S. Coast Guard Academy Cadets Graduated and Commissioned, 1926 Class to Present," mimeographed (New London, Conn.: USCGA, Office of Admissions, n.d. but includes data through 1973), 2 pp.

The scope of Coast Guard activity was increased in 1939 with the incorporation of the Lighthouse Service (formerly in the Department of Commerce) into the Coast Guard, and with the assignment of additional patrol duties associated with the President's national emergency proclamation of that year.[2] However, it was 1940 (fiscal year 1941) before a substantial increase in appropriations for the Coast Guard was made. Appropriations for FY 1941 increased by more than

40 percent, from $26.5 million to $37.3 million.[3] The number of admissions to the Coast Guard Academy nearly doubled from 1938 to 1939, and increased another 20 percent in 1940.

The ground forces in this period were small in number—eighteenth among armies of the world in 1938 (with less than 200,000 men on active duty), according to the U.S. Army Chief of Staff. The massive buildup of the Army did not begin until after passage of the Selective Service Act of September 1940, although expansion of the Army's air arm had begun earlier with Roosevelt's strong encouragement. Enrollments at the Military Academy, however, rose sharply only after Pearl Harbor.

Each service academy was profoundly affected by the war, although the effects were felt somewhat earlier at Annapolis and at New London than they were at West Point. As enrollments increased, the academies brought in additional staff and faculty members. The demand for professional military officers for combat assignments rapidly depleted their numbers at the academies. The result was that within a short time, the majority of the staffs and faculties at the academies was comprised of reserve officers and civilians. (The distinction between reserve officer and civilian was seldom great, since the former typically was a civilian who had been commissioned in the reserves when the war began.)

Just as accelerated programs had been instituted in all of the armed services to mass-produce reserve officers (the "ninety-day wonders," as graduates of the accelerated programs were known), so each service increased the production of regular officers by shortening the program at its academy from four years to three. At the Coast Guard Academy, summer cruises to foreign ports were discontinued. In the academic year, sharp cuts were made in so-called general subjects (principally English, history, and social studies). The three-year program reflected not merely the elimination or shortening of activities previously included, but also a shift in the direction of even greater emphasis than before on immediately utilizable skills, especially those applicable to combat. The Coast Guard had come under the jurisdiction of the Navy for the duration of the war as of August 16, 1941; consequently, Coast Guard cadets were given additional training in naval techniques and procedures. Submachine guns, pistols, and rifles were stored in cadet barracks, ready for use in possible emergencies, including "the unlikely event than an enemy submarine should penetrate the Thames River."[4]

Like the Coast Guard Academy, the Naval Academy discontinued summer cruises. However, by mid-1943 a new feature, flight indoctrination, had been added for all midshipmen in a ten-week course at Jacksonville, Florida. Most of the time in the summer that had been utilized for the cruise was now utilized for academic work; the result was that even with the shortening of the program by a year, less than 10 percent of the four-year curriculum was cut. The program, William Appleman Williams recalls, provided "a very useful education in life as well as some damn tough intellectual discipline. . . ."[5] The atmosphere was rarely collegiate, however; it was martial. A ban was placed on general visitation to the Academy. Armed Marine guards stood watch twenty-four hours a day to make sure no unauthorized persons entered or left "the yard." A tough program of physical training was instituted. All midshipmen were required frequently to run an obstacle course (similar to one in Marine Corps boot camp), to learn the skills of hand-to-hand combat, to jump from a 25-foot tower, and to dive into water through flaming oil or gas (simulating an emergency exit from a disabled ship).[6]

The shortening of the program at West Point from four years to three (beginning in the autumn of 1942) resulted in reductions especially in English, the social sciences, military history, and civil engineering. Those cadets who were taking flight training experienced even greater cuts in their academic work than did their ground-force peers. Furlough time also was cut, and military training was emphasized. Apart from the military calisthenics, including the "fit-to-fight" obstacle course, which were required of all cadets, the nature of training varied depending upon whether the cadet had chosen to enter the Air Corps or one of the ground combat arms upon graduation.

Roughly 60 percent of the first and second (senior and junior) classes sought aviation training when it was first offered in 1942, and met the physical qualifications.[7] The first contingent of air cadets went to various flying schools around the country for training. Subsequently, with the activation of Stewart Field north of West Point, cadets received both basic and advanced flight training there—more than 850 hours of it.[8]

Cadets to be commissioned in the ground forces received training in all types of infantry weapons, in grenade throwing, minelaying, demolitions, artillery, tanks, and military engineering. As first-classmen (seniors) they were given temporary assignments to stateside tactical units in the branch of service into which they were to be commissioned.

Because aviation cadets and those to be commissioned in the ground forces spent much of their time independently of one another, two fairly distinct cadet subcultures developed, despite the efforts of academy authorities to ensure a uniform identification with the corps. One such effort was made in the form of an order that required air cadets to rejoin their "earth-bound" peers in drill field and parade ground exercises. The resulting "picturesque screams of the Air Cadets were very amusing," cadet "Ducrot Pepys" reports.[9]

This wry cadet comment points up another facet of the wartime experience at the service academies that should be mentioned. The lives of cadets and midshipmen during the war did not go on "as usual"; rather, there was much greater emphasis than there had been in the prewar days on learning the "basic necessities" that would enable a junior officer to function effectively in combat. However, life *did* go on. Disrupted as the academy routines were in some respects, in other respects they continued largely unaltered. The wartime diary of "Ducrot Pepys," for instance, is largely a record of cadet concern with "the endless soirées,[10] the eternal inspections, the hoped-for rain [so that parades would be cancelled], the impartial, satanic justice of the Academic P's [professors]," as cadet Roger Hilsman noted.[11]

Likewise, although some extracurricular activities were curtailed, the feverish involvement of cadets and midshipmen in athletics continued largely unabated. Jack Dempsey came to referee a boxing match between the Coast Guard Academy and West Point in 1943 (the Coast Guard cadets won, 6-2). The Army-Navy football game continued to arouse emotions from cadets, midshipmen, and alumni, and from interested servicemen listening to the game by radio in various parts of the world. The corps of cadets could not attend the game when it was played in Annapolis, nor could the regiment of midshipmen attend when it was played at West Point. But spectator support for visitors as well as for home team was maintained by designating a portion of the regiment of midshipmen to cheer for Army when the game was in Annapolis, and a portion of the corps of cadets to cheer for Navy when the game was at West Point.[12] Navy won the game in 1942 and in 1943, but the Doc Blanchard–Glenn Davis combination that would lead the Army team to three undefeated seasons had made its appearance by 1944, and Coach Red Blaik's Army team won 23-7.

Assimilating the Lessons of War

Some critics of the academies argued that "too many academy graduates went on playing the Army-Navy game throughout life." If enthusiasts of the interacademy rivalry on the athletic field found therein the nurturing of group *esprit* and martial virtues, critics argued, on the contrary, "that the long-established traditions and customs of the . . . institutions created attitudes and encouraged behavior that obstructed full and effective interservice cooperation."[13] Military campaigns from Normandy to Okinawa had required that units from the Army, Navy, Marine Corps, Coast Guard, and Air Force (that is, Army Air Corps) work together. Often they had done so with marvelous effectiveness; but sufficient misunderstandings, frictions, and breakdowns in communication had occurred to alert postwar planners to the imperative requirement of improving, and possibly fundamentally altering, the institutional structures through which military policies were formulated and implemented. Unification of the services was the most radical of various proposals that were made; for a time it was one that enjoyed high-level support in each of the services. Within the context of consideration of unification, the question arose as to whether the service academies should be replaced, supplemented, or reformed.

As early as the winter of 1943-44, a board appointed by Secretary of the Navy Frank Knox and headed by Rear Admiral William Pye had considered alternatives such as having midshipmen attend civilian colleges for two years, or attend a joint Army-Navy-Air academy for two years, with the final two or three years of training provided at Annapolis. These alternatives were rejected by the board, however, which favored instead expansion of the physical plant at Annapolis to accommodate a four-year program there for an enlarged body of midshipmen. The latter proposal also won tentative approval in Congress in 1945 over a variety of bills that had been submitted calling for the establishment of additional naval academies in other parts of the country.[14] However, the unification issue and the subsidiary one of a common academy for all services were far from dead. Moreover, Knox's successor as Secretary of the Navy, James Forrestal, although a vigorous opponent of unification, wanted further study of the methods to be used in procuring and educating naval officers. Rear Admiral James Holloway, Jr., then heading the Pacific Fleet Training Command (and later to become Naval Academy Superintendent), was asked by Forrestal to undertake such a study, assisted by a board of officers. The Holloway Plan, as the report of the board became known, called for the maintenance of the program at Annapolis, but emphasized a federally subsidized program of Naval Reserve Officer Training at civilian colleges, some graduates of which could obtain commissions in the regular Navy rather than in the reserves. Congress enacted the plan into law by unanimous vote in 1946.[15]

An analogous board, headed by Karl Compton, president of the Massachusetts Institute of Technology, had been appointed in 1945 by the Secretary of War, but with the more limited mandate of reviewing the curriculum that had been devised at the Military Academy for the postwar transition. As Academy Superintendent Lieutenant General Maxwell Taylor accurately noted in retrospect, "the academic reorientation effected in that period was significant but far from revolutionary as it was sometimes described."[16] With but a few suggestions, which were accepted by Academy officials (as noted below), the Compton Board endorsed the West Point program.

A study of the needs of the Coast Guard Academy in the postwar period also had been undertaken by the Superintendent, Rear Admiral James Pine, in conjunction with a group of Coast Guard officers and the civilian advisory committee. Admiral Russell Waesche, Commandant of the Coast Guard, had asked for a general assessment of the procedures for procuring and educating Coast Guard officers in the postwar years. Noting that the problem of the Coast Guard was complicated by the uncertainty of its future missions (being then under Navy control, but to revert to the Treasury Department in the postwar period), the Pine Committee nonetheless endorsed a program that represented only a modest reform of the prewar Coast Guard Academy curriculum.[17]

Each academy had received the blessings of its parent service for the program to be implemented in the postwar period. However, the services themselves (notably the Army, Navy, and advocates of an independent air force) were becoming increasingly embroiled in the continuing unification controversy. The National Security Act of 1947, creating an independent Air Force linked to the Army and Navy loosely through an office of the Secretary of Defense with (then) severely limited authority, formally resolved the issue in favor of those (such as Forrestal and the Navy) who had opposed full unification. By 1949 Forrestal as the first Secretary of Defense had announced his support of the establishment of a new academy for training air cadets. However, as he told the Service Academy Board which he had appointed early that year,

> Although this legislation will, if enacted by the Congress, provide facilities for the training of a larger number of future officers, this decision will nevertheless leave unanswered many fundamental questions concerning the kind of basic education which career officers of the three Services should receive and the role of the three academies in providing it.[18]

The Service Academy Board was chaired by Robert Stearns, president of the University of Colorado, with General Dwight

Eisenhower, former Chief of Staff of the Army, who now was president of Columbia University, as vice-chairman. Assisted by high-ranking military officers with matters of military training, and panels of educators from each of several academic disciplines with questions of academic curriculum, the board prepared an 82-page report. The creation of a separate Air Force academy along the lines of West Point and Annapolis was supported. Endorsement of existing programs at the latter two institutions was qualified only by a series of recommendations (several of which are discussed below) designed primarily to reduce the isolation of the academies from the civilian academic community.[19]

Even during 1944 and 1945, the first months in which the postwar status of the academies was being considered by the parent services and by the Congress, academy officials could contemplate a return to the four-year peacetime programs relatively confident that the Military Academy, Naval Academy, and Coast Guard Academy would be retained intact. They could be equally confident, however, that there could be no simple reversion to the prewar routines. Changes that were made at the academies in the early postwar period reflected an effort to incorporate the lessons of World War II, to anticipate the demands of coming decades upon the military profession, and to undercut arguments from external critics for even more far-reaching changes.

Interservice Cooperation

Even before encouraged to do so by directives from the Secretary of Defense and recommendations of the Service Academy Board, the Superintendents of the Military Academy and the Naval Academy had initiated measures designed to increase contacts between members of the two institutions and to improve the understanding that cadets and midshipmen had of the sister services. Before assuming command of the Military Academy in the autumn of 1945, Major General Maxwell Taylor stopped by Annapolis to meet his counterpart, Vice Admiral Aubrey Fitch, who had just assumed the superintendency of the Naval Academy. A few weeks later, Taylor returned to Annapolis, accompanied by his Commandant of Cadets and a group of top-ranking cadet officers, to participate in the Naval Academy centennial celebration. Six weeks later, a group of senior midshipmen officers visited West Point. The following year, a regular program was instituted whereby first-classmen of each academy visited the opposite academy for several days in the spring. More intensive interaction between cadets and midshipmen was provided by a two-week program of joint amphibious training (known as CAMID, the acronym for Cadet-Midshipman), begun on an annual basis in the summer of 1946. Further effort to foster improved relations between the two academies was made by the assignment of a few instructors from the Naval Academy to teach temporarily at West Point in exchange for Military Academy instructors to be assigned at Annapolis. However, this plan was abandoned within a few years in favor of one in which a tactical officer from West Point each year would serve at Annapolis in exchange for a company officer from the Naval Academy.[20]

At the Coast Guard Academy, awareness of the activities at least of the Naval Academy had been increased as a result of the close association of Coast Guard officers and Navy officers during the war. However, with the return of the Coast Guard to Treasury control in the postwar period, the Coast Guard Academy was largely ignored by the other two academies as well as by the various boards that had been created to consider possible unification or modification of the academies.

Science and Military Technology

A second facet of the wartime experience that affected the postwar programs at the service academies was the military benefit to be derived from investment in science and technology. Radar, sonar, the proximity fuse, the jet turbine, and atomic weapons were among the important military instruments that had been forged in the feverish scientific activity occasioned by the war. In addition, developments from the First World War, such as the combat use of armored forces, aircraft, and submarines, and from even earlier wars, such as rocketry and mines, had been further refined and modified in World War II. These and related developments in military technology found their way into portions of the military training of cadets and midshipmen in the postwar years, as well as into academic coursework in the physical sciences, military ordnance, and engineering.

The service academies kept ahead of available textbook coverage of several of these developments. For example, Jerry B. Hoag, who taught physics and mathematics at the Coast Guard Academy, previously had been at the University of Chicago, associated with the development of the nuclear reactor. Upon moving to New London, he had set up an accelerator in the basement of his home. The morning after the atomic bomb was dropped on Hiroshima, Hoag assembled a group of Coast Guard Academy cadets, drew a sketch of how an atomic bomb is constructed, and explained how it works. Within less than a year, the McMahon bill on atomic energy had passed, and information about atomic weapons of the sort Coast Guard Academy cadets had been acquiring in the classroom had become highly classified and thereby banned from all classrooms.[21]

At about the same time, the Military Academy introduced a subcourse (within physics) in nuclear physics. The Military Academy also increased the number of classroom hours in electricity by nearly 50 percent, with much of the additional time to be spent by cadets in an electronics laboratory or in the related radar laboratory (the latter included television equipment). The availability beginning in 1945 of a well-equipped electronics laboratory and radar laboratory enabled academy officials to state "without fear of contradiction that the Academy has the best facilities of any institution in the country for practical instruction in elementary electronics."[22]

In a more applied vein, Annapolis became the first academy to create a separate department of aeronautics. Aviators had made a sudden rise to prominence in the Navy. Secretary of the Navy Forrestal was a former naval aviator himself; with his urging, the emphasis on aviation was being felt at all levels. In the summer of 1945, Vice Admiral Aubrey Fitch, who had been Deputy Chief of Naval Operations for Air, was personally selected by Forrestal to become the first aviator in Naval Academy history to serve as its Superintendent. Captain Stuart Ingersoll, selected a few months earlier as Commandant of Midshipmen, also was an aviator and, like Fitch, had gained distinction through his service with fleet aviation during the war. Upon assuming command at Annapolis, Fitch announced that "After this year every man who graduates from the Naval Academy either is going to become an aviator or is going to have a lot of knowledge and a great respect for air power."[26] Ingersoll, working closely with the midshipmen, conveyed the same attitude. He also contributed to Forrestal's efforts to further the cause of carrier aviation forces through frequent trips to Washington to confer with the Navy Secretary. When Ingersoll left Annapolis in the summer of 1947 for Hawaii to assume command of a fleet air wing, he was replaced by Captain Frank Ward, another aviator. When Ward in turn left two years later, still another aviator, Captain Robert Burns Pirie, was named as his successor. Indeed, so closely associated had the appointment become with naval aviation, that of fifteen Commandants of Midshipmen at Annapolis in the period from 1945 to 1977, twelve were naval aviators.

The appointment of Major General Maxwell Taylor as Superintendent at West Point at the end of the war also was symptomatic of changing emphases in military technology and shifts that were occurring in the weight attached to various skills or credentials in the selection of officers for key posts. Taylor had gained a reputation in military circles as the epitome of the soldier-scholar, because he combined military prowess with extensive linguistic competence and broad experience in foreign cultures. He had taught French and Spanish for five years at the Military Academy in the 1920s, followed in the 1930s by a four-year tour where his principal assignment was to live in Japan and learn the language. However, additional assignments during the latter tour included one in China as Colonel "Vinegar Joe" Stilwell's assistant and Japanese expert. Ironically, the Army had transferred Taylor back to the United States shortly after Pearl Harbor to prepare for action in Europe, despite Stilwell's and Taylor's mutual desire to have Taylor remain in Asia, and despite the fact that he had become one of the few officers in the Army who were fluent in Japanese.[24] Fortuitously for his career, his successes in Europe as a combat leader of the newest striking element of the Army—the airborne forces—gained him even greater prominence. Taylor came back to West Point in September 1945 as Superintendent, well known for campaigns in which he had participated from Normandy to the Battle of the Bulge as commander of the 101st Airborne Division. The assistant commander to that division, Brigadier General Gerald Higgins, joined Taylor at West Point in early 1946 as Commandant of Cadets.

If the emphasis on airborne warfare symbolized the beginning of a new era in the Army, the replacement of the horse cavalry with mechanized infantry and armored units symbolized the end of another era. At West Point, the end of the cavalry era was acknowledged in 1947 by the termination of horsemanship, a long-standing element of training that plebes especially had experienced. This meant also the end of polo, and the annual horse show, which had served as a link with the landed gentry of the Hudson Valley. A great horse auction was held on the Cavalry Plain at West Point in 1947, leaving only the Army mules, which would continue to serve as mascots for football games, as imperfect links to the romance of the equestrian past.[25]

Modern Techniques of Leadership and Management

World War II represented not only the marshalling of the physical sciences on behalf of warfare, but also the first extensive application (far beyond the primitive beginnings of World War I) of the behavioral sciences to military requirements. Psychological warfare is the most popularly remembered example of such application. Probably more basic to military needs, however, was the application of behavioral knowledge and insights to tasks such as (1) the selection of individuals for military service; (2) the identification of persons with requisite skills or aptitudes for particular assignments; (3) the testing of service personnel for their ability to perform adequately under various levels of stress; (4) the design of equipment, vehicles, aircraft, and weapons that would best enable service personnel to operate and maintain them; (5) the improvement of training environments; (6) the preparation of individuals for leadership roles with greater awareness of the factors that motivate or demotivate subordinates; (7) the development of techniques appropriate to the management of large complex organizations.

Among the service academies, West Point took the most significant steps in the early postwar period to update approaches to leadership training and to incorporate modern behavioral knowledge into training practices. During the war, the Military Academy, aided by Army psychologists, had experimented with the testing of candidates for mathematics and language aptitude—tests similar to ones that would become standard nationwide. A series of peer ratings, supplemented by ratings of cadets by those in classes senior to them in their company and by the company tactical officers, also was introduced, beginning with the Class of 1944, in an effort to predict aptitude for military service and leadership.[26] Under the impetus of Army Chief of Staff Eisenhower's expressed concern for the quality of training in leadership and personnel management at the Academy, and for having some practical or applied psychology included in the curriculum, the Office of Military Psychology and Leadership was created in 1946, and a course in applied psychology was introduced into the curriculum beginning with the 1946-47 academic year. The new office was headed by an Army lieutenant colonel, but had as its associate director a civilian psychologist with a Ph.D. and twenty-five years of experience in the field.

The new office would continue to suffer, even into the 1970s, from its ambivalent status, ostensibly an academic department but operationally under the control of the Commandant of Cadets (as the officer with primary responsibility for leadership training). Moreover, abstract theories of human behavior were all very well; but for officers who themselves had had leadership experience in combat (as was true of all officers assigned to the Commandant's staff in the early postwar period and of most officers on the academic faculty), the natural tendency was to continue to rely on techniques that had seemed to work for them, and to impart to cadets a strong preference for knowledge acquired through experience rather than that imparted in the theories of psychologists. Furthermore, if knowledge of military leadership was to be acquired in part in the classroom, there were many who thought that it was already provided in appropriate form through the study of "the great captains" of military history, in a course that was taught in the Department of Military Art and Engineering.[27]

Factors such as these inhibited the new office from fulfilling its potential for greatly affecting aspects of the cadet socialization process, such as the plebe system. Modest reforms were made, however. A system was introduced whereby each plebe was assigned to an officer-sponsor whose role was entirely that of making the plebe feel welcomed into the Academy community (for example, through inviting him into the officer's home for meals on weekends), rather than that of serving in a disciplinary capacity.[28]

The Naval Academy also introduced classroom instruction in applied psychology in leadership training during this period, although no separate department of psychology or leadership was created. At the Coast Guard Academy, the entire classroom instruction in leadership consisted of four hours of lectures. Despite prevailing skepticism regarding the usefulness of behavioral theory, the Coast Guard Academy did hire a psychologist in 1947, who in time would play an important role in systematizing academy selection and admission policies, and in helping academy officials to deal with the chronic problem of attrition.[29]

Social Sciences and Humanities

The increasing interest in the behavioral sciences that was displayed during World War II by each of the services, in varying degrees, was closely related to a growing interest in the social sciences and humanities (especially the former). The war had posed a number of challenging social, political, and economic problems for military professionals: working together with persons from other nations in allied operations, coping with problems of supply and industrial mobilization, administering occupied territories. Many military professionals came to recognize that their military education and training had provided skimpy preparation, at best, for coping successfully with such problems. The impending responsibilities of the United States as a superpower, including the postwar occupation duties of military personnel, augmented the argument advanced by a number of military professionals to the effect that subjects such as economics, comparative government, international relations, and foreign languages should be regarded as staples rather than frills in a program of military education.

Of the three service academies, the Military Academy had devoted the most time to the social sciences in the prewar period (with the prodding and guidance of Colonel Herman Beukema); West Point also moved more vigorously than the other two academies in the early postwar period to expand its offerings in the social sciences. In many respects those associated with the social sciences department at West Point, and perhaps to a lesser extent its English department, experienced a period of creative ferment in the early postwar years akin to that felt in the State Department under George Marshall. Beukema's position was strengthened by the assignment in 1947 of George A. Lincoln as a permanent professor and deputy department head. Lincoln, who accepted a reversion to the rank of colonel from brigadier general to accept the position, had served under Marshall in the War Department as Chief of the Strategy and Policy Group of the Operations Division during the war. Then in the early postwar period, when Marshall took over at State, Lincoln served as the War Department representative on the important State-War-Navy Coordinating Committee (SWNCC) that was responsible for coordinating the economic, political, and military policies and actions of the three departments in regard to countries that had experienced enemy occupation—most notably wartorn Europe. Thus Lincoln brought with him not only a wealth of expertise at top policy levels, but also useful continuing links to Washington.

A change of the name of the department was effected at Beukema's initiative the same year of the Lincoln appointment, from "Economics, Government, and History" to "Social Sciences." Although the change did not alter the scope of departmental responsibilities, it was symbolic of high-level acknowledgement of the need for the Academy to stay abreast of modern thinking in the field. At a more substantive level, in the spring of 1948 the social sciences department canvassed third-class cadets to ascertain what level of interest existed in a course on Russian history for those who could validate their previous study of European history and thereby become exempted from that required course. Finding an enthusiastic response to the idea and a sizable number (54) of cadets who could validate the European history course, the department introduced a special section of Russian history, and another in Latin American history.[30] In 1949, largely in response to proposals from Beukema and Lincoln, the Academic Board gave approval to a program of "national security studies." Offerings in social science courses such as "the economics of national security" and "international relations" were at the core of the program. However, approval by the Academic Board doubtless had been facilitated by a description suggesting that the "great captains" course offered in the Department of Military Art and Engineering, the practical military training in the Department of Tactics, the weapons training in the Department of Ordnance, and instruction in virtually every other department from mathematics to hygiene had some contribution to make to "national security." Furthermore, the proposal had the merit of putting the Military Academy on record as promoting "a spirit of interservice teamwork," with attention in the program to such details as

> the precise use of the terms "armed forces" and "Department of Defense" in speaking of broad military matters rather than narrowing the cadet's view to the "Army"; the use of examples drawn from Navy and Air as well as Army experiences in all courses of instruction; and a general climate of service understanding and teamwork in the instructional staff.[31]

Finally, a proposal for a program in "national security studies" had the appeal of being "tough-minded." This was no abstract exegesis on the importance of a liberal arts education, but rather an indication of the knowledge that the American combat leader needed in the complexities of the modern world.

Emphasizing the same national security theme that had gained wide national appeal, the Department of Social Sciences obtained support from both the Academy and the Carnegie Corporation in 1949 to begin an annual student conference on United States affairs (SCUSA). The focus of the conference in 1949 was on U.S. policies in Europe, in 1950 on U.S. policies in the Far East. Each of the other service academies eventually would follow West Point's lead in launching such a conference, bringing student participants from the academies and from civilian colleges into contact with high-ranking policy makers and social science experts. To those military traditionalists who may have been apprehensive about possible adverse effects of the intensive contacts with civilians, the Department of Social Sciences could point to benefits that had been derived from a public relations standpoint. Typical of comments sent to the Academy from civilian participants in the 1950 conference, for instance, were one providing the assurance "that any unfavorable or misguided opinions that we may have had on the Academy were quickly dispelled by the frank sincerity and considerateness of the Cadets," and another indicating that "one of the most valuable things about SCUSA II . . . was the breakdown of the attitude of the civilian conferees toward the so-called 'military mind.'"[32]

The humanities at West Point still suffered from cuts in their portion of the curriculum that had been made during the war. However, Russian was added to Spanish, French, Portuguese, and German as a foreign language option for cadets in the 1945-46 academic year. Moreover, the English department, headed by George R. Stephens, who had been a civilian faculty member at the Naval Academy, and Russell K. Alspach, who joined the faculty with a Ph.D. in English from Pennsylvania, imparted to the cadets a humanistic component that had been largely absent in the previous emphasis on composition and logic rather than literature.[33]

Mathematics retained its predominant role in the Military Academy curriculum, with cadets required to spend far more time in the mathematics classroom than in English, in foreign languages, or in any of the social sciences. Even though the portion of the curriculum devoted to scientific and engineering subjects at the Military Academy was identical to that at the Naval Academy (48 percent of classroom hours), West Point had found more time for the social sciences and humanities (38 percent of the total) than had Annapolis (26 percent).[34]

Although figures from the Coast Guard Academy at the time are imprecise regarding semester hours per subject, it appears that even less time was devoted there to the social sciences and humanities than was true either at West Point or Annapolis, and proportionately more to scientific and engineering subjects.[35] A special board of consultants that had assisted the Coast Guard Academy in designing their postwar curriculum had suggested that there be "adequate" provision for training in the social sciences and humanities; but the emphasis should be on mathematics, the natural sciences, engineering, and professional nautical subjects. As Superintendent Wilfred N. Derby observed in 1948, the goal of the Academy was

> to provide a sound *engineering education* with adequate attention to fundamental mathematical, scientific and engineering knowledge to enable graduate[s] to continue post graduate work in advanced studies at leading universities. General studies are included in the curriculum to the extent generally approved by civilian engineering schools.[36] (Emphasis added.)

At Annapolis, the lack of emphasis on the social sciences and humanities was similarly defended with references to civilian engineering schools. The "bull" subjects (as such subjects were known among midshipmen) were accorded even more emphasis at Annapolis than at MIT in 1948, according to Naval Academy calculations.[37] Such comparisons are difficult to assess, given differences in course titles at different institutions and differing bases for the computation of semester hours. However they may have fared in comparison with civilian engineering schools, it was clear that the Coast Guard Academy and the Naval Academy were largely carrying over into the early postwar period the "trade school" emphasis on teaching practical, applied skills that had characterized instruction in the prewar period. Despite some lingering tendencies in this direction at West Point, officials there in the early postwar years defined their goal as that of laying an educational foundation upon which an officer could build throughout his entire career; they were preparing future generals, not second lieutenants. In contrast, at the Naval Academy explicitly and at the Coast Guard Academy implicitly, the mission was that of producing immediately utilizable junior officers.

Faculty Recruitment and Qualifications

Curricular innovations such as the advanced courses in Russian history and in Latin American history that Beukema and Lincoln had introduced at West Point, or the honors courses in English that Stephens and Alspach had introduced, were possible in part because there were more faculty members available than in the prewar years with postgraduate training, and therefore specialized knowledge. This was true at all three academies, in varying degrees.

It became West Point policy to send all officers who were to be assigned as faculty members first to a civilian university for graduate study (usually to the master's degree). This resulted in an upgrading in faculty qualifications in most departments, although some of them lagged. In chemistry, for example (within a Department of Physics and Chemistry), the Panel on Instruction in Science and Engineering of the Service Academy Board found that as late as the 1949-50 academic year "there is not a single scientist with advanced training in the professional field of the department."[38] Because assignment to the faculty at West Point was generally regarded among Army officers as a stepping-stone to career advancement, selection typically could be made among a large number of highly promising candidates for a position (whatever lack of graduate training they might have at the time of the application). The science and engineering panel of the Service Academy Board noted, however, that although "faculty members are chosen with great care," the "extremely high ratio" of West Point graduates to non-academy graduates (9 to 1) "carries with it the danger of perpetuating methods of teaching and subject matter instead of encouraging needed changes."[39] The Academy responded cautiously. Whereas there had been thirty non-academy graduates on the faculty in the 1948-49 academic year, there were thirty-nine in 1949-50, with plans for adding an additional five the following year.[40]

At Annapolis, a policy was instituted shortly after the end of the war of hiring only those civilian applicants who possessed a master's degree, and of making the Ph.D. a prerequisite to promotion beyond associate professor. However, as of 1949 there were 220 naval and marine officers on the faculty and 180 civilians.[41] The typical academic qualification of the officer-faculty member continued to be the bachelor's degree. The number of non-academy graduates among officers assigned to Annapolis tended to be considerably higher than at West Point; but this was true primarily because, in contrast to the situation in the Army, assignment to the Naval Academy was regarded as detrimental rather than beneficial to career advancement (except for the few key positions). One indication of the low priority accorded by the Navy Bureau of Personnel was the shortness of a tour of duty at the Academy—typically two years, in contrast to the three- or four-year tour at West Point.

The Coast Guard Academy, which had an extremely small nucleus of permanent faculty (a handful of civilians and a slightly greater number of officers), had been able to be reasonably selective in hiring persons with at least the master's degree in the early postwar years. In 1950, the civilian faculty was enlarged to between seven and ten, and the number of permanent officer billets was increased to about a dozen.[42] Several of the permanent faculty, both civilians and officers, had come to the Academy as reservists during the war. In many cases, their acceptance by the Coast Guard "regulars" and even by cadets was tentative. As Stanford Dornbusch has noted, there were a variety of subtle means at the Coast Guard Academy of indicating to the reservist "that although the interlopers may have the same rank, they do not have equal status."[43]

At West Point, the appearance of full integration of reservists or civilians—such as George Stephens, Russell Aspach, Ben Gault, and Wilfred Burton (all in the English Department)—with the predominantly Academy-graduate faculty could be achieved by putting civilians in uniform and giving them field-grade commissions (thereby also increasing the incentives for them to conform to the morés of military behavior rather than to those of the civilian academician). However, the mere absence of the West Point ring on the finger was clue enough to cadets as to who the "outsiders" were; their heavy concentration in English branded that department as marginal to the "real mission" of the Academy. (In order to avoid that stigma of being regarded as a "soft" discipline, Colonels Beukema and Lincoln made a concentrated effort to find highly decorated combat veterans for the social sciences department.)

At Annapolis, the frictions between reserve officers and regulars were far less significant than those between officers (regular or reserve) and civilians. Symbolic of the lack of full integration of civilians into the Naval Academy community was their exclusion, until the 1950s, from commencement exercises.[44] Civilian faculty members were becoming increasingly active in professional meetings, such as those of the American Mathematical Society or the Modern Language Association. They were active, too, in educational symposia at the Naval Academy, where eminent scholars from other institutions were invited to deliver papers and join in discussions on a variety of scholarly topics.[45] However, these sessions served to widen the gap between the academically oriented faculty and the naval officers, many of whom were impatient with their assignments and anxious to return to the fleet. The complaints that civilian faculty members heard from their better midshipman students about the teaching of some of the officers accentuated the feeling that the Athenian component of the academy goals was being sacrificed to Spartan commitments.[46] Many civilian faculty members also resented the almost total dominance of military officers in policy discussions. In 1949, a gesture toward including civilians in matters of academic policy was made through creation of an academic council, in which senior civilian professors were represented along with military officer representatives.[47] However, the policy continued of having all departments headed by a naval officer—often one with no specialized training at all in the discipline of the department.

The perpetuation of traditional prerogatives and status even in the face of an acknowledged need to innovate and modernize the program of academics and military training is illustrative of organizational internal contradictions which would plague the service academies throughout the postwar years. Within each academy, a consensus had developed that the education of prospective officers for the complex demands of the postwar era required a heightened emphasis on the Athenian component of the academy mission. Yet this was a consensus at the intellectual level. Emotionally, the Spartan component of the mission retained its primacy.

PART II

The Change Process

Case Studies in the
Evolution of the Service Academies

4

An Old/New Model: The Air Force Academy in Its Formative Years

The founding and early evolution of the Air Force Academy provide a case study of a rare opportunity for organizational innovation that was only partly exploited. However, both the energetic departures from the seminary-academy model that were made and the anxious determination in other respects to emulate the older academies are instructive. On the one hand, a distinctive reputation for the new academy could be established only through novelty and experimentation. On the other hand, to follow traditional norms and practices was to maintain a sense of security and, because such forms and practices seemed initially to have been accepted by a majority of the members of Congress as well as by the Air Force Chief of Staff, to ensure a measure of legitimacy.

The initial bias of the founders of the Air Force Academy, the vast majority of whom were West Point graduates, was toward replicating the Military Academy with some aeronautical accoutrements. However, severe early problems, including the threat that accreditation would not be granted in time for the first graduation, led to a major reshuffling of personnel and a revision of programs.

The man who emerged as Academic Dean following this first organizational crisis, Robert McDermott, became the driving force for reforms that broke with the seminary-academy tradition in important respects. The totally prescribed curriculum was abandoned, academic majors were instituted, and cadets were given credit for college-level course work taken elsewhere. As the fruits of these reforms began to be realized, the Dean pushed for an even more ambitious plan, whereby the Air Force Academy would be authorized to award master's degrees to selected cadets who could acquire the requisite credits within four years. This proposal put the Air Force on a collision course with the other services; after intense debate, only a compromise was approved, whereby selected academy graduates could go immediately to civilian graduate schools for the degree.

Internally, rapid change had been accompanied by mounting tensions and disputes over both goals and methods. The recurrent theme of this book is illustrated by the struggle between McDermott and a succession of commandants of cadets, although internal frictions were not exclusively along Athenian-Spartan lines. However, the revelation of a major cheating scandal in 1965, followed by another one in 1967 just as Academy officials were complimenting themselves on having taken the necessary corrective measures, came as a devastating blow to all the Academy leadership. The dream of developing an institution geared to meet the challenges of the aerospace age had been realized to a significant extent in a remarkably short time. But Academy officials had to ask themselves if the gains were worth the cost if they came at the expense of fundamental values such as honor.

The Founding

Although proposals by aviation enthusiasts for a separate air force academy can be traced all the way back to the months immediately following the World War I armistice,[1] the necessary precondition for getting executive and congressional support for such an idea came in 1947 with the establishment of an independent Department of Air Force. Less than six weeks later, a bill calling for the establishment of an Air Force Academy at Randolph Field, Texas, was introduced by U.S. Representative Paul Kilday of that state. Kilday's bill, along with numerous others that followed, remained "alive" in the Congress for the next several years. However, not until proposals had been "staffed" through the executive branch was the necessary support on behalf of a single proposal forthcoming. The major stimulus to such support was provided by the report of the Service Academy Board that Secretary of Defense James Forrestal had created in 1949.

The board reported in early April 1949 that "an Air Force Academy should be established without delay. . . ."[2] A week before the report, Forrestal had been succeeded as Secretary of Defense by Louis Johnson, a Washington lawyer who had chaired the Democratic Party's finance committee in 1948.[3] Johnson's former role as director of the company that manufactured B-36 bombers led Air Force protagonists to hope and many Navy and Army advocates to fear that he would support all Air Force programs to the neglect of those of the other services. However, in the case of a proposed Air Force Academy, Johnson disagreed with the board's preliminary recommendations. He took the position that there should be a third Department of Defense academy (the Coast Guard Academy was not included in these deliberations because it was under Treasury control); but the new academy should be one where cadets and midshipmen of all three services would receive part of their training.[4]

In their final report, however, the members of the Service Academy Board explicitly took issue with the alternative that Johnson had advanced, and instead reiterated their recommendation for an academy whose sole purpose would be that of producing qualified officers for the Air Force.[5] The board bolstered its position politically by routing the final report to Secretary Johnson through the three service secretaries, each of whom endorsed the board's recommendations.[6] Thus faced with a united front, Johnson finally gave the report his endorsement in the summer of 1950. By then, however, the attention of Congress as well as of the executive branch was consumed by the Korean War, which had begun in June. It was not until the war was over and General Eisenhower, who had co-chaired the service academy study, had become President that the momentum on behalf of creation of a separate air academy was resumed. Even then, the politics of site selection delayed final approval. However, in March 1954 a bill authorizing creation of an Air Force Academy at a site to be determined by the Secretary of the Air Force, Harold C. Talbott, was passed; the following month it was signed into law by President Eisenhower. In June, Talbott announced that he had selected a site near Colorado Springs from among three possibilities suggested by his Site Selection Commission (the other two being Alton, Illinois, and Lake Geneva, Wisconsin).

The West Point Model

The Air Force had begun tentative planning for an academy in 1948. After a brief study, a board headed by Air Force Vice Chief of Staff General Muir Fairchild recommended that when the new academy was created, it serve to provide the final three years of a five-year program, the first two having been provided by civilian colleges and universities. The Air Force Chief of Staff, General Hoyt Vandenberg, rejected this recommendation, and created a new planning board at the Air University at Maxwell Air Force Base, Alabama, that was to "plan a four-year course of instruction along the lines of the other service academies and without provision for pilot training."[7] Nearly seven years would elapse before the Air Force Academy would actually begin operations, at temporary quarters at Lowry Air Force Base in Denver. The design that emerged from the long years of planning is remarkable, however, not for its novelty or inventiveness but, on the contrary, for its close resemblance to the traditional pattern of training and education at the older academies—especially that at West Point.

To be sure, the program at the new academy was distinctive in its orientation toward the production of what Lieutenant General Hubert R. Harmon, first Superintendent of the Academy, called "air-faring men." Air Force Academy graduates would be "air-minded and thoroughly indoctrinated in all aspects of air operations," Harmon had assured members of the House Armed Services Committee prior to final enactment of the legislation establishing the Academy.[8] Even the emphasis on aviation was less than some critics in Congress and in the press had hoped, however.

With only a fifty-hour dose of aerial navigation training added to placate critics, the program at the Air Force Academy at the outset bore a remarkable similarity to the one at the Military Academy. The system of discipline was patterned after that at West Point. So also was the plebe system, the program of varsity and intramural athletics, and the leadership training that was provided to cadets. There was even an effort to create instant tradition by adapting West Point customs and lore to Air Force Academy use. The first cadet handbook of the Air Force Academy, *Contrails*, written by officers of the Commandant's staff, reflected this effort.[9]

The West Point honor system had received widespread unfavorable publicity only three years before the founding of the Air Force Academy, in a cheating incident which resulted in the dismissal of ninety cadets. Nevertheless, Superintendent Harmon was determined to encourage his officers to "drop a hint here and there about West Point and its honor system" to Air Force cadets shortly after their arrival.[10] There was some thought given by Academy officials to letting the cadets vote on the honor system to be adopted. However, according to official Academy records, "it became apparent quite early during the discussions of this subject that there was a strong unanimity among the cadets for its acceptance. Further, it was believed that the Honor Code was of such importance to the success of the Academy mission that it should be established whether or not there was complete acceptance."[11]

Before becoming Superintendent, Harmon had said that he "would resist every effort to overload the honor system, or to make it a means of enabling the authorities to apprehend pranksters, or of easing their administrative or police responsibilities."[12] However, as the official Air Force Academy history points out, "on a number of these points General Harmon had to compromise after he became the Academy's first Superintendent." The "all right" was used in lieu of direct inspection of rooms at evening bedcheck. A cadet's word of honor was required to ensure that he did not violate the prohibition against drinking, gambling, or narcotics. With these and other practices modelled after the West Point system, the Air Force Academy honor system had become entwined, as it was at West Point, in the system for maintaining order and discipline.

The academic curriculum at the new Academy differed in content from that at West Point and at the other two academies, the major difference being that Air Force cadets were given somewhat greater exposure to the social sciences and humanities than was provided to their counterparts at the older academies. However, like the older academies, the Air Force Academy had a curriculum that was almost totally prescribed. (Cadets were allowed to choose between taking a foreign language and studying aircraft design.) Moreover, the long shadow of the West Point Thayer system was evident in the classroom format utilized at the Air Force Academy. Faculty members were informed that cadets were to be graded on at least half of classroom attendances. In mathematics, the core academic discipline from the seminary-academy tradition, cadets were tested almost daily.[13]

The Founding Fathers

Emulation at the Air Force Academy of West Point practices and traditions becomes understandable when one recognizes that the vast majority of those who held key positions at the new Academy at its founding were graduates of the Military Academy.[14] The West Point dominance prevailed despite the fact that approximately 90 percent of Air Force officers had commissions from other sources.[15]

The Superintendent

Hubert R. Harmon was the son of Millard Fillmore Harmon, who had graduated from West Point in 1880. Two of Hubert's older brothers also attended the Military Academy, Kenneth graduating in 1910 and Millard, Jr., in 1912. Hubert graduated in 1915, attended flying school the following year, and in 1920 had transferred from the Coast Artillery to the Air Service. His service in the interwar period had been marked by assignments that stood him in good stead in career terms, including two tours as White House aide, a brief stint back at the Military Academy as an instructor, and attendance at the Army War College, from which he graduated in 1938. He had attained his first star before Pearl Harbor, the second early in the war, and in

1948 was promoted to Lieutenant General. In December 1949, two days before the Service Academy Board submitted its final report, Air Force Chief of Staff General Hoyt Vandenberg (USMA 1923) named Harmon his special assistant for Air Force Academy matters.

In 1953, Harmon retired—twice—recalled to active duty in each instance by his West Point classmate, now President, Dwight D. Eisenhower. To no one's great surprise, Harmon was selected to be the first Superintendent of the Air Force Academy, assuming his duties in Denver in August 1954. He selected as his chief of staff Colonel Robert R. Gideon, who had graduated from West Point in 1939. As a member of his faculty who was present during the first years of the Air Force Academy recalls of Harmon's goals for the institution, "He just wanted it to be another West Point, and if we had copied West Point exactly, he would have been satisfied and very pleased."[16]

The First Commandant of Cadets and His Staff

The first Commandant of Cadets at the Air Force Academy was Robert Stillman, who had been a noted heavyweight boxer, football player, and All-American lacrosse player as a West Point cadet. After his graduation in 1935, Stillman had returned to the Military Academy each fall until the war to serve as a member of the football coaching staff. Stillman's deputy commandant, Benjamin B. Cassiday, had graduated from West Point in June 1943. Serving under Stillman and Cassiday in a capacity analogous to that of the tactical officer at the Military Academy were air training officers (ATOs), who were to live in barracks with the cadets, maintaining discipline and assisting with military training. All the ATOs were young bachelor officers. Although the plan was to have a spread, with approximately 25 percent West Point graduates, 25 percent Annapolis graduates, and 50 percent of the ATO's selected from other sources, the West Pointers became the dominant element.

The athletic program, in which Stillman had great interest, also was under the direction of a West Pointer. Lieutenant Colonel Robert W. Whitlow had earned his football letter under Coach Red Blaik at Army in the early 1940s, and brought a Blaik-like zeal for athletics to his duties at the Air Force Academy.

The Academic Faculty

Brigadier General Don Z. Zimmerman, who was selected to become the first Dean of Faculty at the Air Force Academy, had had the extraordinary experience of starting as a plebe at West Point in 1925 after having received both a bachelor's and a master's degree from the University of Oregon. After graduating from the Military Academy in 1929, he attended flight school, and received his wings in 1930. He was back in graduate school briefly in the mid-1930s, to acquire training in meteorology at the California Institute of Technology. This training provided the foundation for his authorship of a weather manual which became used in Air Corps flying schools. He served for a semester as an instructor at West Point on the eve of World War II, and then was assigned for a few months of additional schooling at Cal Tech. This was his last academic experience, however, before becoming Dean of the Air Force Academy in 1954.

The Assistant Dean, Colonel Arthur E. Boudreau, was one of the rare members of the Academy staff in a relatively significant administrative position who was not a West Point graduate. Boudreau was a reserve officer, who had been recalled to active duty in 1948 from a position as Dean of the Inter-American College at Coral Gables, Florida. Boudreau had worked closely with General Harmon as project officer and director of the Curriculum Group of the Air Force Academy Planning Board for several years before joining the Academy staff.

Boudreau had pushed hard during the planning stages for a faculty that was a mix of civilians and military officers. At first, both General Harmon and Colonel Zimmerman accepted the idea.[17] However, after an impassioned counterargument by a young colonel, Robert McDermott, Harmon became persuaded to follow the West Point rather than the Annapolis practice in the selection of faculty. Faculty members would be exclusively military officers, selected according to the criteria that (1) all of those selected would be officers with "superior" to "outstanding" effectiveness reports, (2) at least one-fourth of them would be service academy graduates, and (3) an effort would be made to recruit officers with outstanding war records or those who in other ways had become famous.[18] Although a strict adherence to the second of these criteria would have resulted in a faculty composed largely of officers commissioned from sources other than the service academies, in fact the key faculty slots initially were filled primarily with officers who had graduated from West Point or had served there, or both. Seven departments had been established by the time the first cadets entered in the Air Force Academy in 1955. Four of these were headed by West Point graduates, and two of the remaining three by officers who had taught at West Point. In addition to the department heads, nearly all officers who received appointments as professors were graduates of the Military Academy or had taught there, or both.

Turbulence at Takeoff

When the Air Force Academy moved from the planning to the operational stage, problems latent in the organizational design became manifest almost immediately, aggravated by clashes of disparate personalities. For example, the prescribed curriculum, despite the years of planning that had gone into it, was a jerry-built structure, the enormity of which reflected the inability of planners to reconcile their differences other than by including virtually everything that was proposed. In a detailed analysis of the development of the curriculum, William T. Woodyard, a member of the original faculty and later Dean of Faculty, confesses that he has "not been able to determine the rationale behind the thinking of the planning group that led them to believe that the Air Force cadets could complete a four-year curriculum that was from 50 to 85 percent heavier than the four-year curricula of civilian universities."[19]

General Zimmerman, the Dean, attempted unsuccessfully to create an organizing rationale for the curriculum after the cumbersome program already had been created. The Eisenhower administration, which had come into office in 1953, had given considerable emphasis to strategic air power in promulgating the doctrine of "massive retaliation"—the threat of nuclear reprisal to be used to contain Communist expansion. Zimmerman, seizing upon this element of strategy in which air power played the dominant role, saw possibilities for infusing all coursework with a sense of applied relevance; in practice, however, an intellectually coherent rationale for the program was still missing.

Zimmerman's cold war orientation to the curriculum was a doctrinal approach to reconciling the inherent conflict between Athens and Sparta in the Academy mission. The policy body within which Spartan-Athenian differences were supposed to be reconciled was the Academy Board (modelled along the lines of the Academic Board at West Point). The composition of the board differed from its counterpart at West Point in the important respects of representing some, not all, heads of academic departments, and in according representation not only to the Commandant of Cadets but also to three of his ranking subordinates as well as to the director of athletics.

Far from serving to alleviate tensions among the various interests represented at the Academy, the board became a key arena of struggle which often aggravated internal frictions. At the board's first meeting in the spring of 1955, for example, a heated debate had taken place over the weight to be assigned to various factors in selecting from among candidates qualified for admission to the Academy. The Dean's spokesman for the academic department heads, while acknowledging the relevance of a candidate's extracurricular activities in high school, argued fervently for an emphasis upon academic criteria. However, the director of athletics, bolstered by the strong interest which Secretary of the Air Force Talbott had expressed in developing a nationally ranked football team at the Academy, had insisted that outstanding athletes ought to be recruited even if doing so meant that the recruits' College Board scores or high school academic records were below the Academy's standard. The debate was resolved in favor of making such allowances; but the struggle was far from over. By midyear, Air Force Headquarters had begun to receive complaints from members of Congress whose candidates for admission to the Academy had been rejected in favor of others whose academic qualifications were much weaker. In turn, the new Secretary of the Air Force, Donald A. Quarles, had demanded that the Academy Superintendent investigate the situation.[20]

It was in this context that the Academy received a visit from Manning M. Pattillo, Jr., a consultant to the North Central Association of Colleges and Secondary Schools, who had been asked to meet with Academy officials in preparation for an accreditation examination. Normally, accreditation of an institution was not granted until at least three classes had graduated. However, urged by Academy officials to make an exception, the association had agreed to undertake a review. (Having published a catalog which announced that the bachelor's degree would be awarded upon graduation, Academy officials found themselves under special pressure to meet accreditation requirements.)

Pattillo arrived on a day when General Harmon was in Chicago, discussing plans for permanent-site construction with an architectural firm. However, meetings were arranged with other members of the cadre. Toward the end of his visit, Pattillo shared with some of the officers some reservations which he had regarding the potential of the Academy for accreditation. There were areas of great strength, including plans for a superb physical plant and prospects of a very able student body. But there were glaring deficiencies. Many of the faculty and even the Superintendent and the Dean seemed to have been selected without much regard for academic qualifications. Moreover, the concern at the Academy with developing the strongest possible football team was disturbing.[21] (Among the sources of Pattillo's dismay, apparently, was a briefing by the director of athletics that began with a chart announcing that "The mission of the Air Force Academy is to beat Army and Navy in football.")[22]

A letter from Pattillo to Harmon detailing his observations and concerns evoked no reply. However, the Air Force Deputy Chief of Staff for Personnel and his assistant contacted Pattillo. Disturbed not only by Pattillo's assessment but also by a report from the governor of Colorado that all was not well at the Academy, the two officials flew to Denver to meet with Harmon.[23] Harmon rejected their recommendation that he relieve General Zimmerman from his post as Dean. However, the Air Force reassigned Zimmerman to the Pentagon in Air Force Planning. Harmon himself now assumed the position of Dean in addition to that of Superintendent, while a search began for a replacement for Zimmerman. McDermott became secretary of the faculty, and in that capacity became de facto the man upon whom Harmon came to rely most heavily in the effort to reform the academic program. The emphasis was on meeting the requirements necessary for accreditation.

As Lieutenant General James Briggs recalls of his perspective as head of an Academy curriculum review board appointed by the Air Force Chief of Staff:

> We *had* to be accredited, or the cadets couldn't have a degree when they graduated, and we were going to get that Class of 1959 with degrees if it was the last thing we did. So Bob McDermott and I, and the Commandant, General Stillman—oh, gosh, we went to Chicago [where the North Central Association was headquartered] a thousand times, and anything they wanted we'd do, anything that was within possibility, to get that accreditation.[24]

Perhaps the most important recommendation which the Curriculum Review Board made was that Colonel McDermott be appointed as Zimmerman's replacement as Dean. Learning that he had been selected, McDermott initially replied that he simply could not accept, for a variety of reasons including his desire to get back to regular line duty with the Air Force. However, Briggs, notified that he was to come to the Academy as Superintendent to replace Harmon, who was terminally ill with cancer, rebutted each of McDermott's arguments for rejecting the assignment as Dean. He even assured McDermott that the Air Force was changing his date of rank for full colonel to 1943, the year of McDermott's graduation from West Point, thereby giving him seniority over all other colonels on the faculty!

Attempting one final caveat, McDermott said, "Well, I'm not going to accept this unless I can organize this faculty to get the job done." Brigg's reply was terse: "You're going to accept this because you're in the Air Force, whether you like it or not. The only alternative is to get out of the Air Force. But," he added, "you go ahead and organize it the way you think it should be. . . ."[25]

In late July, the ceremony at which command of the Air Force Academy passed from General Harmon to General Briggs was held at Lowry Air Force Base in a driving rainstorm. Harmon was in pain and had developed a chronic cough. It rained four and a half inches in just over half an hour; but the ceremony continued. Briggs recalls that he was "scared to death."

> Thunder and lightning were all over the place, and the kids were out there with their guns, and the lightning arrestors—the rifles had the bayonets on. I was sure we were going to kill about ten cadets out there, if we didn't kill General Harmon in the pouring rain. But it came out all right. It didn't happen. General Harmon retired and went to San Antonio on the first of August, and I got into the job up to my ears.[26]

New Directions

Major General James E. Briggs had close connections to West Point and its traditions. He had been cadet First Captain at the Military Academy, graduating 31st of 261 members of the Class of 1928. Two years later he gained pilot's wings. After ten years in various assignments as a junior officer in the Air Corps, Briggs had returned to the Military Academy for a two-year tour with the Department of Mathematics, one of the most hidebound in its adherence to the methods of the Thayer system. Brigg's son, James, Jr., had entered West Point in 1950 and had graduated with the Class of 1954.

Yet Briggs found some Military Academy practices silly and counterproductive, such as the plebe system (which the cadre of air training officers at the Air Force Academy had emulated). Moreover, Briggs was too much of a tough-minded pragmatist to let sentimental attachments to tradition impede his judgment as to what was required at this critical stage in the infancy of the Air Force Academy. Still, some practices persisted despite Brigg's better judgment and his efforts to eliminate or reform them. For example, he saw no reason why prospective Air Force officers should train with M-1 rifles, but his efforts to provoke the Board of Visitors into raising questions that would lead the Air Force to approve a change got nowhere. Likewise, he was unable to make the reforms that he thought were essential in the indoctrination of "doolies" (freshmen).[27]

On other, more substantive issues, Briggs met daily—sometimes several times a day—with his key subordinates: the Dean, the Commandant of Cadets, the Registrar, and the Superintendent's Chief of

Staff. "All of my first years spent there, we were in trouble, real trouble," Briggs has noted. In that crisis atmosphere, discussions often became heated—especially those involving McDermott, the Dean, and Stillman, the Commandant. Both McDermott and Stillman were "strong characters," Briggs recalls, "fighting for that ten minutes of cadet time, without too much mutual respect between them. They respected each other as successful people, but not in their own areas of responsibility."[28]

From Stillman's perspective, the Air Force Academy was first and foremost a military institution, and to attempt to transform it into a high-powered academic institution was bound to undermine the primary mission. Even several years after his reassignment from the Academy, Stillman (by then a major general) continued to criticize what he saw as the mistaken trend at the Air Force Academy toward overemphasis on academics. "American taxpayers," he said, "did not build service academies to compete with Massachusetts Institute of Technology, but to train military leaders of tomorrow." Describing the current generation of American youth as weak, Stillman warned that "if we have to butt heads with a relatively primitive country we are going to be in trouble."[29]

McDermott, in turn, argued that "even if the Academy had 'the best airmanship and athletic programs that man could devise,' the failure to offer a four-year accredited baccalaureate degree program would not attract the quality of students needed and desired by the Academy."[30] Briggs shared McDermott's view. As he observed: "I don't believe Stillman and Henry Sullivan [Stillman's successor as Commandant in 1958] really believed in teaching academic work, and they couldn't see why all the furor about being accredited. I was hellbent that we were going to be accredited."[31]

McDermott tended to have the upper hand in policy debates with his adversaries, not only because Briggs shared McDermott's sense of urgency regarding accreditation, but also because McDermott proved a more skillful and tenacious advocate of his views than did Stillman or his successor, Sullivan. Such skill and tenacity were illustrated in the discussion of the Superintendent's suggestion that the reassignment of some courses that had been under the Commandant's jurisdiction to the Dean would serve to clarify lines of responsibility.

> Of course General Stillman and [later] General Sullivan didn't like it at all. The Dean liked it. The Dean was a much better salesman. He prepared his lessons, and in front of every presentation you could just tell he knew

> what he was talking about, and that he had thought it out. He had an answer to every objection, and of course the commandants got up and fumbled and stumbled around . . . although they were fine, fine officers, they were in pretty fast company with that academic side over there.[32]

McDermott, Briggs recognized, was an "academic whiz." Thirty-six years old as the 1956-57 academic year began, McDermott was one of the youngest "bird colonels" in the Air Force. His secondary schooling at the Boston Latin School had provided him with the foundations of a classical education. Indeed, he had had four years of Latin, four of French, four of German; and at graduation, he had won one of two prizes for excellence in music. (The other was awarded to Leonard Bernstein.)[33]

Then he had spent two years at Norwich University, making the Dean's list in his studies there, before entering the Military Academy at West Point in 1939. When he returned to West Point in 1950 as a faculty member in the Department of Social Sciences, he soon became identified by the department head, Colonel Herman Beukema, and his associate, Colonel George Lincoln, as one of their most talented instructors. In endorsing one of McDermott's annual effectiveness reports during that period, Beukema had written:

> In my 25 years of service at the Military Academy I have had no officer in this department surpassing Colonel McDermott in quantity and quality of his input, independence and balance of judgment, initiative, devotion to duty, and loyalty to superiors. His future should be brilliant.[34]

It was Beukema who, in 1954, had suggested to Harmon that he consider bringing McDermott to the Air Force Academy as an assistant dean. The occasion for the Beukema-Harmon conversation had been a dinner at the White House with their West Point classmate, President Eisenhower. Harmon had followed Beukema's advice, getting McDermott's orders to a much coveted flying assignment changed, appointing him head of the Department of Economics with additional duties as Vice Dean.[35] Now in 1956 as Dean, McDermott drew upon his past experience and training in an effort to reform the academic program. The heart of the problem lay in the prescribed curriculum, and was compounded by the unquestioning commitment of several of the Military Academy graduates on the faculty to perpetuation of traditional practices.

Validation and "Enrichment"

The prescribed curriculum typically had been particularly distasteful for cadets who, like McDermott, had come to an academy after a year or more of college. Such cadets were likely not only to be assigned to a number of courses in which they had no real interest, but also to have to repeat coursework taken elsewhere. In 1956, 103 students were admitted to the Air Force Academy with previous college experience, 28 of them with two or more years.

The mathematics department had experimented in 1955-56 with accelerated coursework for students who were able to demonstrate they could handle it. In 1956-57, accelerated coursework was added also in chemistry, English, and graphics.[36] However, McDermott had in mind a more general and basic reform. In a memorandum to all professors in December 1956, he proposed (1) to give cadets credit for courses in the prescribed curriculum taken at other colleges, and to allow these cadets to take electives in lieu of courses validated; (2) to allow gifted students to carry extra courses, over and above the prescribed courses and/or semester-hour load; (3) to broaden the curriculum in the social sciences and humanities and to deepen the curriculum in the sciences through substitutes and extra course offerings.[37]

McDermott noted that he was "concerned that a West Point *rigor mortis* might set in at the professor level to kill the faculty interest in these proposals. I hope that this anxiety is unfounded. . . ."[38] His knowledge of the West Point system was only partly at the root of his concern. McDermott, as well as Briggs, had become convinced that a number of those who held academic positions either were not qualified to provide the leadership needed for their deparments or were too recalcitrant to adapt to the needs of the Academy for innovation.[39] Fortunately, there were some who had proven their mettle, including most notably Archie Higdon, head of the mathematics department, and Peter Moody, head of the English department. Higdon and Moody often disagreed with one another and with McDermott.

> Higdon had strong opinions, strong views that science and engineering [were] going to be the big need of the Air Force in a technological environment. . . . He articulated his position very well, and of course Pete Moody, on the other hand, was equally as forceful in arguing that the future soldier-scholar would have to be an articulate man and appreciate the humanities in order to be a successful leader of men.[40]

Despite their differences—indeed, partly because of the clarity and logic with which these differences were articulated in policy discussions—Higdon and Moody had become valuable and influential members of the initial cadre of senior faculty members, and would continue to be influential in shaping the evolution of the Academy in the years ahead. Many others who had held top posts, however, were eased out—they "would simply disappear, without any official explanation of what happened," Moody himself recalls.[41] Before the end of the 1956-57 academic year, professors of physics, engineering drawing, philosophy, law, and geography and the assistant dean had left the Academy for assignments elsewhere or had resigned from the service.

Administrative Reorganization

The turnover of faculty provided an opportunity for reorganizing the departmental structure. Some departments had combined disciplines in temporary marriages that were not intended to endure. Foreign languages had been joined with English at first, because no language instructors had been assigned yet; this presented no problem initially, since cadets were not to be exposed to language instruction during their first year.[42] Other conglomerates, however, followed the makeshift arrangements that had characterized many departments at West Point.

A related concern was the representation on the Academy Board, which had only five academic members. With reorganization, combinations such as "English and foreign languages," and "chemistry, physics, and electrical engineering" were broken into separate departments. The head of each department was represented on a faculty council, chaired by the Dean. The heads of the four divisions and the Dean were represented on the Academy Board. Thus, curricular decisions were made or approved by the Faculty Council, and then forwarded to the Academy Board, where, McDermott observes, "I figured [they] would get rubber stamped."[43]

The Overload Dilemma

The changes that were made to enrich the curriculum and to restructure the academic departments represented more than an effort to provide the institution with the academic respectability it needed for accreditation. More fundamentally, the goal was to create an environment in which intellectual development could flourish. But given the regimentation and discipline which the Air Force Academy had adopted, following the West Point model, could such a goal be realized? McDermott had stated the issue vividly to the Superintendent in a memorandum in October 1956:

> If you schedule a man's activities six days a week and half of a Sunday, you have reached the ultimate in discipline. You are producing the perfect follower. . . . Leaders develop from a system where a man has many opportunities to solve problems, make decisions, and assume responsibility for the decisions he makes. He has to have time to think, time to sit and time to reflect. Furthermore . . . a cadet is a human being. . . . As a citizen he has the right to think about, read about, hear about and discuss the Suez problem, the school integration problem, the World Series, the Elections, Marilyn Monroe, and Elvis Presley. We have no right to isolate him mentally for four years, but we are doing just that by the simple device of not giving him any time to pursue his own interests. If he takes the time, he does so at the risk of failure in one or more programs.[44]

Ironically, the program of academic enrichment that McDermott had pushed through, with its provision for taking elective courses on an overload basis, added to the demands on cadets' time. Even without overload, the demands were severe. The Curriculum Review Board had succeeded in getting the 140 semester-hour program that cadets had faced during the initial year reduced to 129 semester hours by 1956-57. However, by 1960-61, the academic curriculum was up to 146 hours. Not included were an additional 40 to 50 hours of required military airmanship studies and physical training during the regular academic year. Thus the military-academic program had fluctuated between about 180 to 190 hours, in contrast to 120 to 130 for a typical liberal arts college and perhaps 140 hours for a four-year engineering school.[45] McDermott produced evidence that 21 of the 28 key players on the varsity football team participated in the academic enrichment program, 18 of them on an overload basis. Of 46 first-classmen who held key leadership positions in the cadet wing, 37 were participating in the enrichment program, most of them on an overload basis.[46]

However, producing evidence that even otherwise-busy cadets were actively seeking academic enrichment experiences, thereby filling their crowded schedules still further, was different from claiming that a solution had been found to the problem of providing cadets with "time to think . . . time to reflect." The fact was that cadets with "overloads" as well as those with "normal" course loads often felt hassled by competing demands on their severely limited time, and by conflicting signals as to the goals toward which their prime energies should be directed. Cadet athletes, for example, felt cross-pressured. Ben Martin, who had been brought to the Academy in 1958 as head football coach, has described the chronic problem:

> . . . I know from my own experience with cadets during their on-season in football, that they feel a great deal more pressure from other areas than they do from me as their head football coach. In other words—we are going to play Notre Dame—I don't put any pressure on them, but somebody in one of the departments may, about writing a five-page paper and reading five hundred pages and "do it by nine o'clock tomorrow morning." I don't put that kind of pressure on my players. So I would say, yes, there was and there continues to be time suspenses [sic] on performance in the intellectual area, and in the military area too.[47]

In short, even with the word "overload" dropped from official usage (as it was in 1961 in favor of "extra electives"),[48] a fundamental dilemma remained. How could intellectual, physical, and emotional demands be imposed so as to challenge and foster development, rather than overwhelming cadets or leading them to make only Pavlovian responses to cues of the system in the order of their intensity?

The dilemma remained unresolved. It was nonetheless, as Coach Martin has noted, a time of "unique challenge." In the academic realm, it was a period of creative ferment. Fine arts, sociology, group dynamics, astrodynamics, and space technology were among the courses introduced during these years that were completely novel to a service academy curriculum.[49] (The Dean also attempted to get a course in comparative religion initiated, but the proposal was defeated, apparently in part because some of his critics saw it as an insidious means of promoting his Roman Catholicism.)[50] Moreover, the academic reforms that came with the early years of the enrichment program opened doors that were bound to lead to other changes.

Fruition

On Labor Day weekend 1958, the move of Academy personnel from the temporary site at Lowry Air Force Base to the permanent quarters near Colorado Springs was made. The following April, the long-awaited good news came. Accreditation had been granted—and in time for the graduation of the first class that had entered the Academy. It was a time of triumph for members of the Class of 1959 and their families. Yet in some respects it was an even more exhilarating time for those members of the staff and faculty who had come to refer to themselves as the group who "got there before classes started." They were the "founding fathers," who had been working furiously since those first difficult days and nights at Lowry, struggling to develop an effective curriculum, agonizing over the design of classroom and laboratory facilities, debating priorities among a host of competing demands. It had been, as one of this group recalls, "the most exciting period in my career."[51] Another notes, "There was excitement and pride too in beginning this new institution. We wanted the [Air Force] cadets to be smarter, stronger, better educated, better in every way than their counterparts at USMA and USNA." He adds, "Perhaps too much attention was paid to continual comparisons made with other military academies and with civilian schools. And perhaps it was wrong to strive to be the best in everything."[52]

But the comparisons invariably were made. From the vantage point of Colorado Springs, at least, they provided clear evidence of a remarkable success story. June Week 1959 was a historic moment. Officials were being commissioned into the Air Force from its own academy for the first time. Although short on tradition, the Air Force and its young academy were rapidly creating it, and memorializing their prophets and saints: Billy Mitchell, Hap Arnold, Muir Fairchild, Hoyt Vandenberg, Hubert Harmon. With the widow present for each of these leaders, buildings named in their honor were dedicated. In this setting, the Class of 1959, resplendent in Hollywood-designed uniforms, were graduated. Even the diplomas were modernistic—aluminum rather than parchment—"a little twist," the Superintendent observed, "to identify the age."[53]

Accelerating Change and Mounting Resistance

Once the sanctity of the prescribed curriculum had been violated, modifications to introduce further opportunities for students flowed naturally. Validation of coursework taken elsewhere and provision of limited electives had led almost inexorably to opportunities for students to put together a combination of required courses and electives that would constitute an academic "major." Majors became available in 1957-58 in three fields, which by the following year had been redesignated and expanded to include basic sciences, engineering sciences, English, and public affairs.

McDermott agreed with recommendations of the Curriculum Review Board and of the Board of Visitors that no additional important changes should be made at least until a four-year cycle had been completed. However, with validation and enrichment, a number of cadets were accumulating more credit hours than were required for the baccalaureate degree. Some were taking advanced work which in credit hours would be the equivalent of master's degree requirements at civilian institutions. The obvious question was: why should not additional incentives for such advanced work be provided by awarding the master's degree to qualified students? Such reasoning seemed particularly compelling in the light of a severe shortage of Air Force officers with postgraduate degrees.

These considerations prompted McDermott to seek executive and congressional authorization and the accreditation required for awarding the advanced degree. Indeed, it became his passion. The earlier reforms were beginning to bear fruit. The credentials of the faculty were becoming more impressive, so that by 1962-63, 97 percent would have advanced degrees and 20 percent the Ph.D. Cadets were "thirsting" for the enrichment program, as McDermott had predicted, with more than three-fourths of the cadet wing participating by the 1959-60 academic year. Moreover, mean test scores from the Graduate Record Examinations showed the first group of graduates from the Air Force Academy ranking second among students from 187 institutions of higher education in the natural sciences area, second also in the social sciences, twenty-first in the humanities, and second overall.[54] The program in engineering was of such excellence that the Engineers Council for Professional Development soon would authorize the Academy to graduate its engineering majors as "bachelors of science in engineering science."[55]

However, the path that had to be travelled if authorization for awarding the master's degree was to be attained was one filled with more obstacles than those that had temporarily impeded progress toward obtaining accreditation as a baccalaureate-awarding institution. The essential first convert to the cause had to be the Superintendent, Major General William Sebastian Stone, who had succeeded General Briggs in August 1959. Stone had entered West Point at the age of twenty during the first year of the Depression, graduating in 1934. He was one of the exceptional military officers during the prewar period to receive postgraduate training, obtaining an M.S. in meteorology from Cal Tech in 1937. Like McDermott, Stone was one of "Hermie Beukema's boys," with experience on the social sciences faculty at West Point. Stone had two such tours of duty, the first from 1940 to 1942, and the second from 1947 to 1950. The latter tour gave him the opportunity to earn a second master's degree during off-duty hours, in economics from Columbia University. Upon completion of the second faculty tour at West Point, Stone was selected to attend the National War College. From 1953 to 1956, Stone had been director of personnel planning in the Office of the Deputy Chief of Staff for Personnel at Air Force Headquarters. In that capacity he had been deeply involved in the curriculum planning prior to the opening of the Air

Force Academy as well as in the selection of faculty—including the selection of McDermott (at Harmon's urging, following Beukema's suggestion, as indicated above). Moreover, Stone had served on the Curriculum Review Board, which, early in 1956, had set the stage for the reform of the initial curriculum and the selection of McDermott as Dean.

Thus, Stone came to the job of Superintendent with a knowledge of service academy programs in general and the curriculum of the Air Force Academy in particular, as well as with extensive previous association with McDermott. With Stone's essential backing, the master's degree proposal received the endorsement of the Academy Board shortly after Briggs took office. Plans were made for launching master's degree programs in two fields: astronautics (tailored after a comparable program at MIT), and public policy (emphasizing economics and political science).[56]

In March 1960, McDermott and Stone—primarily the former—shared their ideas regarding the master's program with key officials at the other three academies at the annual service academy superintendents conference. The proposal was greeted with considerable skepticism. Lieutenant General Garrison Davidson, Superintendent of the Military Academy, voiced a concern that others would express as well in ensuing months of discussion. Would the proposal be seized upon by those who in the past had argued that the academies ought to be transformed from undergraduate into graduate institutions? McDermott acknowledged the possibility, but thought that the Service Academy Board had put this issue to rest. The director of admissions at the Naval Academy, William Shields, raised another issue that also would recur in subsequent discussions. Did a military academy provide the intellectual environment appropriate to postgraduate education? It was not simply a question of whether cadets or midshipmen could accumulate the hours requisite for an advanced degree in four years, Shields argued; such an approach missed the essential difference in purpose between undergraduate and postgraduate education.[57]

Although clearly there was no enthusiasm for the master's degree plan at the other academies, their support did not then seem essential. A heartening endorsement of the proposal "in principle" was provided the following month by the Board of Visitors to the Air Force Academy by an 11-0 vote. Since the board included members of Congress, it was with some confidence that Academy officials forwarded to Air Force Headquarters for approval a draft of a bill which would provide legislative authorization for the degree programs. Air Force readiness to offer their support was undercut by an expression of opinion from President Eisenhower, however. While noting that he had "no objection in principle" to the proposal of which he had been informed, he recommended that "a study be made with the faculties of the three service academies with a view to taking uniform action among all three in this regard." To be avoided, the President emphasized, was a "cheapening of academic degrees by careless legislation."[58]

A meeting of representatives of the three academies in an effort to satisfy the President's concerns proved fruitless. Indeed, McDermott interpreted follow-up memoranda from West Point and Annapolis detailing their objections as evidence of efforts to "sabotage" the proposal. His twenty-seven page rebuttal included five appendices documenting the accomplishments of the enrichment program and the readiness of Air Force Academy cadets to participate in a master's degree program.[59]

The Defense Department remained noncommittal, however. At the Navy's request, the American Council on Education was asked to review the feasibility of the proposal. The ACE referred the matter to their Committee on Relationships of Higher Education to the Federal Government. Speaking first to the committee at their meeting in April 1961, McDermott repeated all the arguments that he had used in the past as to why a master's degree program at the Air Force Academy was not only feasible but highly desirable. He was followed by Brigadier General William W. Bessell, Jr., Dean of the Academic Board at West Point, who testified in opposition to the proposal.

Bessell, the son of an Army colonel, had graduated from West Point in 1920, the year McDermott was born. By 1950, when Bessell was completing thirty years of military service as one of the most senior members of the Military Academy faculty, McDermott was just reporting for duty at West Point as a junior faculty member. Thus, in April 1961, when Bessell explained to the committee why the McDermott proposal conflicted with the very essence of the service academy mission, he could do so with the conviction that his depth of understanding of that mission far exceeded that of his junior colleague.

The committee met in executive session to discuss this extraordinary debate, but made their decision public. It was a bland, almost noncommittal decision, scarcely strengthened by the announcement that the committee's action was being taken on a unanimous basis. The committee noted that,

> while it views with sympathy and even enthusiasm the efforts of Academy leaders to stimulate and challenge students by a program of enrichment,

> and is impressed with the quality of work being done by the Academy, the Committee is not in position to endorse the recommended solution at this time.[60]

Privately, the committee told McDermott that if the Department of Defense wanted the endorsement of the American Council on Education, DOD was going to have to endorse the program themselves; the ACE did not want to have to referee among the services.[61]

McDermott was angry. Continuing to fight for the program, he enlisted powerful support from General Curtis LeMay, who, although not an academy graduate himself (and previously hostile to the creation of an Air Force Academy), could agree that "this is the best goddam academy in the country," and could say to other members of a general officers advisory committee to the Academy that "everyone around this table is going to support that master's degree program."[62]

However, West Point and the Army continued to be uncompromising in their opposition to the program. Navy officials continued to express their disapproval, but took the position that they would not stand in the way of legislation that permitted the Air Force Academy to launch a master's degree program. Without unified support, however, the program was dead, even though Stone's successor as Superintendent (Major General Robert H. Warren, who took command in July 1962) would continue the futile effort to persuade the Defense Department to back the program.

In consultation with senior faculty members, McDermott now modified his goals, seeking to enable selected students to obtain the master's degree at civilian institutions after graduation from the Air Force Academy, with graduate credit granted for some of the advanced work which they had taken at the Academy. Within a few years, cooperative master's programs of this sort had been arranged with eight civilian graduate schools in a variety of fields. These were supplemented by programs that enabled a small number of Academy graduates each year to go directly to law school or to medical school. Such programs were relatively short-lived, however. Beginning with the Class of 1976, the opportunities for gaining master's degrees through a cooperative program had ended; and beginning with the Class of 1978 (which entered the Academy in 1974), the opportunities to go directly on to law school or to medical school had been eliminated.

From the outset of the period of academic innovation at the Air Force Academy, there had been both internal and external critics who had argued that to shift additional emphasis to the academic component of the Academy's program would undermine the essential fabric of military discipline and commitment to a military career that were the mainstays of the Academy mission. By the late 1960s, and even more so in the early 1970s, the views of such critics had become official policy.

The Falcons Come Home to Roost

Despite the many reforms that had been implemented in a short period of time (and in part because of them, paradoxically), McDermott's support among his faculty in the mid-1960s was far from total. On the one hand, there were officers on the faculty who shared the view that was prevalent on "the Commandant's side of the terrazzo" that a service academy exists first and foremost to instill discipline and to provide military training, and only secondarily to provide a college education. As the emphasis on academies increased through enrichment offerings, through access to postgraduate degree programs, and even more so with the introduction of "majors for all," such officers tended to perceive a relaxation of the standards of discipline in the classroom which they found distressing. (Cadets were coming to class with shoes not shined to the proper gloss, for example. They were also growing careless about prefacing their remarks to their instructors with "Sir," and generally seemed to be adopting many of the habits of civilian college students.)[63]

On the other hand, there were other faculty members wholly supportive of the goal of transforming the Academy into an institution that justifiably could boast of its academic excellence, but who nonetheless objected to McDermott's tendency (as they saw it) to make all key academic policy decisions himself, in consultation with a few trusted advisors, rather than decentralizing the academic policy process. Even among those faculty members who accepted the need for strong, centralized leadership in this period of rapid transition, there were many who wondered if at this point in the mid-1960s, McDermott wasn't pushing too far, too fast.

Whatever differences in outlook individual faculty members may have had with the Dean, however, were mild in most cases in comparison to the differences in viewpoint that characterized the faculty, led by the Dean, on the one hand and the AOCs (Air Officers Commanding, who were assigned to supervise cadet companies), led by the Commandant of Cadets, on the other hand. The "terrazzo gap"[64] on many days was a chasm.

The Seawell-Stone Interregnum

However, in Brigadier General William T. Seawell, who assumed the post in June 1961, General Stone found a Commandant who could work harmoniously with the Dean. Seawell shared the Superintendent's and the Dean's concern about eliminating the hazing and petty harassment that had become characteristic of the approach to discipline in the cadet wing, and also shared their concern with establishing more sensible priorities.

In the spring of 1961, just before Seawell's arrival at the Academy, Stone and McDermott had instituted an organizational change that paved the way for the reforms in discipline and military training that Seawell would oversee. Officers who had been serving as faculty members were assigned to each of the twenty-four cadet squadrons to assist the Air Officers Commanding. Moreover, Colonel Victor Ferrari, who had taught psychology and more recently had been serving as assistant to the Dean, was assigned as Wing Air Officer Commanding, one of the key positions under the Commandant of Cadets, with special responsibility for supervising the practical leadership training of cadets.[65]

When Seawell took over, his first concern was reform of the fourth-class ("Doolie") system. The slogan which he tried to impart to the first-classmen who had the most direct contact with the incoming cadets, was "teach and train, do not hinder or harass." One important and immediate result of the program was that the retention rate during the first year was higher than it had been in the past. Whereas the stresses of the first summer alone had accounted for the attrition of as much as 10 percent of incoming classes, in the summer of 1961 that rate was cut by over 50 percent.[66]

The positive leadership approach was extended also to the regular academic year, and to relationships between the AOCs and the cadets, as well as to those between upperclassmen and doolies. AOCs were urged to exercise more restraint than they had in the past in supervising the cadet chain of command, letting the cadet leaders learn from mistakes rather than "hovering over them" to ensure that no mistakes occurred.[67]

However, Seawell had been Commandant only for a year when Stone was reassigned from his position as Superintendent. The following February, Seawell cut short his own tour as Commandant and gave up a promising Air Force career to join Pan American Airways (becoming president and chairman of the board). Their successors tended to restore the mode of discipline that had been characteristic before the Stone-Seawell interregnum. The new Superintendent, Major General Robert H. Warren, was at age forty-four the youngest man to have been appointed to the post, having graduated from West Point only a year ahead of Seawell. As a journalist-guest of the Academy observed, the new Superintendent was "not the jocular type"; rather, he had a "MacArthurian mein." The simple explanation of why he had been selected as Superintendent in preference to 167 other available generals in the Air Force was that the Chief of Staff, "Gen. Curtis LeMay wanted him."[68]

The new Commandant, Brigadier General Robert W. Strong, Jr., was Warren's West Point classmate. Strong's father was a retired Army general who had graduated from West Point in the celebrated class of 1915, with Eisenhower, Bradley, Herman Beukema, Hubert Harmon, and others. Young Strong had returned to West Point after World War II as a tactical officer. Most of his subsequent assignments were with the Strategic Air Command until his tour at the Air Force Academy.

"Still vividly remembering his days at West Point," a cadet interviewer reported, Strong expressed the conviction that his appointment as Commandant was "'one of the greatest honors' that a brigadier general can have."[69] He told his interviewer that he felt that one of the important aspects of his job was to "bug" cadets, in order "to make [cadets] perform to our maximum ability." The one goal that cadets should pursue at the Academy, Strong asserted, was that of achievement—and the three keys to achievement were "character, high standards, and the will to win. . . ."[70]

Honor Scandal and the Search for Diagnosis

The Superintendent, the Dean, the director of athletics, and probably every individual member of the faculty and the staff, each in his own way, shared the belief in "the three keys to achievement" that Strong had cited. In January 1965, however, such aphorisms took on a ring of hollow mockery. Evidence of widespread cheating on examinations was uncovered. An informal investigation revealed that blatant violations of the honor code had been occurring for nearly two years. "Ringleaders" had been stealing examinations, making them available to other cadets for prices ranging from $2.50 for a quiz to $25.00 for final examinations.[71]

Two of the fourth-classmen who reported the honor violations revealed that they had been offered examinations "for a price" during the final examination period in December, and had been threatened with bodily harm if they revealed the offer.[72] Moreover, as the preliminary investigation of a possible cheating ring continued, an unregistered pistol was found in the room of the cadet whom Academy officials believed had first stolen the examinations.

In this context of potential violence, an assistant dean called McDermott aside and revealed to him a previously undisclosed incident associated with the 1951 honor scandal at West Point. The assistant dean had been one of the West Point cadets in 1951 who, learning of honor violations, had reported them. Shortly after revealing what he knew, he had been attacked from behind while walking along a road behind the cadet barracks by several cadets who were involved in the cheating, and had been badly beaten. Knocked unconscious, he had never learned the identity of his assailants. McDermott, shocked at this disclosure of violence associated with the 1951 incident, and recalling in addition an as-yet unsolved mysterious disappearance of a cadet shortly before the 1951 scandal broke, immediately telephoned the Superintendent.[73]

Horrified by these revelations, General Warren called Washington for assistance from the Office of Special Investigations of the Inspector General of the Air Force. Over the next several weeks, many cadets and faculty members would become dismayed at the techniques used in interrogating cadets suspected of having information regarding cheating, and in searching their rooms. However, apprehension of the "ringleaders" (some of whom had threatened other cadets) and their removal to Lowry Air Force Base for further interrogation at least had the positive effect of alleviating the fear that physical violence would occur.

In March, the Secretary of the Air Force announced that he had accepted the resignations of 109 cadets, all of them implicated in violations of the cadet honor code. To investigate more thoroughly the underlying causes of the incident, he appointed a special advisory committee, headed by retired General Thomas D. White, who had served as Chief of Staff of the Air Force during the years 1957-1961. Members of the committee interviewed more than two hundred witnesses, ranging from present and former cadets, and present and former members of the staff and faculty, to all living former Superintendents and Commandants, and the present Superintendent, Commandant, Dean, and director of athletics. The Superintendent and Commandant at the Military Academy and at the Naval Academy also were interviewed, as were some civilian psychiatrists and other specialists.

The report of the White Committee, submitted in May, contained a number of findings that read eerily like the report on the 1951 honor scandal at West Point. Identical examinations were being administered on alternate days (leading to "Monday-Tuesday" cheating, as the committee described it). Frequent quizzes rewarded rote memorization in many courses. Graduation in order of merit reinforced the competitive emphasis on grades even on daily quizzes. The formality of many instructors added to the anxieties which cadets experienced regarding academics. Moreover, the distinction between the rules of discipline and those applicable to the honor code often were blurred.

Furthermore, as at West Point in the earlier scandal, the committee found that a growing emphasis on "big-time" football had led athletes to develop team loyalties that conflicted with their loyalties to the Academy as a whole. The committee established that 44 percent of the 105 cadets who had been cheating had come to the Academy as recruited athletes.[74] Moreover, as McDermott learned to his consternation (independently from the White Committee investigation), there had been a recurrence of recruitment abuses, similar to those that had plagued the Academy in its first year. Some athletes had been admitted in lieu of nonathletes who, on the basis of test scores, should have had preference. McDermott discovered the scope of this problem with the aid of a conscience-stricken secretary in the registrar's office who had secretly marked admission records of candidates whose scores had been changed. A phone call from McDermott to former Superintendent Stone, who now was Deputy Chief of Staff for Personnel of the Air Force, triggered a separate, more narrowly focused investigation.[75]

In addition to these specific factors that had proven to be a corrupting influence on the climate at the Academy, the White Committee identified some more general institutional factors that had led to an erosion of the foundations of a system of honor. In the first place, the committee noted, there was "confusion over the Academy's mission among officers and cadets, leading to conflicts of loyalty and purpose among the institution's various departments. . . ." It was a problem with dimensions beyond the "terrazzo gap," manifested in "an uneasy union" of academics, athletics, and military training.[76]

The committee identified as a second institutional weakness "the Academy's concern with the symbols of achievement, sometimes at the expense of reality. . . ." Whether it be zeal for winning football games or anxiety to demonstrate how cadet performance on the Graduate Record Examinations compared with the performance of students from prestigious civilian institutions, the committee questioned whether the concern with "image" had become excessive.[77]

Third, the committee found that "rigid and centralized leadership policies by the authorities, coupled with cadet failure to maintain Wing morale in the face of these policies, helped to create an atmosphere of disaffection in which cheating was more likely to occur." The Class of 1965, the committee observed, had "lived under three Commandants, each with a different view of leadership training and discipline." The results were deleterious, especially in squadrons where AOCs simply provided unthinking compliance with the Commandant's demands for "tightening up" the discipline. Cheating tended to be especially prevalent among cadets in squadrons where the AOCs were regarded as "uncommunicative, preoccupied with standardization, or overly concerned with petty detail."[78] This observation was confirmed by Dr. Alan L. Morgenstern, a psychiatrist asked to offer his independent analysis of the situation in the wake of the honor scandal. "Cadets view AOCs as unintelligent and possessed of 'checklist minds,'" Morgenstern reported;

> it is unsettling to see a theme of contempt for the AOC. Although outwardly respectful, cadets are appalled to see a major crawl under a bed to look for dust or to encounter a captain whose leadership centers around a book of regulations; they wonder if commissioned service is what they imagined.[79]

Aftermath of Scandal: The Search for Cures

In March, even before the committee had completed its study, the *Denver Post* reported that Generals Warren and Strong were being reassigned from the Academy. The report was denied at the time, but in April, the reassignment of both the Superintendent and the Commandant was formally announced, to become effective in July and June, respectively. (Under the circumstances, the reassignments had the appearance of being punitive; however, Warren would soon make his third star, Strong his second.) It fell to their successors, Major General Thomas S. Moorman as Superintendent and Brigadier General Louis T. Seith as Commandant, to attempt to effect the remedial measures proposed by the White Committee.

The White Committee had unearthed some disturbing symptoms of malaise at the Air Force Academy which, had they been explored in depth, might have provided the justification for a major redefinition of Academy goals, procedures, and priorities. However, despite its various critical observations, the committee had provided reassurance that they had found that "the Academy's program is fundamentally sound." The committee dismissed as "misguided" the notion that "the cheating episode reflects either a major institutional collapse or requires a major overhaul of the Academy's program. . . ." Rather, the committee concluded, "We are convinced that stability is essential at the Academy, and that changes affecting cadet life may have a harmful effect if too abruptly made."[80]

It was to the latter cues that Academy officials responded. There had been staff and faculty members in testimony before the committee who had contended that the institutional problems of the Academy were extremely deep-rooted. However, those who so testified were branded as "finks" for sharing criticism with "outsiders."[81] Moreover, two years later, when a congressional investigating subcommittee would attempt to obtain the transcript of the testimony received by the White Committee, they would be informed that it had been destroyed.[82]

The reform efforts of Academy officials proceeded on the assumption that the institution was fundamentally healthy. Some additional security was needed at exam time, with locks on doors to academic buildings and to elevators and filing cabinets. Athletes needed to be integrated more fully into the cadet wing, and all cadets needed additional indoctrination on the rationale for a system of honor. However, these actions could and would be accomplished with dispatch.

In addition, there was an effort to increase the sense of organizational identification. In the operation of the cadet wing, the pendulum swung back from the Strong emphasis on centralization in the direction of the Seawell philosophy of letting cadets take primary responsibility. A program of summer field training, designed to familiarize cadets with counterinsurgency techniques and to instill in them "individual initiative and self-reliance," was instituted. In 1966, optional programs were provided to volunteers from the first and second classes who sought experiences such as parachute training, underwater demolition training, jungle operations training, or a number of others. Lest football players believe that they had been stigmatized by the recent honor scandal, the Superintendent, General Moorman, took a personal interest in the team unsurpassed by any of his predecessors.[83]

In April 1966, the Secretary of the Air Force, Harold Brown, wrote to Moorman commending him on the success that the Academy had had in implementing the recommendations of the White Committee. Brown noted that both he and the Chief of Staff, General John P. McConnell, "agree with your insight that the most beneficial outcome over the long pull probably will be the intangible gains of self-study which the Academy has undergone. As you mention . . . the time has now come to put the cheating incident behind us and move on. . . ."[84]

In March 1967, however, still another rash of honor violations was uncovered at the Air Force Academy. Although on a smaller scale (46 cadet resignations) than the 1965 incident, the recurrence was sufficient to shatter the illusion that remedial actions had been effective. However, Academy officials continued to espouse the "few rotten apples in the barrel" thesis that officials at the Military Academy had found attractive in 1951 and that had seemed plausible to Air Force officials in 1965 in the light of the crass thievery and hustling of exams that had characterized the behavior of ringleaders.[85] Evidence adduced in 1967 included the concentration of cheating among the members of one class who lived in two squadrons of the cadet wing. Furthermore, it seemed obvious to Academy officials that some cadets had been unduly swayed by the temper of protest against authority and by the attitude of "do your own thing" that prevailed on college campuses. McDermott told a congressional investigating subcommittee that one of the cliques of honor violators in the 1967 incident had called themselves the "cool group . . . symbolized by kooky music, long hair, and a few other things."[86]

That the "few rotten apples in the barrel" thesis was inadequate as an explanation of the phenomenon that had been occurring, however, was suggested by the number of resignations because of honor violations that had occurred *independently* of the 1965 and 1967 incidents—more than three hundred from the classes of 1965 through 1968. One cadet in twelve that had entered with the Class of 1965 had resigned because of honor violations; the same ratio held for the Classes of 1967 and 1968. For the Class of 1966, the ratio was even higher—one cadet in eight.[87]

Thermidor

Despite the tenuousness of the rotten apples thesis, Academy officials were correct in sensing the effects of the changing mood among American youth upon cadets. Currents of social change were sweeping American society in the late 1960s. Domestic protest against American involvement in the war in Southeast Asia was rising, and civilian campuses were in ferment. Among students at the service academies, there was much ambivalence. Attitudes vacillated between kinship with and estrangement from civilian peers, even as cadets and midshipmen found themselves now accepting, now rejecting the rationale for the pervasive rules and regulations which defined their lifestyle, circumscribed their behavior, and decreed that they soon might be called upon to give their lives to the cause that had become so widely questioned.

To officials at the Air Force Academy, with traditional military values under heavy attack in the society at large, it was a time to defend and sustain, not depart from, those values. The need was especially acute in the light of the threat to institutional legitimacy that the honor scandal represented. It would be difficult to exaggerate the shattering impact that the revelation of widespread violations of the cadet honor system in 1965, followed by a renewed outbreak of violations only two years later, had upon those closely associated with the Air Force Academy. For over a decade, Academy officials had been struggling to develop an institution that would embody the enduring virtues of character building that had characterized the traditional seminary-academy, while building a forward-looking program of education and training that was appropriate to the needs of the space age. Despite heated internal debates, despite sharp differences among members of the staff and faculty regarding priorities that ought to be established in the program, there was nearly universal pride in the accomplishments—dazzling accomplishments, some thought—that had been recorded in so short a time.

But suddenly the record had been blemished. Suddenly officers who had accepted with unquestioning enthusiasm the wisdom— indeed, the necessity—of being bold and experimental in the development of a curriculum, of expanding the cadets' intellectual horizons through permitting increasing numbers of electives, began to have doubts.

This is not to say that fundamental realignments occurred among competing "mission elements," as they were termed at the Academy. Interpretations of the nature of the calamity that had overtaken the Academy and of its root causes tended to reflect the orientations and biases that participants had developed over the preceding years of internal debate. Those on the staff and faculty who had felt that McDermott and the academic "whiz kids" were pushing too hard, too fast, tended to see academic pressure as a fundamental corrupting influence that had contributed to the breakdown in the honor system. In contrast, those who had viewed the Commandant and the AOCs as being little more than out-of-place Marine Corps drill sergeants continued to believe that the effects of such "drill sergeant behavior" on cadet attitudes and morale were counterproductive. Those who long had lamented the emphasis at the Academy on "big-time" football continued to do so.

However, there was a general desire to avoid further institutional crises, and a general rallying to the watchword "stability." The cues that Academy officials received from their task environment—Air

Force Headquarters, the Board of Visitors, the Congress—reinforced the predisposition toward the maintenance of the status quo. For example, the Board of Visitors to the Academy in the spring of 1967 had assured Academy officials that,

> If anything, the incidents of cheating seem to have strengthened and consolidated the belief in the Code. To the cadets, the Code has become a bond between themselves and the entire military heritage.[88]

Likewise, the special congressional subcommittee investigating the three DOD academies in the aftermath of the second honor scandal at the Air Force Academy underscored the continuing relevance of traditional military virtues and training practices. The subcommittee was "convinced that the honor codes as administered at the respective service academies are enthusiastically endorsed and supported by the cadets and midshipmen themselves." Moreover, the subcommittee praised the academies for their plebe systems, voiced concern about the curriculum only in a suggestion that "the professional and military training aspect of the curriculum may not receive its proper share of attention," and strongly supported the emphasis in the academy programs on intercollegiate athletics.[89]

Thus, the Air Force Academy entered a period in which the emphasis was on stability rather than innovation. There continued to be incremental program developments, of course. But the organizational milieu had been decidedly altered. Symbolic of the transition that was occurring from one phase in the Academy's history to another was the resignation from the Academy and from the service in 1968 of Brigadier General Robert McDermott.[90] The innovations from the earlier phase of its history, however, already had become a major stimulus to change at the other three academies.

5

Agonizing Reappraisal: The Davidson Years at West Point

The Military Academy, the Coast Guard Academy, and the Naval Academy each came to adopt reforms similar in intent if not identical in form to those that had been instituted at the Air Force Academy. However, it would be an over-simplification to attribute the changes that eventually occurred at the older academies uniquely to the stimulus that the Air Force Academy provided.

It is true that innovations at the Air Force Academy were threatening to long-cherished assumptions which had served to justify the maintenance of the status quo at the older academies. Moreover, the modernistic organizational image which Brigadier General McDermott and other officials at the new academy consciously worked to create stood in sharp contrast to the image of romantic traditionalism that the older academies had cultivated. With a new institution competing for a limited pool of young men able to meet academy mental, physical, and emotional standards, the older academies could ill afford to discount the possibility that a more modernistic image (implying also a substantive modernization to accord with the image) would enhance the recruitment appeal.

Yet the older academies were subjected to pressures for change in addition to those generated by a new and innovative competitor. The professional requirements that parent services imposed were in flux in the late 1950s and early 1960s (with a new sense of urgency for developing technological expertise after the Soviet launching of *Sputnik I* in the fall of 1957). Critics of the academies, such as Vice Admiral Hyman Rickover, began to receive receptive hearings in Congress, which in turn made the academies more receptive to the criticisms. Moreover, advocates of change emerged from within the academies themselves—most visibly and influentially in the form of Superintendents newly assigned to the academies and convinced for a variety of reasons that they must become the instruments of change, but also to some extent in faculty or staff members who saw in a new Superintendent the possible champion of causes that they had long repressed.

The experiences of the Coast Guard Academy and of the Naval Academy are described in subsequent chapters. Here we focus on West Point, which in the mid-1950s had been a bastion of the status quo. However, with the arrival in 1956 of Lieutenant General Garrison Davidson, an important period of ferment began. A case study of this period allows us to investigate a number of questions about the process of organizational change.

For example, to what extent is a period of organizational reform predictable in terms of the key reformer's predisposition, as evidenced by previous career actions or experiences? Given Davidson's many ties to the Military Academy and its traditions, there was little in his background to suggest that his appointment as Superintendent would bring other than a perpetuation of existing practices. In fact, however, he arrived with some definite ideas about practices and structures that needed change, ideas which were bolstered by careful analysis that was initiated shortly after his arrival.

What leadership strategies are likely to prove effective in mobilizing support for reform? Davidson's use of internal study committees and his canvassing of the opinions of alumni and other senior military personnel provide interesting examples of a relatively successful strategy.

How important are personal friendships and animosities among key organizational decision makers in determining the outcome of particular policy issues? Long-standing frictions between Davidson and the football coach, his warm earlier association with the Commandant of Cadets, and his lack of enthusiasm for the personality of the Protestant chaplain were clearly relevant to the outcomes of policy discussions regarding athletics, military discipline and training, and chaplaincy selection. In academics, the most important area of reform during the Davidson years, the dynamics of change were more complicated. Personalities were important, as were earlier friendships. But so also were the traditions of the Academic Board, the relative power which various departments wielded because of the share of the curriculum which they controlled, the access which Davidson had to external authorities, and yet the veto power, in effect, which the Army Chief of Staff was able to exercise.

Finally, when an innovative leader like Davidson departs from an organization, what is the likelihood that changes which he instituted will be sustained, and what factors help to explain the relative durability or fragility of such innovation? The directions in which Davidson had led the Academy during his tenure were not reversed under his successor; but neither was the momentum for change fully sustained. The cues from the Army Chief of Staff, who had been less than fully supportive of Davidson's academic reform efforts, provided external pressure after his departure for stability rather than further change. It seems clear that General William Westmoreland was selected as Davidson's successor to provide such stability and to "restore the balance" between academics and military training.

In short, the experience of the Military Academy during these years holds interest both for the organizational dynamics that resulted in new departures, and for those that thwarted change and preserved traditional practices.

West Point in the Mid-50s

Subsequent to the superintendency of Lieutenant General Maxwell Taylor, there had been no significant programmatic changes at West Point to the mid-1950s, although the modest reforms begun in the early postwar period were continued. For example, increasing numbers of prospective faculty members were being sent to civilian graduate schools for study before assignment to the Academy (although the all-military faculty concept remained unquestioned). Limited experimentation with specialized courses for students with advanced preparation continued, especially in the social sciences under Colonels Herman Beukema and George Lincoln, and in English under Colonels George Stephens and Russell Alspach.

However, as noted approvingly by a board of Academy officers appointed by the Superintendent in 1953 to review the curriculum and to consider explicitly the possibility of offering cadets a choice among major fields of study, "the current USMA curriculum is, for all intents and purposes, completely prescribed." Cadets were afforded a choice only among foreign languages (French, Spanish, German, Portuguese, or Russian). In a wry bit of Orwellian logic, the board, headed by Colonel James W. Green, Jr., of the electricity department, argued away the compulsory nature of the curriculum. "It should not be forgotten," the committee observed, "that he [the cadet] has exercised a complete freedom of choice in his decision to come to West Point at all—he chose the prescribed curriculum."[1]

The rationale invoked by the Green Board on behalf of perpetuation of a prescribed curriculum was the traditional one, with two key elements. First, because the Academy was preparing its students for a common professional career, it was possible to identify a single "best" program of studies and practical experience that would provide the requisite preparation. Secondly, there was the traditional concern that any significant departure from a uniformly prescribed program would carry the risk of destroying the sense of organic unity among the corps of cadets.

The Dean, the Academic Board, and the Superintendent found the reasoning of the Green Board persuasive. Priorities in education and training therefore continued to be those of the traditional seminary-academy, with its paramount emphasis on the development of "character"—that is, the qualities of courage, loyalty, and discipline that were the hallmarks of the reliable military leader. "The entire system of instruction, training, athletics, discipline, and administration is geared to this goal [character development]," the Green Board had noted proudly.[2] And they were right. The emphasis on instilling the martial virtues—especially discipline—began the day a new cadet entered the Point, extended into every facet of his cadet life, and continued through to graduation.

The selection of officers to be assigned to the staff or faculty at West Point tended by the mid-1950s to be made from among those who had compiled enviable combat records in Korea. The group included nationally publicized heroes, such as "Iron Mike" Michaelis, who had assumed command of an infantry regiment early in 1950 as a lieutenant colonel and by the spring of 1951, at the age of 38, had been promoted to brigadier general and had been awarded the highly coveted Distinguished Service Cross. Michaelis, a graduate of the Class of 1936, returned to West Point as Commandant of Cadets in 1952. Less renowned among the general public but no less so among cadets were such officers as Lieutenant General Blackshear Bryan, who came to the Academy as Superintendent in 1954 from Korea, having commanded the 24th Infantry Division during the months of intense fighting in 1951.

If his officer-superiors and his instructors provided a reservoir of recent wartime experience on which the cadet could draw and to which he could look for inspiration, there also was inspiration to be derived from the occasional visits of even more legendary figures such as General Douglas MacArthur. MacArthur had returned from Korea in 1951 a martyred hero. Cadets were told by their instructors that President Truman had been fully within his rights in removing a military commander whose actions had been insubordinate. However, dismissal from his command could not diminish the luster associated with MacArthur's fifty-two years of distinguished military service. On the contrary, the principle that MacArthur had espoused in his insubordinate speeches and letters, "There is no substitute for victory," attained the kind of venerated status at West Point that had been accorded at Annapolis to the dying words of Captain James Lawrence: "Don't Give Up the Ship." Far from "fading away," as MacArthur had told the joint session of Congress he was doing at the conclusion of his speech to them upon his return home in 1951, MacArthur's spiritual and physical presence were felt at West Point even more intensely throughout the 1950s than they had been in the recent past.

Cadets were inspired also by the knowledge that the man who had fired MacArthur had yielded the White House to another West Pointer, Dwight D. Eisenhower (USMA 1915). The omnipresence of revered alumni, combined with abundant cues which Academy officials provided, effectively emphasized the priorities among organizational values; first Spartan, then Athenian ones. There was little evidence of a desire for major changes among cadets. To the contrary, the evidence was that the cadet who made it through the ordeal of "beast barracks" followed by four years of severe regimentation was likely to take a fierce pride in the maintenance of the system as it was. A survey of the Class of 1954 that was conducted by the *Pointer*, the cadet magazine, a few months before the respondents were to graduate, found that approximately eight cadets in ten would make the same decision to enter West Point as plebes if they had it to do over again. Moreover, two-thirds of the respondents indicated that they would like to return to the Academy as instructors or as tactical officers.[3]

The attractiveness of the Military Academy to prospective applicants, however, seemed to be declining. In 1954, the Academy filled only 67 percent of available vacancies, continuing a downward trend since the first year of the Korean War, when 82 percent of available vacancies had been filled. Moreover, the recruitment base was changing, with heavier reliance than in the past on sons of military personnel and upon young men from working-class families. Whereas nearly a third of the corps of cadets on the eve of World War I were sons of men in civilian professions, such as medicine and law, by the eve of World War II the percentage had declined to 20 percent; by the mid-fifties it was down to 10 percent. In contrast, membership in the corps recruited from sons of skilled workers had fluctuated between 5 and 10 percent from the turn of the century until 1950, when it began to rise slightly, to about 17 percent in 1953.[4] Roughly one cadet in five in the mid-1950s was an "Army brat." These statistics were sufficient to prod the Academy to devote additional time, energy, and resources to recruitment and to public relations.[5] However, major change in the substantive programs of the Academy would await new leadership.

Selection of Garrison Davidson as Superintendent

If one looks to career background for clues, the appointment of Major General Garrison Davidson as Superintendent of the Military Academy, to assume command beginning in July 1956, suggested that the status quo would be maintained rather than challenged. Indeed, there is no evidence that those who appointed Davidson to the post (most notably the Army Chief of Staff, General Maxwell Taylor) thought otherwise, or that they prompted him to initiate changes.[6] Davidson had had an exceptionally long and close relationship with West Point: as cadet, as instructor, and (like Lieutenant General Blackshear Bryan, his predecessor as Superintendent) as football coach. His affection for the institution was such that even after he retired, to reside 3,000 miles away in California, his wife would chide him good-naturedly, "You've never really left West Point." "She's right," he would confess, "spiritually I never have."[7]

Gar Davidson had been admitted to West Point as a plebe in 1923, the first alternate for a vacancy that became open to him when the principal appointee, a close friend, was discovered to be color-blind. Davidson's father, who had fought as a National Guardsman in the Spanish-American War, and revered the Military Academy, had never pressured Gar to apply, but was intensely proud when his son was accepted as a cadet. Young Davidson was delighted also, the more so because the family income could not have stretched to permit him to go on to college otherwise. He marvelled then, as he would do so frequently in the future, at the equality that was fostered among classmates.

> It impressed me that the day we entered all identity with the past was lost, except for our names, that we received the same clothes, the same food, the same income, the same allowance, the same everything. From that day on, we each stood on our own two feet and our progress depended solely on our individual ability. It permitted a graduate from a high school on the lower east side on New York [such as Davidson] to compete successfully.[8]

Davidson excelled as a cadet. His military aptitude so impressed Academy officials that at the outset of his first-class year he was selected to lead the indoctrination and training of the incoming plebe class (as "king of the beasts"), and to become the number two cadet in command assignments during the academic year. He was a starting member of the football team, served as a member of the twelve-man honor committee, and graduated twenty-second in order of merit (based primarily upon academic performance) in a class of 203.

From an engineering assignment after graduation, Davidson was able to return each fall to West Point on temporary duty as an assistant football coach. Beginning in 1930, he was assigned to the coaching staff on a full twelve-month basis. For various periods, these duties placed him also in the classroom as an instructor in natural and experimental philosophy (later termed "mechanics"), at the side of the Superintendent as his aide, and in contact with a company of cadets as their tactical officer.

Earl "Red" Blaik, who had graduated seven years ahead of Davidson from West Point and was serving with him as an assistant coach, was the man some persons thought should be elevated to the top spot when the head coach was reassigned in 1932. However, following the traditional policy of filling the position with an active-duty officer, Davidson was made head coach. He was, he reports, "flabbergasted, [but] I was overjoyed as never before. . . . Football was my love, really my vocation." Blaik stayed on briefly as Davidson's assistant, but then resigned abruptly to become head coach at Dartmouth.[9]

During his five years as head football coach, Davidson came to regard his staff "almost as a family." Marriage and the birth of his first two children added to the special fondness that he would come to feel for these years. With war imminent, however, Davidson and his coaching staff were disbursed to new assignments. Combat soon took its toll. Four of the "football family" died; two others became amputees. But four of the coaching staff who in the 1930s had been company grade officers emerged from the war as general officers, testimony to outstanding performance. These included Davidson himself, on whose shoulders General George Patton pinned a pair of his own stars at the close of the Sicilian campaign.[10]

Davidson served in stateside engineering and staff assignments in the early postwar period, until the Korean War broke out and he became Assistant Division Commander of the 24th Division in Korea. The latter role placed him in command of a variety of task forces, fighting first the North Koreans and then the Chinese.

In 1954, having returned to a stateside assignment, he received a phone call from General Matthew Ridgway (who had succeeded MacArthur as Commander of United Nations forces in Korea, and now was Army Chief of Staff). Ridgway had two important positions to fill: the Superintendency at West Point, and the Commandant's post at the Command and General Staff College at Fort Leavenworth. Davidson was told that he and his old coaching colleague from the 1920s,

Blackshear Bryan, had been the key candidates under consideration for each of the jobs; Ridgway had decided to assign Bryan to West Point and Davidson to Leavenworth. Two years later, when Bryan was promoted to lieutenant general and was to be reassigned from West Point, Davidson wrote to General Maxwell Taylor, who had succeeded Ridgway as Chief of Staff. He had a long devotion to the Military Academy, Davidson explained, and although he had never before requested a particular assignment, he would like to be considered for the Superintendency. In March 1956 Davidson learned that he had been selected.[11]

Davidson's First Year as Superintendent

Davidson himself recalls little influence of the Leavenworth experience upon his approach to his Academy responsibilities. However, as Ivan Birrer's assessment suggests, at the very least the earlier command provided evidence of a penchant for organizational reform which Davidson would display again at West Point. From a perspective of thirty years of service as education advisor to a long succession of Command and General Staff College commandants, Birrer has identified changes which Davidson initiated in the mid-1950s as having determined the fundamental thrust of the Leavenworth program for another twenty years.[12]

However, Davidson moved cautiously during his first year as Superintendent, as he had in his first months at Leavenworth, getting acquainted with his staff and faculty and sizing up the situation. General Bryan had assured Davidson that the Academy was "in fine shape."

> It is not perfect, but I think that the plans and policies now in existence are sound and can be defended easily in the event of any attack [from critics]. The morale of the Corps is excellent; they work hard, and they are a credit to West Point. They are well-disciplined, and they know well that if they step off base they pay for their sin.[13]

However, the recruitment problem had become urgent. The number of unfilled vacancies for cadet appointments to West Point had been growing since early in the Korean War. The opening of the Air Force Academy in 1955 compounded recruitment difficulties, because from its first year the new academy attracted greater numbers of applicants than did its older competitors. Bryan had taken a number of steps designed to combat the problem: cadet speaking tours; films, television programs, and even a cartoon book on West Point; contacts with high school counsellors and educators and with Boy Scout groups.

Davidson expanded on this recruitment-publicity effort in various ways: (1) the pool of potentially acceptable candidates was expanded by shifting from exclusive reliance on College Entrance Examination Board tests to a "whole man" selection procedure which supplemented academic achievement tests with evidence of character, leadership potential, and physical fitness; (2) increased effort was made to appeal to Academy alumni to assist in the search for and encouragement of prospective applicants; (3) an office of registrar and admissions was created with an admissions information branch as one of its principal components; (4) a registrar and director of admissions was appointed, with tenure on the same status as permanent professors; (5) each member of Congress received a letter from the Superintendent encouraging the congressman to permit the Academy's Academic Board to designate the principal appointee from among four candidates nominated for each vacancy (thereby in effect giving the Academy a much larger pool of applicants from which to choose); (6) an effort was made—unsuccessful during Davidson's superintendency—to persuade Congress to revise the statutes so as to base the number of appointments to the Academy annually on the average annual strength of the Academy during the preceding year rather than on the anticipated strength of the Academy at the date of entrance of the new class (thus increasing the number admitted).[14]

Although time pressures before his assumption of command had precluded preparation of a firm "plan of action," Davidson had arrived with a number of concerns beyond the physical-plant modernization and the recruitment matters conveyed to him by Bryan. Recollections from his own cadet and staff experiences at West Point were the genesis of one set of concerns. (Could not increased academic challenges be provided for cadets who were especially bright or who entered the Academy with previous college experience? Why were the subjects of "military art" and engineering married in a single department?) Other concerns stemmed from impressions that Davidson had gained in later years as an interested alumnus who kept track of developments at the Academy through the Association of Graduates and through informal contacts. (Should not the study of ordnance and of engineering be coordinated through a single academic department, with new emphasis on the latest developments in electronics? Had the factors that had caused the breakdown in the honor system in 1951 been eliminated or corrected?)[15]

By the end of his first year as Superintendent, Davidson had formulated not only a set of goals but also a strategy for achieving the goals. In terms of insight that they provide into the institutional prerequisites for successful reform at the Academy, the steps that Davidson took to pave the way for change are as interesting as the substance of the changes that eventually were effected. The strategy combined the use of extensive staff studies, an approach to problem solving and reform that had a well-established tradition in military (as well as civilian) bureaucracy, with the use of surveys of the opinions of relevant constituencies, a technique that was relatively novel for a military commander to employ. The strategy had the dual advantage of developing a massive informational base from which to proceed, while disarming numerous potential opponents of change by the flattery of seeking their advice and by posing questions that encouraged respondents to consider problems or deficiencies that the Superintendent wished to remedy.

In the spring of 1957, a questionnaire was administered to all members of that year's graduating class, seeking their opinions regarding virtually all facets of their cadet experience. Concurrently, a 24-page questionnaire, supplemented by a 14-page booklet for open-ended comments, was mailed to some 14,000 graduates from the USMA classes of 1900-1954. A few months later, a similar survey instrument was mailed to a sample of Army officers with commissions from sources other than the Academy, including (1) a random sample of officers of all grades on active duty; (2) all active and retired three- and four-star generals; (3) all active and a small group of retired major generals.[16]

Even before the data from these surveys were fully tabulated and analyzed, the Superintendent had appointed special study groups and committees to prepare analyses of various problem areas and to submit recommendations for change. A board of three senior officers was appointed to study the honor system. Still another, under the direction of the commander of the Second Regiment of Cadets, Colonel Richard Stilwell, was to study the system of measuring military aptitude among cadets and developing military leadership. Another, under the Assistant Commandant, Colonel Julian Ewell, was charged with determining the qualities and attributes that military officers would require in the years 1968-1978. Once the Ewell Committee had made its report (January 1958), follow-on studies would be launched of the historical evolution of the academic curriculum from the founding of the Academy to the end of World War II, and of the present curriculum and future trends.

From Analysis to Action

Never in the history of the Military Academy had such a substantial portion of its staff, faculty, student body, and alumni been engaged in contributing to a critical review of Academy programs. Having raised the consciousness of those whose support he needed, Davidson began to enact a program of reform, sometimes armed with studies or surveys that demonstrated the need for change, sometimes in response purely to his own instincts and observations.

Honor and Athletics

The honor system, linchpin of the Academy tradition and reputation, had been at the top of Davidson's list for critical review. When the board that he appointed to review the honor system reported that there were "no important deficiencies" in the system *per se* (that is, in the methods used to teach cadets the honor code and in procedures used to operate and enforce the system), Davidson accepted the finding. Yet, he had felt at the time of the 1951 honor scandal that the root cause of the incident had been "loss of perspective of a few at West Point because of overemphasis [on] football."[17] Consequently, he felt that it was imperative to ensure that there was no recurrence of this loss of perspective.

As a former football player and coach himself, naturally Davidson believed in maintaining an effective football program at West Point. More importantly, however, he believed in the value derived by all cadets from participating in physical training and athletics, whether the latter were in varsity or intramural competition, and whether the sport be football, swimming, or something else.[18] On the eve of the 1957 football season, Davidson issued a formal statement of the principles that were to govern the intercollegiate athletic program of the Military Academy. The statement emphasized that the program was designed

> to provide opportunity for all cadets representing the Military Academy in athletic contests, and to prepare properly so that they can represent the Academy with distinction WHILE CONTINUING THEIR PROPER DEVELOPMENT TOWARD BECOMING CAREER OFFICERS IN THE REGULAR ARMY.[19]

The effort to encourage other sports rather than maintain an extremely disproportionate allocation of attention and resources to football was bound to put Davidson on a collision course with his old colleague, now head football coach, Red Blaik—however much the two of them might try to avoid a showdown.[20] Yet it was hardly a propitious time for Davidson to be challenging the claims of football for attention and resources at the Academy. Blaik, his own reputation and the fortunes of his team near rock bottom following the dismissal of nearly the entire football team (including reserves) in the 1951 honor scandal, had done a dramatic job of rebuilding.

Nevertheless, Davidson was explicit about the implications of his August policy statement for athletic recruitment. First, coaches in all sports were expected to recruit young men who not only had demonstrated superior athletic ability, but who also had the qualities of character, intelligence, and leadership ability that were desired in career officers. Secondly, among the vacancies to the Academy that were earmarked for recruitment of athletes, less preponderance than in the past was to be given to football.

Blaik argued that "equality for all sports" would be a splendid ideal "if there were enough appointments to go around. But since there were not, football must be first established on a solid footing, if for no other reason than self-preservation." Football, after all, paid the bills—"for eighteen years we had underwritten the school's entire intercollegiate program," Blaik noted. Furthermore, the Army football team had to compete against major teams in schools with recruitment programs that made the one by West Point look puny in comparison. Whereas in most sports Army could adjust its schedule from year to year, depending upon the caliber of manpower available, big-time college football required one to schedule several years in advance—and Army was facing some tough teams in the years ahead.[21]

Davidson was unpersuaded by Blaik's argument. Although Blaik had insisted that thirty-five was the absolute minimum number of football appointments that were needed in the incoming class of cadets, Davidson permitted Blaik only a guaranteed minimum of twenty-eight. Other available appointments were to be allocated equitably among other sports.

Still unreconciled to Davidson's athletic recruitment policies, Blaik again found himself stymied when Army received a bid to play a post-season game in the Cotton Bowl and Davidson announced that the Military Academy could not accept. Early in the 1958 season Pete Dawkins, the team captain, had asked Blaik if the team could go to a bowl game if they compiled an undefeated season. Blaik "promised him I would not oppose the wishes of the squad" (but apparently did not consult the Superintendent at this stage). Undefeated after victories over such powerful teams as Notre Dame and Rice, and tied only by Pitt, in November the Army team received an invitation to play in the Cotton Bowl. Blaik convened the Academy Athletic Board, which he headed, and found them unanimously in support of accepting the invitation; Blaik himself thought it proper to abstain from voting.[22]

Davidson recognized that a decision to decline the invitation would be an unpopular one. However, at issue was a challenge to the philosophy that he was trying to promote, and football had to be kept in perspective with the primary mission of the Academy. "In my opinion," he noted in explaining his decision, "it is not in the best interests of the accomplishment of our mission to participate in post-season football bowl games."[23] In January 1959, Blaik resigned to become vice-president of AVCO Corporation. The announcement was made without warning, as his resignation from Davidson's coaching staff in 1934 had been. (Blaik reports that even his own family was caught by surprise at his resignation in 1959.)[24] Davidson expressed his regret at the resignation privately to Blaik and publicly to the press and to alumni, noting that "Red is the finest football coach in the country."[25]

There is no doubt that Davidson was genuine in the admiration that he expressed for Blaik's skill as a coach. Still, the resignation proved fortuitous in Davidson's efforts to realize a more balanced athletic program. Whereas Blaik had served concurrently as football coach, director of athletics, and chairman of the Athletic Board, Davidson was now able to fill the positions with three separate individuals. Moreover, in moving earlier to elevate the position of Director of Physical Education (occupied by still a fourth individual) to one with tenure, comparable to that of a permanent professor, Davidson was underscoring his desire to have the program of athletics and physical training serve all cadets, and not just a select elite.

Military Discipline and Training

The realm of military discipline and training was one which Davidson could approach without arousing old animosity, as he had with Blaik in athletics. Curiously, however, again the key figure with whom Davidson had to deal was one he knew from earlier experiences in football. Brigadier General John Throckmorton, Commandant of Cadets from the spring of 1956 through the summer of 1959, had played B squad (junior varsity) football as a cadet when Davidson was coach. Throckmorton had been appointed while Bryan was still Superintendent, although General Bryan had courteously sought Davidson's concurrence in the appointment.[26] Davidson was delighted to have Throckmorton as his Commandant, because he brought to the role an unusual combination of credentials.[27] On the one hand, he had compiled an enviable record of performance in combat and in the Pentagon (each in its own way vital to success in a military career). On the other hand, Throckmorton had had an exceptionally varied and extensive series of assignments at West Point.

Rising to the temporary grade of lieutenant colonel less than eight years following his 1935 graduation from West Point, Throckmorton had served throughout much of World War II as Assistant G-3, First Army, in combat in Europe and in the Pacific. After the Korean War began, Throckmorton had been given command of a regimental combat team, earning a Distinguished Service Cross, a Silver Star, and an oak leaf cluster to the Legion of Merit that he had won in World War II. From Korea, he had been assigned to Washington to serve as aide to the Army Chief of Staff, General J. Lawton Collins. Interspersed with these assignments in the field and in Washington, Throckmorton had had three return tours to the Military Academy. He had served briefly as an instructor in chemistry in 1941-42. Then after the war, he had returned during the Maxwell Taylor superintendency to serve first as S-3 in the tactical department and then as commanding officer of the First Regiment of the Corps of Cadets. In 1955, after a brief tour in the Office of the Secretary of Defense, Throckmorton had returned to West Point again, this time as Chief of Staff to the Superintendent, General Bryan. He was made Commandant in April 1956.

Throckmorton worked hand in glove with Davidson in the design of reforms of the program of military training and discipline. An early policy decision, of modest significance in retrospect but radical to Academy personnel at that time, was the introduction of "Project Equality" in the assignment of cadets to companies. For years, the single criterion applied had been height; cadets of the same size were assigned to the same company. The result was that there was no more than an inch or two variation in height among cadets in any given company; therefore, each company presented a uniform appearance when marching in parade. Indeed, appearance on the parade ground was the prime—perhaps the only—rationale for assignment by height. Over the years, most cadets had come to enjoy the usually friendly rivalries that developed between those in the short companies ("runts") and those in the tall companies ("flankers"). Yet Throckmorton and Davidson decided that the competitive disparities among companies (for example, in intramural athletic competition) inherent in such a system, and the benefits to be gained by applying other criteria in the assignment process, outweighed the inevitable cost associated with any departure from customary practice. Thus, beginning with the class that entered the Academy in the summer of 1957, assignments to companies were made in such a way as to attempt to gain an equitable distribution of cadets according to scholastic ability, physical ability, leadership potential, and height.[28]

The summer of 1957 was also one in which some initial steps were taken in the direction of smoothing the transition from cadet to junior officer, an improvement that both Throckmorton and Davidson thought necessary. Cadets of the upper two classes who were making trips to military installations around the country were granted the same privileges regarding access to beer and liquor that were available to junior officers. Beginning in September, the system of authorizations and privileges for first-classmen was expanded, granting them privileges that for previous classes had been doled out gradually as graduation day came closer (for example, opportunities to go outside the Military Academy gate on afternoons or weekends). In return for expanded privileges, first-classmen were asked to assume increased responsibilities for running the corps, taking on duties such as inspection that previously had been performed by the company tactical officers. Cadets also were alerted that privileges that had been granted would be withdrawn in the event of substandard performance in any realm, ranging from their leadership responsibilities to academics.

Greater familiarity with the realities of life as a junior officer was accorded also by a change made in the summer training for second-classmen. Throughout most of the post–World War II period, second-class summer had been used to provide the cadets with an orientation to the various arms and branches of service. A few days aboard ship with the Navy would be followed by a brief stopover at the Transportation Center at Fort Eustis, then on to the Quartermaster School at Fort Lee. A brief tour of the Helicopter School at Fort Rucker would be followed by extensive exposure to combined-arms training at the Infantry School at Fort Benning. Air Force indoctrination would be provided at Maxwell Air Force Base, followed by a tour of the Engineer School at Fort Belvoir, a look at the Signal Corps at Fort Monmouth, and then back to West Point. In terms of actual training accomplished, it was, as General Throckmorton has put it and as most cadets who participated in the training realized, "an expensive boondoggle."[29] Thus, beginning in the summer of 1958, the orientation trip was reduced from seven weeks to two. For the remainder of the summer, one-third of the second-classmen were assigned to assist first-classmen in the conduct of "Beast Barracks." The remainder of the second class were assigned as assistant platoon leaders to regular Army units with the 2nd Infantry Division, the 101st Airborne Division, and the 82nd Airborne Division.

The Academy training and disciplinary practices that had been most at variance with those in the regular Army units were those associated with plebe year. Within plebe year, the eight weeks that the plebe experienced from the time he entered during the first week in July were by far the most grueling. It was true that the initial training for new cadets had elements in common with the basic training that Army recruits experienced: the physical conditioning, the introduction to drill, familiarization with the rifle, the denial of most privileges and opportunities for liberty or relaxation. However, the initiation to cadet life was basic training increased several-fold in intensity. Hazing, in the sense of outright physical abuse, had been proscribed at the Academy since congressional legislation in 1901. Yet such proscription did not prevent some of the upperclassmen who were charged with putting new plebes through their paces from subjecting their charges to emotional torment and making demands upon them for performance that would bring them to the point of physical and mental exhaustion. It was an ordeal that had become termed "Beast Barracks" with reason. The new plebe learned that regardless of the status which he may have enjoyed as a high school class president or an all-state athlete, or, in a few cases, as a combat veteran, he was now a "beast" until he proved himself by surviving the severe test of new-cadet indoctrination, training, and discipline.

Throckmorton's recollections of his own experience as a plebe were not fond ones, and he was determined "to see to it that the plebes were 'led' rather than stomped on all the time. . . ."[30] However, a review of the existing system proved inconclusive, except to reveal that little had changed since the MacArthur superintendency, when customs associated with the system first had been codified.[31]

Morale and Moral Leadership

The quest of Throckmorton and Davidson jointly to revitalize cadet morale and to upgrade the process by which leadership and character were developed extended even to the Sunday chapel service. The Protestant chaplain of the Korean War period, Frank Pulley, had been replaced in 1954 by a younger man. The selection process had been rather casual. A mimeographed letter describing the job and announcing the salary ($5,200 per year) had been widely disseminated among clergymen. The Commandant of Cadets, two faculty members, and Chaplain Pulley reviewed applications, after which a selected few of the applicants were invited to the Academy in the spring of 1954 as "guest preachers." The Reverend George Bean, then serving at Saint Mark's Episcopal Church in Richmond, Virginia, was selected after this first-hand appraisal. The deciding factor, Bean believes, was the favorable impression that the Commandant developed while sitting next to Bean's wife at lunch, after which he was heard to remark, "That is just the kind of chaplain's wife we really ought to have at West Point."[32]

Although the selection of Bean maintained a tradition that had been broken only a handful of times since the founding of the Military Academy in 1802, in the appointment of Episcopalians to the post, Bean quickly proved to be decidedly different from his predecessor in his approach to the chaplaincy. Not only did his relative youth and informal style enable more cadets to identify with Bean than had been able to do with Pulley, but also his sermons sounded themes at variance with those with which Protestant churchgoers at West Point had become familiar. Bean's baccalaureate for the Class of 1955, for example, focused on Paul's entreaty to the Romans to "be not conformed to this world." "Perhaps no greater temptation awaits you as you go off to Fort Benning and Moore Air Force Base," Bean told the graduates, "than temptation to conform, to follow the line of least resistance, to become one more cog in the mass production line of American culture." He urged his audience to resist this temptation, so that they could become "part of the solution" to life's problems, rather than "part of the problem."[33]

Whatever inspiration and challenge Bean may have provided, especially in contrast to his predecessors, the fact of compulsory chapel remained; thus, it would be fatuous to describe the attitudes of cadets toward the Sunday chapel as typically enthusiastic. On the contrary, nearly 40 percent of respondents to a cadet-conducted survey of the Class of 1954 had indicated that they would not attend chapel weekly if attendance were voluntary.[34] When Academy officials dared to raise the issue of compulsory chapel in their own surveys of first-classmen, starting with the Davidson-initiated survey in 1957, they found fewer than 50 percent of the classes of 1957, 1958, 1959, and 1960 agreeing that "chapel attendance at West Point should continue to be compulsory."[35]

Not long after his arrival at West Point, Bean himself had become an opponent of compulsory chapel. To make worship compulsory was a contradiction. Moreover, it was evident to him that "compulsory chapel was doing far more harm than good . . . there was a strong feeling of opposition in the Corps and I know of a number of cadets whose relationship to God was blunted because they felt so strongly on this point."[36]

Bean's protestations to Academy officials, however, got nowhere. The rationale given was that "since religion was not taught anywhere within the curriculum that a basic part of a cadet's moral and religious education would be provided by what happened to him in chapel on Sunday morning. It was also said that since everything else at West Point was involuntary, something as important as chapel should not be made an exception."[37] As Throckmorton notes, "I don't recall much if any *serious* discussion about making [chapel attendance] voluntary. As a matter of fact, I seem to recall we resisted strongly efforts by the ACLU [American Civil Liberties Union] to do just that."[38]

Davidson never established the close rapport with Bean which Bryan, the former Superintendent, had maintained. From Davidson's perspective, Bean was "a nice guy," but lacked the qualities desirable for the chaplain's role. "He was more a companion than a 'father.' "[39] Well publicized criticism of West Point by the Military Chaplain's Association in the spring of 1958 for the alleged "calculated and unwarranted discrimination against other denominations" may have provided an incentive for considering a non-Episcopalian replacement for Bean.[40] In Bean's view, however, the allegation was but a thinly veiled effort to persuade the Academy to hire military chaplains in lieu of civilians.[41] Moreover, Davidson has discounted the impact of the criticism on his thinking. It was one often heard in the past; but to the best of his knowledge, there had been no bar to the appointment of non-Episcopalians. However, the fact that he was an Episcopalian himself and the grandson of an Episcopalian minister put him in a good position, he thought, to break with the traditional pattern in the next appointment.[42]

The key consideration in the minds of both Davidson and Throckmorton, who joined in the search for a replacement to Bean, was that "Protestant cadets needed someone like Monsignor Moore [the senior Catholic chaplain].[43] They found the man they wanted in Theodore C. Speers, then pastor of the Central Presbyterian Church on Park Avenue in New York. He was, Davidson recalls, "made to order"—a big man, once hammer-throwing champion at Cambridge University, who had served in World War I as an artillery corporal and retained a fondness for the military, and possessed a booming voice. His appointment, Davidson contends, was "the best decision I made at West Point."[44]

Academic Reform

The appointment of a new chaplain may be recalled by Davidson as his "best decision"; but it was not the initiative that had the most enduring impact on the Military Academy. The latter came in the realm of academics. Because many of Davidson's ideas encountered stiff resistance, by no means were his plans for upgrading the academic program fulfilled completely during his superintendency. However, a period of academic ferment had begun at West Point during the Davidson years, with many of the proposals that were thwarted or "studied to death" during those years actually implemented under Davidson's successors.

Davidson could effect some academic reforms by decree; and he proceeded to do so not long after becoming Superintendent. For example, he announced that henceforth all officers who were appointed to permanent positions (as professors) would be expected to obtain their doctorate within a reasonable time after appointment if they did not hold the degree already. Moreover, a program of sabbatical leaves was instituted to encourage professors to maintain their intellectual development. Between sabbaticals, professors were to take an assignment of up to three months outside the Academy studying current and probable future requirements of the Army that were relevant to the West Point educational mission.[45] Informally, Davidson urged all his officers to write in professional journals.

Reform of the academic curriculum, however, required the assent of the Academic Board, the venerable institution that had managed to frustrate, if not entirely to thwart, such earlier reform-minded superintendents as Douglas MacArthur. In the early stages of Davidson's reform efforts, it appeared that his carefully formulated strategy of building a foundation of staff studies and surveys would enable him to avoid entirely the frustrations that predecessors had experienced.

The Ewell Board, charged with determining the qualities and attributes that Army officers would require to cope successfully with the challenges of the period 1968-1978, had been energetic in their exploration of alternative ideas and perspectives. They had sought the advice of military officers whose careers had been associated with new developments in military technology, such as Lieutenant General Leslie Groves (West Point Class of November 1918), who had headed the Manhattan Project and was serving as vice-president of Remington Rand. Discussions with civilian educators had brought committee members up-to-date on many recent developments in civilian institutions of higher education. Visits to the Air Force Academy had given committee members a chance to discuss first-hand with members of the Air Force Academy staff and faculty their reactions to innovations in "enrichment" and in academic organization that McDermott as Dean was beginning to put into effect.[46]

The heart of the Ewell Board's conclusions lay in their assessment that

> [The officer of the future] will require a wealth of knowledge and a capacity for applying that knowledge beyond any requirements of the past. Character, leadership qualities, intellectual capacity, and a fairly fixed body of knowledge are no longer enough. The inroads of physical science and political science into the military realm demand military leaders who are well based in these areas and who have the intellectual curiosity, the initiative, and the quality of creative thinking which will enable them to expand their base of knowledge in a flexible manner, and apply it to ever-changing situations.[47]

Following on the heels of the Ewell Board, a subcommittee chaired by Lieutenant Colonel Cranston E. Covell conducted a review of the historical evolution of the Military Academy curriculum. A parallel subcommittee, chaired by Colonel John Jannarone (who would become Dean of the Academic Board in 1965) assessed the current curriculum. The reports of the two subcommittees, supplemented by the broad guidance that had been provided by the Ewell Board, became the basis for the synthesis attempted by an eight-man evaluation committee headed by Colonel Walter J. Renfroe, Jr., professor of foreign languages.

In 1954, Renfroe had been a member of the Green Board, whose recommendations in favor of the retention of the prescribed curriculum had served to maintain the status quo. However, General Davidson had let the committees that he had appointed know that he felt that change was imperative if the Academy were to carry out its mission of preparing apprentice officers for the challenging demands of the future. He had mobilized approximately thirty officers from the staff and faculty on the various committees whose studies were to provide a solid analytical framework within which to design needed reform. Graduates of the Academy had been asked to send in their suggestions for needed changes. As Davidson had informed the Army Chief of Staff in his annual report for 1957, this would "represent the most complete survey of our curriculum that has been made in well over half a century. It will be a monumental task but one whose fruit I believe will be well worth the effort."[48]

In short, it was almost unthinkable that the Evaluation Committee, charged with synthesizing the findings of the other committees and presenting recommendations, would follow the Green Board precedent and essentially endorse the status quo. Nevertheless, the report of the Evaluation Committee was cautious in tone. The committee began by recommending "that the curriculum of the Military

Academy be an essentially standard one. . . ." The modest qualification to this recommendation, however, was significant in the light of the deep roots of the prescribed curriculum: the committee added that there should be "increased opportunities for accelerated and advanced work throughout [the cadet's four years] and at least two elective possibilities in the last two years."[49]

Some of the other proposals of the Evaluation Committee would move the existing program beyond its seminary-academy limitations. For example, the committee called for the establishment of a fine arts committee "charged with the responsibility for developing and monitoring a positive program designed to stimulate the interest of cadets and expand their knowledge and appreciation of the fine arts." They urged that "wider use be made of essay type questions in examinations and laboratory reports," and that "less emphasis be placed on daily grading and less frequent reviews be scheduled in upperclass sections."[50]

However, most of the forty-six recommendations of the committee were either platitudinous in nature (for example, departments should review their teaching methods constantly), or were merely calls for tinkering with the existing structure of the program—shortening the time devoted to a course here, lengthening the time devoted to another course there. (Because all cadets followed the same program, it had been possible to have courses ranging in length from a few weeks to one or two semesters, rather than have them all uniformly run the full semester or two.)

The report of the Evaluation Committee did not represent as thorough a mandate for academic reform as General Davidson had hoped to receive. However, a possibility for expanding the mandate lay in the review of the committee's recommendations which Davidson had asked for from a group of civilian educators and senior military officers.

The Curriculum Review Board consisted entirely of persons familiar with and sympathetic to the Academy. Generals Alfred Gruenther (Davidson's brother-in-law) and Anthony McAuliffe had graduated from West Point in 1919 and retained close contacts at the Academy. Troy H. Middleton, president of the Louisiana State University, although not a West Point graduate, had retired from the military in the grade of lieutenant-general. Moreover, he had served as a member of the Board of Visitors to the Military Academy in 1951, 1952, and 1953. John M. Kemper, headmaster at Phillips Academy, Andover, Massachusetts, was a classmate of Brigadier General Throckmorton (USMA 1935) and had served in the Army through World War II, resigning in the grade of lieutenant colonel in 1948. The other three members of the Curriculum Review Board were also members of the Board of Visitors to the Military Academy concurrently.

Support by the Curriculum Review Board of Davidson's ideas for academic reform could be important in at least two ways. First, its conclusions could be used to exert leverage on members of the Academic Board who might prove resistant to change. Secondly, a favorable response in the Curriculum Review Board to Davidson's views virtually assured support in the Board of Visitors.

In a meeting with the Curriculum Review Board at West Point in December 1958, Davidson told the members that "the recommendations of the local Evaluation Committee had not gone far enough in proposing modifications to our curriculum." Davidson then made his own recommendations, which, in their report in March 1959, the Curriculum Review Board substantially endorsed. Thus, with the support of the Curriculum Review Board, Davidson was able to forward a series of recommendations to the Academic Board. The key objectives of the recommendations were to:

> (1) Provide opportunity through electives for each cadet to go deeper in an area of his own choosing and for which he has a natural liking and aptitude.
> (2) Increase the opportunities for acceleration for the brighter cadets in accordance with their abilities.
> (3) Expand the basic scientific content of the curriculum, particularly with regard to instruction in nuclear physics and the solar system.
> (4) Provide for increased emphasis on the social sciences and humanities, including the communication skills.
> (5) Decrease the vocational instruction during the academic year to the minimum essential.[51]

The Academic Board 1959-60

The struggle for academic reform now was in the key arena: the Academic Board. (See Table 2.) Although members of the board in 1959-60 were, with the exception of the Commandant, men in their fifties or sixties, as hidebound board members in earlier decades had been, they lacked the kind of marathon service on the board that had been characteristic at the turn of the century. Nine of the 1959-60 board members had been among those designated to fill permanent professor positions created in the postwar reorganization during Maxwell Taylor's superintendency. Like Taylor himself, they had returned to the Military Academy cognizant of the need to update cadet education and training, and bringing with them valuable wartime observations and experiences.

Table 2 The Academic Board, U.S. Military Academy, 1959–60

Position	Board Member	Age as of 1-1-60	Reported USMA This Assignment	Highest Academic Degree	Undergraduate Education
Superintendent	Garrison H. Davidson	55	1956	BS	USMA 1927
Dean of the Academic Board	William W. Bessell	58	1946	BS	USMA 1920
Head, Dept. of Electricity	Boyd W. Bartlett	62	1942	PhD	USMA 1919
Head, Dept. of Physics and Chemistry	Edward C. Gillette	59	1946	MS	USMA 1920
Head, Dept. of Military Topography & Graphics	Lawrence E. Schick	62	1946	BS	USMA 1920
Head, Dept. of Law	Charles W. West	61	1943	LLB	USMA 1920
Head, Dept. of English	George R. Stephens		1946	PhD	Princeton 1921
Head, Dept. of Foreign Langs.	Charles J. Barrett	59	1947	BS	USMA 1922
Head, Dept. of Military Art & Engineering	Vincent J. Esposito	59	1947	BS	USMA 1925
Head, Dept. of Mathematics	Charles P. Nicholas	56	1948	BS	USMA 1925
Head, Dept. of Mechanics	Elvin R. Heiberg	55	1949	Dipl. HE	USMA 1926
Head, Dept. of Ordnance	John D. Billingsley	55	1951	MBA	USMA 1928
Head, Dept. of Social Sciences	George A. Lincoln	52	1947	MA	USMA 1929
Head, Dept. of Military Hygiene	Philip W. Mallory	50	1959	MD	Texas 1930
Commandant of Cadets	Charles W. G. Rich	49	1959	BS	USMA 1935

Yet the Academic Board represented a group with deep institutional ties. The vast majority of board members (including Davidson, who chaired the meetings) had been nurtured in the Thayer system as cadets. Bonds to the Academy which had been forged as cadets had been supplemented in many cases by tours of duty at West Point in the 1920s or 1930s, before the return to the Academy faculty permanently in the World War II or postwar period. The only two board members who were not graduates of the Military Academy were the heads of two of the departments with the least prestige and influence: the Departments of English and Military Hygiene. Yet each of these men had direct family ties to West Point. George Stephens's daughter had married a Military Academy graduate, and Philip Mallory's son was a cadet in the Class of 1961, concurrent with his father's service on the Academic Board.

Of course, the "old school" ties among board members did not necessarily work to General Davidson's disadvantage. As a West Point graduate himself, with extensive subsequent experience on the Academy staff, he could and did appeal to fellow graduates in terms of the necessity for the Academy to adapt to a changing world in order to maintain the West Point standards of excellence. Moreover, he could and did reiterate his own dedication to Academy traditions. It was a Burkeian approach. Or, as he preferred to say, it was the approach his great-great-great-grandfather, Chief Justice John Marshall, might have used.[52] Modest reforms made now could avert the necessity for radical alterations of the Academy program—perhaps imposed from without—at a later date.

Beyond the general appeal of a common alma mater, Davidson had had close previous association with several members of the board. His cadet days had overlapped with those of Esposito, Nicholas, Heiberg, Billingsley, and Lincoln. Schick and West had been instructors at West Point in the 1930s when Davidson was there as a coach. And there were family ties: one of Davidson's sons had married Bessell's daughter.

However, the nature and composition of the Academic Board also served to limit whatever authority Davidson might otherwise have been able to wield by virtue of his superior rank and position. His leverage was reduced, in the first place, by the general recognition that he would be leaving the Military Academy soon for another assignment (although, as noted below, he volunteered to stay on for an additional four-year tour), while all the academic members of the board were assigned to West Point permanently. Moreover, although junior in grade to Davidson, most of the board were senior in age and in date of graduation from West Point. Four board members, including the Dean, had graduated from West Point in 1920. This was also the class of General Lyman Louis Lemnitzer, who had been Army Vice Chief of Staff during the period when the first curriculum studies under Davidson had been initiated (1957-59) and had taken over from General Maxwell Taylor as Chief of Staff in mid-1959. It was to Lemnitzer that any changes that Davidson was able to persuade the Academic Board to endorse would have to go for final approval.

A further factor reducing the probability of major change was the fact that each board member was a department head, jealously cognizant of the diminution of his domain that might result from modification of the curriculum. Although board members had one vote each, the authority with which they were able to speak about the curriculum as a whole was dependent to some extent upon the share of it over which they presided. Thus, the heads of two of the traditionally important departments, mathematics and military topography and graphics, and of the newly established "empire," social sciences, were more likely to be able to speak with authority of the cadet's academic needs and performance than were the heads of departments whose claims on cadet time were light, such as military hygiene, law, or ordnance. (See Table 3.)

Mathematics in effect had two voices on its behalf, the current department head and his predecessor, now Dean. Like the claims made on behalf of the traditional coursework in drawing represented in military topography and graphics, those made on behalf of the need of cadets for such extensive exposure to mathematics would come under increasing challenge by other board members in coming years. However, the senior members of the Department of Mathematics would continue to evoke not only the pragmatic arguments on behalf of mathematics as a skill useful in other disciplines, but also the traditional, almost mystical, claims about the value of mathematical training as "mental calisthenics."[53]

If in the champions of mathematics, Davidson faced zealous argument and fervent belief in the primacy accorded to this discipline, in other board members Davidson confronted men confident of their own abilities and judgment. Some had an intensity of academic preparation that Davidson himself could not claim. Philip Mallory, of course, had an M.D. degree. Boyd Bartlett had both a master's and a Ph.D. degree in physics from Columbia University, and had engaged in postdoctoral study in Munich. George Stephens, a Princeton undergraduate, had a master's and a Ph.D. degree from the University of Pennsylvania. Although the remainder of the board could not boast impressive academic credentials, collectively they represented a wealth of knowledge and experience, including high-level assignments in Washington during and just after World War II.

Table 3 Relative Importance of Military Academy Departments in Terms of Their Claims on Cadet Time During the Academic Year 1959–60

Claim on Cadet Time	Department	Contact Hours Over Four Academic Years
Heavy	Mathematics	424
	Military Topography and Graphics (MT&G)	360
	Social Sciences	321
Medium	Military Art and Engineering (MA&E)	282
	Mechanics	282
	Physics and Chemistry	262
	Electricity	247
Light	Foreign Languages	214
	English	180
	Ordnance	141
	Law	90
	Military Hygiene	26
Special	Tactics, including Military Psychology and Leadership (MP&L)	103 MP&L plus 473 other hours of training not normally considered in discussions of the academic program

Data are from Brig. Gen. William W. Bessell, Dean of the Academic Board, "The Modified USMA Curriculum," *Assembly* 19 (Fall 1960): 14, Table 1; and USMA, Stilwell Committee, "Superintendent's Curriculum Study: Report on Military and Physical Education Components," mimeographed (West Point, N.Y.: USMA, 26 Jan. 1960), Annex B.

The "Great Debate"

Throughout the autumn of 1959, the Academic Board agonized over possible changes in the curriculum. At stake was not only the preservation of local "fiefdoms" against possible intrusion, but also, in the minds of many board members, the preservation of the integrity of the Thayer system. It was, an observer close to the events of the time recalls, an era of "the Great Debate."[54] As General Davidson recalls, the confrontations that occurred in meetings of the Academic Board were "intense to say the least—but friendly. Blood was drawn but the wounded lived."[55]

The "Great Debate" climaxed in February 1960 with a meeting that ran for five successive days.[56] The results, not surprisingly, represented a compromise between proponents and opponents of change. Some of the changes which the board resisted were ones which later boards would find palatable. For example, Davidson's proposal that the Office of Military Psychology and Leadership be shifted from the Commandant's domain to that of the Dean, as an academic department, was rejected in 1960 but put into effect in the 1970s. Similarly, contrary to the recommendation of the Evaluation Committee that a single engineering department be organized, combining the ordnance department with the military engineering component of the Department of Military Art and Engineering, Ordnance was retained intact and MA&E remained an integral unit. In 1969, however, the merger was effected with the creation of a Department of Engineering. A new Department of History also was created at that time, incorporating "history of military art" components from the former Department of Military Arts and Engineering along with the history section of the Department of Social Sciences.

The internal politics that had been at work also were evident in some of the minor changes that the board endorsed. For instance, the name of the Department of Military Topography and Graphics (direct descendant of the nineteenth-century Department of Drawing) was to be changed to the Department of Earth, Space, and Graphic Sciences, with new emphasis placed on astronomy and astronautics in coursework offered. Davidson, who viewed much of the subject matter of MT&G as largely irrelevant to the needs of modern Army officers, observed that the department had "cut me off at the pass by hastily modernizing and reorganizing its course and changing the title of the department to be more descriptive of the new content. . ."[57]

The most notable breakthrough in decisions by the Academic Board came in the expansion of existing opportunities for students with advanced preparation to validate introductory courses and take accelerated course work instead, and in limited inroads on the prescribed curriculum through the provision of electives. Specifically, one elective would be available to cadets during each semester of their senior year. The idea of permitting cadets to develop an academic major was rejected (as it would be repeatedly into the 1970s). However, the electives to be made available were to be grouped into two broad areas of concentration: science-engineering-mathematics, and social sciences–humanities.

One final issue remained to be resolved by the Academic Board. Should one or two additional electives be provided by shifting the two-semester sequence in civil engineering during first-class year from a prescribed basis to an elective one, in whole or in part (one semester)? Those who argued for retention of at least one semester of the sequence on a prescribed basis argued that "actually there is no engineering as such" in the coursework which the cadet receives during his first three years, only courses which lay the foundation for engineering. It is in his final year, they contended, that the civil engineering course provides a "synthesis" of the various engineering-related subjects that had been introduced earlier. Thus, such a course was "indispensible as a finale" in the curriculum.[58]

Those who favored making the civil engineering sequence totally optional (a group which included the Superintendent) argued, to the contrary, that it still would be possible for the cadet to elect the sequence in civil engineering. Because such a course would be elective, the entire academic program of first-class year "would be challenging and hence more fruitful since part of it would be subjects of the cadet's preference."[59]

When the meeting was adjourned, Davidson submitted the recommendations of the Academic Board to the Army Chief of Staff, General Lyman Lemnitzer, writing in longhand an extensive justification of the various proposed changes. Davidson presented both the views of the majority of the board and his own views on the civil engineering issue. In June, just before he left West Point for his new assignment, Davidson received Lemnitzer's reply. The proposals in general had been approved. However, one semester of the civil engineering sequence would continue to be required of all cadets.[60]

The Changing of the Guard

In November 1959, Davidson had written an informal "Dear Lem" letter to the Chief of Staff, offering to stay on at the Military Academy for an additional four years "to supervise the execution of the plan for the revision of the curriculum." He noted that:

> A standard question from the continual flow of official visitors who come thru here is a query as to the tenure of the Superintendent. Without exception, the educators, particularly the college presidents, express surprise that a short military tour of duty has been applied to this unique position and observe that their institutions would be at a great disadvantage under similar circumstances.[61]

Because the superintendency was only a two-star billet, the expectation had been that an officer who received his third star while Superintendent (as Davidson had) would be reassigned to a three-star billet. (Bryan, the only officer in the history of the Military Academy appointed to the superintendency *after* promotion to lieutenant general, had been reassigned after only a year and a half—apparently because of President Eisenhower's insistence that a three-star general ought not to occupy the post.) In his letter to Lemnitzer, Davidson offered to accept a reduction to the rank of major general in order to continue as Superintendent. As he described the offer to his old friend and brother-in-law, General Al Gruenther, "I guess I'm crazy but my vocation is that strong."[62]

It was not until six months after Davidson had made the offer to Lemnitzer, and approximately two months after the recommendations regarding curriculum revision had been forwarded for action, that Davidson received his reply. It was a brief, handwritten "Dear Gar" note from Lemnitzer, flattering—but to Davidson, disappointing:

> The purpose of this note is to let you know on an *eyes only* basis, that I have recommended, and Secretary [Wilbur] Brucker has approved, your assignment as the next Commanding General of the Seventh Army of Europe, effective upon Frank Farrell's retirement on 30 June 60.
>
> This is in recognition of your outstanding ability, your splendid record and the fine job you have been doing. My congratulations to you.[63]

June Week was the traditionally festive occasion at West Point in 1960, spirits dampened only by intermittent rain and losses to Navy in year-end competition in lacrosse, baseball, track, and tennis. President Eisenhower was on hand for a reunion with the Class of 1915, and to plant a "Thayer Memorial Tree," to replace the one that Sylvanus Thayer had planted when he was Academy Superintendent. Thayer's memory was invoked also by the graduation speaker, Army Chief of Staff General Lemnitzer. Thayer, Lemnitzer told the graduating class,

> is the man who, more than any other single individual, is responsible for what West Point has been able to contribute to the Army and to the Nation. The merit of his concept has been proved throughout the years.
>
> . . . Remember, that of all the qualities you possess as an officer, the most important is to be a leader. The qualities that distinguish an officer from other men are courage, initiative, will power, and knowledge. These qualities have been required in the past and the advent of nuclear weapons and great technical developments have not changed the situation in the slightest degree.[64]

A few weeks later, Lieutenant General Garrison Holt Davidson left West Point for Germany, to assume the new duties for which General Lemnitzer had selected him. In his place arrived a man ten years his junior in age. At forty-six, Major General William Childs Westmoreland was the youngest Superintendent of the Military Academy since Maxwell Taylor, whose protégé in many respects Westmoreland had become. Westmoreland had served Taylor as secretary to the General Staff in the mid-1950s, when Taylor was Army Chief of Staff. Then Taylor had personally selected Westmoreland to be commanding general of the 101st Airborne when the division was reactivated, the assignment Westmoreland had had just before becóming Superintendent.

Within five years, Westmoreland's name would be a household word. By then he would be commander of the U.S. Military Assistance Command in Vietnam, overseeing the beginnings of the direct American troop involvement in the war beginning in 1965. He would be dubbed "Supersoldier" by *Newsday* that year, and selected as *Time* magazine's Man of the Year. Lyndon Johnson's approval of his performance would lead the President to name Westmoreland Army Chief of Staff in 1968.

Yet the laurels would not be added without costs as well. As David Halberstam has observed of Westmoreland:

> . . . this war would stain him as it stained everything else. As many of his countrymen came to doubt the war, they would come to doubt him; as so many of the civilians who had helped plan the war bailed out on it, thinking

it unwinnable and not worth the cost, Westmoreland, his name somehow attached to it more than anything else in his career, the men he commanded still serving there, could not let go, and public antagonism would center on him. Even the men who had once praised his sense of duty, his caution, his decency, turned cool.[65]

When he assumed the West Point superintendency in 1960, however, the Eagle Scout image which many observers have identified with Westmoreland was still unblemished.[66] Although not widely known outside military circles, Westmoreland was well known within them, as a "water walker," a "hard charger" whose rise to stardom had been meteoric. He had commanded a division at the age of twenty-nine. He had been promoted to brigadier general at thirty-eight (the same age at which MacArthur had made it), and to major general at forty-two, the youngest two-star general in the Army. He was especially well known at West Point, not only for the dazzling pattern of success since his graduation from the Academy, but also because he had been cadet First Captain in the Class of 1936—the Academy keeps track of those who attain this top command as cadets.

Yet one would be on shaky ground to assume that the selection of this relatively young general officer, still full of ambition and on the rise, reflected a desire on the part of the Army Chief of Staff and his Pentagon associates for acceleration of the ferment that had been generated at the Military Academy under Davidson. It is much more likely that the selection of Westmoreland reflected a concern lest the balance at West Point between Sparta and Athens tip too far in the direction of the latter. "Spartan" values in the American military profession had evolved in the twentieth century along a line described by Morris Janowitz, from the officer who saw himself as a heroic, charismatic leader to one who essentially was a manager, while retaining an affinity for heroic-leader style especially in troop command.[67] Westmoreland in many respects was the culmination of the trend as of 1960, with a Harvard Business School zeal for efficiency combined with a gung ho paratrooper manner with subordinates.

Thus, it is not surprising that the two programs which were introduced by Westmoreland as Superintendent that had the most impact on the Academy were (1) a campaign to improve administrative efficiency, introduced with the slogan "West Point Points the Way in Post Efficiency" and (2) a course, introduced during the cadet summer program at Camp Buckner during the beginning of third-class year, in "Recondo" training (the acronym combined reconnaisance and commando, descriptive of training similar to that provided at the Army

Ranger School). In his previous assignment, at Fort Campbell with the 101st Airborne Division, he also had introduced a program of Recondo training, and also had launched an efficiency campaign, termed "Overdrive."

To emphasize Westmoreland's concern with management techniques and his enthusiasm for the rough-and-ready military training that was provided by the Recondo program is not to suggest that academics were stifled or that they stagnated for want of attention during his tenure as Superintendent. An officer from the Coast Guard Academy who attended every conference of academy superintendents from the first one in 1958 through those in the early 1970s recalls that Westmoreland stood out by comparison with the superintendents from the other academies, making incisive observations about various aspects of academy programs.[68] In the academic realm at West Point, Westmoreland saw to it that the curriculum designed under Davidson was put into effect. Important new facilities were provided, such as an analog computer laboratory for ordnance instruction, a computing center for general academic use, a free flight laboratory for testing small arms and for demonstrating ballistic principles, a subcritical reactor, and a remodelled language laboratory. Construction was begun on a new library. There was experimentation with televised instruction. New elective courses were developed and made available to cadets.

Just before the change of command, Davidson had written Westmoreland that he was arriving at the Military Academy "at a very interesting time."

> The changes that are about to be made inaugurate a new era at West Point. I see them as only the first step in an evolutionary process that must be kept at a reasonable pace if it is to keep up with the probable demands of the forseeable future. I visualize the curriculum of the Academy a generation hence as being considerably different from that of today.[69]

No doubt Westmoreland considered the pace of change during his three years as Superintendent "reasonable." But despite the additions to programs and facilities noted above, Davidson's reforms had been cut back rather than expanded in at least one important respect. As noted earlier, the Army Chief of Staff had approved a curriculum that enabled cadets to take three electives during first-class year. He had sided with the majority of the Academic Board that one semester of the first-class engineering sequence should remain required, with one semester elective, against the expressed preferences of Davidson and the minority of the board who favored having both semesters of the engineering sequence optional. However, as of 1961-62, by action of the Academic Board, the briefly optional engineering course became required again, reducing the number of choices of electives for cadets to two. The Dean of the Academic Board, Brigadier General William Bessell, explained that the original plan—

> to provide a third elective—had been accepted by the Academic Board with great reluctance and genuine doubt as to its advisability. The Board has now decided that the improved coverage during First Class year by the Department of Ordnance and by the Department of Military Art and Engineering last year provided a professional background that every cadet should have.[70]

Several years later, the board would change its collective mind once again, this time in the direction of expanding the opportunities for options (in 1966-67). But in the early 1960s, the majority of the Academic Board took pride in what they saw as a prudently updated Thayer system.

Moreover, in Southeast Asia, American military involvement was expanding, with the likelihood of a major combat role for U.S. forces increasing. President Kennedy had urged the military to readiness for a counterinsurgency mission. West Point had been a step ahead of the President with its new Recondo training, and had responded to subsequent cues from Washington with lectures on counterinsurgency, compulsory for all personnel, and with hastily assembled displays in the library on guerrilla warfare.

MacArthur came to West Point in 1962 to accept the Sylvanus Thayer Award, and to deliver his stirring farewell speech. He was the link to the Academy's past greatness, and its traditional values. Westmoreland, as Superintendent, was at his side. In style, outlook, and appeal, Westmoreland contrasted sharply with MacArthur. Yet, in his own way, Westmoreland almost perfectly symbolized the current era, and was the link to the years immediately ahead.

Westmoreland returned to West Point a few months after he left the superintendency for a new assignment, to make a farewell speech of his own, of sorts. He was on his way to Vietnam to assume command of the American forces there, and returned to speak to cadets of the Class of 1964—the class that had entered the Academy when he had begun his tour as Superintendent. It was, as David Halberstam later observed, "an almost classic exposition of the can-do philosophy," and in capsule form it captured the message that he had conveyed to the corps during his years as Superintendent. "In my view the positive approach is the key to success" Westmoreland said, in stressing to cadets their leadership responsibilities,

> and it's the one that has a strong influence over people. Men welcome leadership. They like action and they relish accomplishment . . . speculation, knowledge is not the chief aim of man—it is action . . . all mankind feel themselves weak, beset with infirmities, and surrounded with danger. They want above all things a leader with the boldness, decision, and energy that with shame, they do not find in themselves. He then who would command among his fellows, must tell them more in energy of will than in power of intellect. He has to have both, . . . but energy of will is more important.[71]

6

Organizational Resuscitation: The Coast Guard Academy under Leamy, Evans, and Smith

Concurrently with the ferment at West Point under Davidson, the Coast Guard Academy at New London entered a period of desperately needed change. Like Thayer at West Point more than a century earlier, Frank Leamy assumed command of the Coast Guard Academy in 1957 when the institution was tormented by internal discontent and was regarded in Washington as being in a grave state of disrepair. Just as Thayer came to be known as the father of the Military Academy for the vital reforms which he implemented as Superintendent, so Leamy would become regarded as the father of the modern Coast Guard Academy for his accomplishments. Unlike Thayer, however, Leamy would have less than three years, rather than sixteen, within which to launch a program of change. Fortunately for the Coast Guard Academy, the momentum for reform which was begun under Leamy was sustained by his successors, Stephen Hadley Evans and Willard Smith.

Leamy, Evans, and Smith proved to be effective not only in recognizing and acting upon needed changes, but also in creating an atmosphere which encouraged their subordinates to generate new ideas which facilitated reform. The eight years that span the command of these three men are of considerable interest for the insight they provide into the change process. Paradoxically, however, these years also are of interest as ones in which latent conflicts between the Spartan and the Athenian commitments of the Coast Guard Academy were brought to the fore.

New London in the 1950s: Distress Signals

By the late 1940s and early 1950s, the Coast Guard Academy had grown considerably. Whereas there had been fewer than two hundred cadets at New London on the eve of World War II, with graduating classes of about twenty, the corps in the early postwar years had increased to four to five hundred cadets, with graduating classes ranging in size from fifty-two to ninety-nine. Unlike the Military and Naval academies, however, the Coast Guard Academy had not yet passed the important threshold of growth beyond which it is impractical for all members of the academy community to get acquainted with one another personally. The Coast Guard Academy retained an intimacy, and took pride in this distinctive quality.

Part of the communal self-image was a derivative of that of the parent organization; members of the Coast Guard, although nearly 30,000 in number by the mid-1950s, liked to refer to their organization as "a family." This was a reasonable myth, not only because of the contrast between the still modest size of the Coast Guard and that of the much larger Army, Navy, and Air Force, but also because most Coast Guard units, like the Academy at New London, were still small enough for retention of a familial approach to interpersonal relations.

The contrasts between the prewar and the postwar Academy were mainly physical in nature. For example, the red-brick buildings that had been built in the 1930s remained the center of Academy activity. However, they had become overcrowded by the 1950s, and the hastily constructed wooden buildings that had been added during World War II, primarily to accommodate an influx of reservists, were becoming fire hazards. The configuration of the waterfront had been altered by World War II construction, and the most prominent feature along the Thames now was a 295-foot bark sailing vessel, the *Eagle,* which had been acquired as a prize of war from Germany at the end of World War II. It is also true that only a few familiar faces from the prewar staff and faculty remained in the postwar years. Most staff and faculty members served on brief tours of duty, of course, and even the relative handful of permanent cadre mostly were newcomers.

Still, an officer who had graduated from New London before World War II would have felt completely at home in the Academy of the late 1940s and early-to-mid-1950s. The routines remained essentially unchanged. The customs, rituals, and expectations were the same. As an officer who had graduated before World War II observed upon returning to the Coast Guard Academy as an instructor in the mid-1950s, "there was still hazing but it was undercover."[1] "Swab year" continued to be a survival test which combined physical stresses somewhat more temperate than those of Marine Corps boot camp with the emotional ones of an English boarding school.

All four years for the cadet were ones in which the emphasis was on discipline and on the acquisition of skills that would be immediately applicable to the problems that a junior Coast Guard officer might face. In this important sense, much of the training provided each academic year was preparatory to applications that were provided by the annual summer cruise. This is not to say that academic coursework was devoid of challenge. As in the prewar years, there were some talented instructors who succeeded in making the subjects they taught intellectually challenging. But even when they had the inclination, instructors had little time for experimentation and in-depth pursuit of their subjects, confronted as they were with having to teach as many as five subjects (fifteen to twenty hours) per week, with additional duties assigned as assistants in coaching, supervisors of extracurricular activities, or officers in charge of inspecting the cadet barracks.

Two of the academic departments were devoted entirely to what were considered "professional" subjects. One of these provided instruction in seamanship and navigation, as well as having responsibility for the planning and conduct of the summer training cruise and for aviation training. The other provided instruction in communications, ordnance and gunnery, small arms, and military law.[2] Within a totally prescribed curriculum, roughly a third of regular class hours and more than a third of laboratory periods were devoted to "professional studies." Upon completion of his four years at the Academy, the cadet would have experienced not only the intensive apprenticeship provided by the annual summer cruise, but also a program during the academic years that included: *Seamanship:* Elementary, Regular, Advanced I & II, Service Seamanship; *Navigation:* Nautical Astronomy, Celestial Navigation, Coastwise Navigation, Advanced Navigation I, II, III; *Ordnance and Gunnery;* I, II, III, IV, V; *Communications:* I, II, III; *Law:* I, II, III (military law, admiralty law, maritime law enforcement); *Damage Control;* and *Administration and Leadership.*[3]

Considered to be "applied science and engineering," as distinguished from "professional" courses, were subjects such as basic machines, power engineering, and ship construction. These and other courses in science, engineering, and mathematics occupied much of the remainder of the curriculum. The Department of Humanities presided over a course in world history, one in government, one in "economics and shipping problems," another in management, a two-year sequence in psychology of leadership, and a two-year sequence in composition, public speaking, and literature.

Even the meager exposure that cadets obtained to the liberal arts was subject to critical and sometimes repressive scrutiny by Academy or Coast Guard officials. In the McCarthy era of the early 1950s, the senior professor on the Academy faculty, who headed the Department of Humanities, received a letter of censure from the Commandant of the Coast Guard (a letter that the professor was convinced had been sent at the suggestion of the Academy Superintendent). The professor's "misdeed" was that of having expressed his agreement in a public discussion with the policy position of a Far East specialist who had come under attack by McCarthy and his colleagues.[4]

The number of cadets who were willing or able to endure four years of an essentially trade school program in a highly restrictive environment had proven to be only a minority of those who sought and obtained entry to the Academy. Fifty-four percent of those who had been sworn in with the Class of 1953 had graduated. Otherwise, no class since the one that had graduated in 1946 (formally designated the Class of 1947) had graduated with more than 50 percent of the young men that had entered that class at the Academy as swabs.[5] A management specialist on duty in the Coast Guard Commandant's office in the late 1940s had made the argument, which many people found persuasive, that the service would be well advised to close down the Academy and rely instead on civilian colleges and universities as the source of commissioned officers.[6] This proposal by a reserve officer had been rejected by the Commandant and other Academy graduates who dominated the key positions in the Coast Guard hierarchy, however. When Admiral Alfred C. Richmond became Commandant of the Coast Guard in 1954, having served in Coast Guard Headquarters continuously since 1945 (most recently as Assistant Commandant), he reviewed the attrition problem at the Academy, but concluded that

> it was not appreciably greater than before the War and was accepted as the penalty of a strictly regimented life, coupled with a difficult technical curriculum. Actually, I found that our attrition was not really much worse than that of any good engineering school, if compared with 4 year graduates.[7]

If attrition during the cadet years was "acceptable" to Coast Guard authorities, the high rate of attrition among junior officers was not. An alarming number of Coast Guard Academy graduates were doing the compulsory three years commissioned service after graduation and then resigning. Ironically, the distinction that the Coast Guard Academy had enjoyed as the only service academy at the time to have attained accreditation as an engineering school also enhanced the prospects of its graduates to find civilian employment. Shortly after Richmond became Commandant, word reached him that

> morale at the Academy was bad—first classmen were advising underclassmen that the Coast Guard so needed officers that they could slough off, more or less get away with murder, graduate, and do three years, resign, and get a good job on the outside.[8]

Richmond made a special trip to the Academy to speak to the corps of cadets, fielding questions from them until nearly midnight. Although he was convinced that the meeting with the cadets had "cleared the air a great deal,"[9] the problems of low morale and high attrition lingered. During rare participation by middle-grade officers at a meeting of the Academic Board (1955-56) that had been called to review the curriculum, one of those officers had suggested his view that " 'morale, motivation and leadership' were much more in need of attention than curriculum." The Superintendent, Rear Admiral Raymond J. Mauerman, responded to the suggestion by appointing a committee of middle-grade officers to assess the problem. The committee spent several months talking to cadets, administering confidential questionnaires, tapping the views of staff and faculty members, and visiting West Point and Annapolis to survey the situation there. Their final report presented "a rather seamy and dismal picture of the then state of affairs at the Academy," with emphasis on serious weaknesses that had been detected in cadet leadership training and problems which cadets experienced in barracks life. The Superintendent, shocked by the report, ordered it classified. An officer at the Academy managed to get a copy surreptitiously to the Coast Guard Commandant, however, who in turn made sure that the officer whom he had selected to succeed Mauerman as Superintendent read it.[10]

When top Academy leadership changed hands in the summer of 1957, the ceremony occurred in the wake of graduation exercises for only 61 of the 198 young men who had entered the Academy as new cadets four years earlier. Admiral Richmond had selected the new Superintendent, Rear Admiral Frank A. Leamy, personally. Although he did not find it essential or desirable to provide Leamy with extensive guidance, he did call Leamy's attention to the extensive "gripe session" that the Commandant had held with cadets; moreover, Leamy was given a copy of the classified committee report, with its dismal picture of the state of affairs at the Academy. "Keep a close watch on the mood of the cadets," Richmond advised Leamy, and "run a tight ship."[11]

The Leamy Era

The latter advice was superfluous for a man of Leamy's disposition and style. Frank Leamy had a reputation throughout the Coast Guard as a strict disciplinarian, even a martinet. Yet as the officer who served under Leamy as Commandant of Cadets recalls, Leamy set the example in the maintenance of high standards, and maintained an "unlimited enthusiasm" that inspired those around him.

Leamy's appointment as Superintendent proved to be fortuitous for the Academy. The fact was that there were few candidates from which the selection could have been made. At any one time in the postwar years, there were only twelve to eighteen flag-rank officers in the Coast Guard. They were the incumbents of the key billets: the area and district commanders, the Coast Guard Commandant and his deputies who headed the various functional offices in Coast Guard Headquarters, and the Academy Superintendent. With the exception of the Commandant's post, which typically was a four-year assignment, rarely were assignments to the key billets made for more than three years. Thus, when a flag officer retired or was reassigned, the post that he vacated was likely to be filled by an officer newly promoted to flag rank or by one of the handful of flag-rank officers who were nearing the completion of a three-year tour in another billet.[12]

Among the few flag-rank officers who might be considered when a vacancy in the post of Academy Superintendent was approaching, the choice often could be limited further on the basis of the social graces possessed by the candidate and his spouse. The role of Superintendent included the hosting of numerous social functions for cadets, their parents, members of the staff and faculty, and distinguished visitors; the ability of the Superintendent's wife to fulfill the responsibilities of hostess successfully was deemed important. Further narrowing the field was the custom of selecting as Superintendent an officer nearing mandatory retirement.

The selection of Leamy, who was fifty-seven years old and had a charming wife, was no exception to the customary practice. The vigor and aggressiveness with which Leamy approached the job, howver, made it clear that he did not view his responsibilities primarily in terms of presiding over social pleasantries. Nor did he view the Academy merely as a preretirement "pasturing ground," as it appeared his immediate predecessors had done. The superintendency was, for Leamy, a new challenge, a new opportunity to add further accomplishments to a career that had been broad, varied, and illustrious.

Shortly after graduating from West Philadelphia High School toward the end of World War I, Leamy had enlisted in the Army, serving in the tank corps until his honorable discharge in 1919. The following year he had enrolled in the University of Delaware, transferring a year later to Temple University. In 1922, he gained admission to the Coast Guard Academy by passing the competitive entrance examination.[13]

Even as a cadet, Leamy's leadership abilities were evident. He became vice-president of his class, president of the athletic association, associate editor of the cadet yearbook, humor editor of another cadet publication, and a company commander. Moreover, it was as a cadet that Leamy won the first of what would be several awards during his career for heroism. In this instance, Leamy assisted in the rescue of two women from drowning in the Potomac River, an act for which he received the commendation of the Commandant of the Coast Guard.

After graduating from what was then a two-year Academy program, Leamy served first with a Coast Guard destroyer force and then with the Navy as a gunnery and small arms observer. In 1929 he returned to the Academy for a brief tour of duty as an instructor and assistant football coach. Willard Smith, who was a cadet at the time, recalls that Leamy did not much like the assignment, but approached it with his characteristic energy and dedication. "He was the strictest officer there," Smith recalls; "the cadets were scared to death of him."[14]

After subsequent service aboard destroyers and cruising cutters, Leamy was selected as the technical advisor for a motion picture that was being produced to tell the "story of the Coast Guard." It was while serving in this capacity that he received orders to take charge of shore rescue operations of passengers from the cruise ship *Morro Castle*, which had burned off the New Jersey coast. In 1935, the adventuresome Leamy began flight training at the Naval Air Station in Pensacola, Florida. Designated a Coast Guard Aviator, Leamy, only two years after he received his wings, became one of the rare individuals to be awarded the Distinguished Flying Cross in peacetime. The award was made in recognition of the skill and heroism that Leamy displayed in piloting his amphibious plane to a landing in darkness in the turbulent seas sixty miles offshore where a seriously injured crewman was removed from a trawler and transported by Leamy back to Boston for hospitalization. The following year, Leamy gained additional distinction for piloting the first flight of the Coast Guard flying boats on a transcontinental flight from the West to the East coast across the Rocky Mountains.

On the eve of World War II, Leamy was assigned as Chief of the Aviation Division at Coast Guard Headquarters in Washington. In 1943, however, at his own request, he was detached for duty overseas. He participated in the initial assaults on South France. Then after assuming command of a 22,000-ton attack transport, Leamy proceeded with his unit to the Pacific, where, as commander of a task group, he led his force in the initial assault on Okinawa. The war in the Pacific ended, with Leamy engaged in preparations for the planned amphibious landings on the Japanese mainland.

In the post–World War II period Leamy served again in Coast

Guard Headquarters. Then he took command first of the 8th Coast Guard District and subsequently of the 9th Coast Guard District. It was while serving in the latter capacity that Leamy learned of his appointment as Academy Superintendent.

Top Priority: Discipline and Morale

Leamy's first concern as Superintendent was the restoration of military discipline and morale. As recalled by Willard Smith, then a captain who had been selected to become Commandant of Cadets at the Academy, Leamy's concern was communicated even before his actual assumption of command. Shortly after Smith and Leamy each learned that they had been selected for their respective positions at the Academy, they met in one of the hallways of Coast Guard Headquarters in Washington. "Every time [Admiral Leamy] met you, he inspected you," Smith recalls. "You felt he did. . . ."

> "Smith," [Leamy] said, "I understand you're going up to the Academy as Commandant of Cadets. . . . Well, I want you to know that I haven't asked for you."
>
> I thought, "Well, that's a great way to get started." [Admiral Leamy] said, "But I do have one question I want to ask you. . . . Do you think that the Coast Guard Academy should be a military academy?"
>
> I said, "Well, I don't think that there can be any question about that. Under the law it's a military academy, and we have a military organization. So if you're going to have a school, it should be a military academy."
>
> He said, "That's all I want to know, because that's what it's going to be!"[15]

Smith and Leamy became collaborators in the promotion of one of the first of a series of changes at the Academy during the Leamy years. Before that time, the Commandant of Cadets had only one full-time commissioned officer to assist him. Ostensibly the Commandant of Cadets and Assistant Commandant were aided by "company officers"; but the latter were faculty members, carrying out responsibilities as company officers as additional duties. Coast Guard Headquarters was reluctant to commit the additional resources that would be necessary for the assignment of additional full-time officers to the Academy. However, Smith argued, with Leamy's strong support, that the Commandant's staff had not grown commensurate with the growth of the Academy as a whole in the postwar period. Moreover, the examples of full-time company officer systems at West Point and Annapolis could be cited. The devastating committee report on cadet morale, which Mauerman had classified but which had reached Coast Guard Headquarters, recommended the adoption of such a system at New London. Leamy and Smith each had made his own visit to West Point and Annapolis, and had registered their own favorable impressions of such a system.[16] Persuaded by the Smith-Leamy argument, Coast Guard Headquarters agreed to the assignment of one commissioned officer to each of the four cadet companies.

There was a cadet chain of command that was given responsibility for conducting drill and parades, for maintaining order in the barracks, and for some coordination of cadet activities. With the assignment of commissioned company officers, cadet officers were able to turn to them for guidance. Moreover, the company officers assisted the Commandant of Cadets in carrying out inspections, in counselling cadets, and especially in attempting to cultivate a sense of "company esprit."[17]

In their concern for generating esprit, the newly appointed company officers became missionaries on behalf of a change in the organization of the intramural athletic program. They had been sent by Captain Smith, the Commandant of Cadets, to study the West Point tactical officer system. (Smith felt that it provided a somewhat better model than did the company officer system at Annapolis.) Upon their return to New London, the company officers proposed that intramural athletic competition, which traditionally had pitted teams against one another on the basis of class (first-classmen versus second-classmen, etc.), be organized instead by company teams. Smith strongly agreed with the proposal. With Leamy's enthusiastic concurrence, the new system was put into effect—but not without expressions of shock from some of the regular officers on the faculty at the intermingling of swabs and upperclassmen on the same teams.[18]

Nels Nitchman, who had stepped down as varsity head football coach, was appointed to direct the intramural program. Nitchman, in turn, encouraged the cadets themselves to assume the maximum responsibility for all facets of the competition, from the scheduling of games, to coaching and officiating. Cadets who were not participating in a varsity sport in a given season were required to participate in an intramural sport. For the vast majority of cadets this particular requirement was welcomed rather than resented. The intramural program became enormously popular, as it had at West Point, providing the stimulus to cadet morale that Leamy, Smith, Nitchman, and the others who helped to launch the program had hoped it would.

The Hiring of Otto Graham

Leamy's efforts to improve esprit de corps led him also to seek to improve the program of varsity athletics. His coup de theatre in this regard was the hiring of Otto Graham, the former star quarterback for the Cleveland Browns, as the Academy head football coach and later director of athletics. Leamy had served in Cleveland as commander of the 9th Coast Guard District in the three years prior to becoming Academy Superintendent. Although he had not known Graham personally, he had become well acquainted with Graham's close friend, George Steinbrenner, who was in the shipping business there (and later became the owner of the New York Yankees). Through Steinbrenner, Leamy had contacted Graham in 1957 to arrange a visit to the Academy on the possibility that Graham, by then retired from professional football, would consider coaching Academy cadets.

Despite the bleak picture New London presented on the rainy days of the visit, Graham was favorably impressed, finding in Leamy a kindred spirit. Both believed "in giving 100% effort, and being disappointed if we are on the short end. . . ." But, Graham notes, neither subscribed to the "win at all cost" philosophy.[19] Leamy, in turn, left no stone unturned in making the job appealing to Graham, even succeeding in persuading the Coast Guard to grant Graham a direct commission at the grade of commander.

The Graham appointment gave a powerful boost to morale. So also did Leamy's emphasis on discipline. Because his reputation as a disciplinarian had preceded him, cadets had been terrified of Leamy at first. However, increasingly they came to respect and to like him, the more so as they found him demanding of his staff and faculty the kinds of high standards that he demanded of cadets. (To the dismay of commissioned officers but to the sheer delight of cadets, Leamy instituted an inspection-in-ranks of all his commissioned officers at the end of the workday every Friday.)[20] Moreover, cadets typically responded favorably to Leamy's demands on them because their participation in the process of running the corps, and even in the evaluation of leadership development (through peer ratings), was increased.[21]

Initiation of a Conference of Academy Superintendents

Admiral Leamy was seldom at a loss for new ideas of his own regarding various facets of the Academy program that might be improved. Yet he was aware of some of the experimental programs that the Air Force Academy was developing, as well as many traditional features of the Military and Naval academies which he admired. The idea of an institutionalized forum for an exchange of ideas among the academies appealed to him. In the fall of 1957, after his own visits to West Point and Annapolis, Leamy asked two of his officers to draft a letter for him to each of the superintendents of the other three service academies inviting them to the Coast Guard Academy in the spring of the following year for a conference to discuss their mutual problems and programs.[22]

Leamy's letter was sent at a time when each of the various superintendents had reason to be receptive to a proposal for a conference of this sort. The Air Force Academy, still in its temporary site at Lowry Field, was in the throes of building a faculty and developing a program that could ensure it accreditation. Thus, the Air Force Academy Superintendent had reason to believe that he could learn from the experiences of the older established academies. The Military Academy under General Davidson was engaged in an intensive period of self-study, and was actively seeking information about programs at the other academies and at other educational institutions. Naval Academy officials, like those at West Point, were curious about the rumblings heard from Denver that significant new departures from traditional academy practices were being considered.

The advantages of a conference and an exchange of ideas among the four academies, however, were potentially the greatest for the Coast Guard Academy. The existence of the Coast Guard Academy scarcely had been acknowledged in the past by the older academies. A conference of academy superintendents afforded not only a potentially valuable source of ideas but also the potential opportunity for recognition of the Coast Guard Academy as a member of "the club."

The significance of the first conference of academy superintendents, which was held at New London April 18-19, 1958, lay less in the substantive conclusions derived from the discussions (which were recorded in a bland report)[23] than in the stimulus which the discussions provided for critical organizational introspection at each academy, and in the precedent that had been established for further exchange at a variety of levels. Each superintendent had brought with him his key staff personnel (most notably the commandant of cadets or midshipmen, the director of athletics, the director of admissions, the academic dean). Subsequently, informal communications linkages would be maintained among the four academies through each of these various staff channels. Moreover, some linkages already existing would be broadened. For example, the program of student exchange visits which West Point and Annapolis had instituted, bringing cadets to the Naval Academy and midshipmen to the Military Academy for a few days each year, was extended to include the Coast Guard and Air Force academies.

Concern for Academic Reform

For the Coast Guard Academy, the conference served as an important stimulus to change, especially in the realm of academics. In 1959, Leamy brought the Coast Guard Academy into closer alignment structurally with the other academies by formalizing the position of Dean, appointing Captain Al Lawrence, the senior professor at the Academy, to the post.

Lawrence had done his undergraduate work at Dartmouth College and had received a master's degree from Harvard on the eve of the Depression. He taught humanities for eight years on the faculty of MIT before leaving to become one of the handful of permanent faculty members at the Coast Guard Academy in 1937, serving there ever since. Chester Dimick, who had taught at the Academy since 1906 (when it was still the School of Instruction of the Revenue-Marine, located at Arundel Cove, Maryland), was acting informally as Dean when Lawrence joined the faculty. When Dimick retired in 1946, Lawrence succeeded him as the unofficial Dean.

When Leamy, in 1959, was able to persuade Coast Guard Headquarters to make the position of Dean official, the act had ramifications beyond those of legitimizing existing routines. An academic division was formally created under the Dean, a cadet administration division under the Commandant of Cadets, and an athletic division under the director of athletics.[24] Lines of organizational authority and communication were thereby clarified (although the sharpening of the distinctions among the various domains had some troublesome implications for the future, to which we shall turn in the conclusion of this chapter). The academic component of the organizational mission was given new emphasis, and the Dean's central role in academic policy making was underscored.

After the 1958 superintendents conference, Leamy had directed Lawrence to "get something going" in the way of academic enrichment.[25] The first ventures at reforming the curriculum were modest, at best. "Enrichment" courses were instituted as two-credit electives available during second-class year on an overload basis, initially in the French language and then also in nuclear physics and advanced mathematics. (A foreign language had been required of all Coast Guard Academy cadets prior to World War II, but the requirement had been dropped when the postwar curriculum was devised.) However, continuing review of the curriculum was prompted not only by the arguments that McDermott, the Air Force Academy Dean, had advanced so eloquently at the superintendents conference, but also by external challenges to the adequacy of the Coast Guard Academy.

An important challenge of this sort was provided by a threat of withdrawal of accreditation by the Engineer's Council for Professional Development (ECPD) that was impending when Leamy assumed command as Superintendent. (The Academy would remain accredited by the New England Association of Colleges and Secondary Schools, but would lose its accreditation as an engineering school.) For years Academy officials had prided themselves on providing a solid engineering education and on having met ECPD accreditation since 1939 (with the exception of suspension during the war due to the accelerated nature of the program). In 1953, however, the evaluation committee of the ECPD raised the minimum standards for accreditation substantially. On the next accreditation of the Academy in 1955, the inspection committee was able to give only a conditional approval of the Academy program. The problem, as the inspection team saw it, was not one of lack of effort, nor perhaps even of talent among Academy staff and faculty. Rather, the perplexing question that the inspection committee raised was whether, given the mission orientation of the Academy, it was feasible to maintain a program that could satisfy the standards of ECPD accreditation.

The ECPD, reflecting the advice of its inspection committee, gave the Coast Guard Academy reaccreditation for only two years, rather than for the usual five years. Academy Superintendent Raymond Mauerman reported to the Commandant of the Coast Guard that he was dubious that ECPD criteria could be satisfied for longer-term accreditation.

In March 1957, the ECPD inspection team visited the Academy again. Their report, however, was not rendered until October, by which time Admiral Leamy had assumed command. As Mauerman had anticipated, the results of the latest inspection were negative; ECPD accreditation was withdrawn. The inspection committee expressed its doubts, as it had in 1955, as to the ability of the Coast Guard Academy to continue their program of apprenticeship-oriented "professional" studies while also developing a program of studies that would meet ECPD criteria. "Furthermore," the committee noted, "it does not appear feasible to develop a strong engineering faculty without research and with the necessity for rotating personnel who are primarily service officers."[26]

The fact is that in the final months before accreditation was withdrawn, Academy and Coast Guard officials had come to the conclusion that they ought not attempt to meet the ECPD requirements. As noted above, Admiral Richmond, the Commandant of the Coast Guard, had become convinced that the engineering accreditation of the Academy contributed to the problem of high attrition among Academy graduates. In effect, many Academy graduates were using their engineering credentials to land attractive civilian engineering jobs after serving only the compulsory three years as Coast Guard officers. Captain Lawrence, the senior professor at the Adademy, shared this view and had expressed it vigorously to the Academy Superintendent (first Mauerman, then Leamy).[27]

Nonetheless, the ECPD report established a basis for challenging the heavy emphasis in the Academy program on the so-called professional subjects (seamanship, navigation, and the like), and eventually would contribute to the decision to upgrade faculty academic qualifications. An additional stimulus to academic reappraisal had been provided just three weeks before the withdrawal of accreditation by the ECPD. Rarely had the confidence of American society been so shaken as by the dramatic announcement that the Soviet Union had achieved a successful launching of an earth satellite, *Sputnik I*, on October 4, 1957. At the Coast Guard Academy, as at thousands of other educational institutions throughout the United States, *Sputnik I* represented a triumph in the application of scientific and technological skills that called into question the adequacy of American educational programs.[28]

By 1959 Leamy and his faculty were exploring reforms more far-reaching than the initial enrichment electives which had been provided. Lawrence, now formally Dean, selected several faculty members below the level of department head to constitute a committee to take a fresh look at the curriculum. After several weeks of study, the committee submitted a number of recommendations. The most significant of these called upon the Academy: (1) to enable cadets who entered the Academy with advanced preparation in a particular subject area to validate as much as the first two years of coursework in that area by examination; (2) to institute a system of electives for cadets during their final two years at the Academy; (3) to give emphasis in teaching to the development of analytical skills, as distinguished from an emphasis on memorization and on what the committee termed "the hardware for academic courses"; (4) to move the teaching of professional subjects (seamanship, etc.) from the academic year to the summer as much as possible. Although, as Lawrence reported at the annual conference of superintendents in the spring of 1960, these were "wonderful ideas, and I think everybody here agrees with them," they were rejected. The Academic Council, consisting of department heads, agreed only with the prefatory statement by the study committee of their subordinates that "the present curriculum [is] basic and sound. . . ."[29]

The action by the Academic Council revealed that the status quo was not easily altered. Indeed, when Admiral Leamy retired in February 1960, his tour as Superintendent cut short by a heart attack, the program of education and training at the Academy retained most of its traditional features: a totally prescribed curriculum, supplemented only by a few overload electives; a swab system that provided an excruciating rite of passage to cadet life; a system of military discipline that if anything had become more intensive (although a few additional privileges had been provided also); an academic program that was heavily oriented toward the applied, professional subjects, with all coursework prescribed except for the few electives that had become available on an overload basis.

Yet the unprecedented ovation that the corps of cadets gave to Leamy at his departure from the Academy reflected their realization, which the staff and faculty strongly shared, that Leamy not only had "saved the Academy from complete deterioration" but also had opened its doors to a new era.[30] Leamy's boundless energy, his enthusiasm, his ceaseless quest for better modes of operation, and his receptivity to new ideas had encouraged many of his subordinates to entertain ideas for reform which previously had lain dormant. A momentum had been established, which his immediate successors would maintain.

The Academy Under Hadley Evans

The new Superintendent, Rear Admiral Stephen Hadley Evans, first had entered the Coast Guard Academy as a cadet in 1924, the year Leamy had graduated.[31] After his own graduation in 1927, Evans had ten years of seagoing assignments before returning to the Coast Guard Academy as an instructor in the Department of Humanities. For four and a half years Evans taught a course in the history of Western civilization as well as one in English composition, with collateral duties as Tactics Officer (Assistant Commandant of Cadets). Shortly after Pearl Harbor, Evans became responsible for the training of the hundreds of research officer candidates who were assigned to the Academy. His subsequent wartime duties included service with convoy escorts in the North Atlantic and later with a Coast Guard training station preparing crew for action in the South Pacific.

A Headquarters tour in the early post–World War II period gave

Evans the opportunity to finish writing the history of the Coast Guard which he had begun before the war,[32] thereby becoming one of the rare published authors in the Coast Guard. Evans acquired additional distinction a few years later by being selected to attend the National War College. (At the time of his appointment as Academy Superintendent, he was the only one who assumed that post as a graduate of any of the nation's war colleges.) In 1954, after completing the War College, Evans began a two-year tour at Coast Guard Headquarters in Washington, D.C., first as Assistant Chief of Operations, then as Deputy Chief of Staff. After a promotion to rear admiral, Evans was given command of the 14th Coast Guard District with headquarters in Hawaii. It was while serving in that post that he received orders to return to the Academy to assume command as Superintendent.

Evans, like Leamy, had been hand-picked for the Superintendent's position by Admiral Richmond, the Commandant of the Coast Guard. No special instructions were given to Evans, except the general advice that Richmond had given to Leamy: "Run a tight ship."[33] This Evans, like Leamy, could be expected to do. If his reputation as a disciplinarian did not generate quite the same awe and trepidation service-wide that Leamy's reputation had produced, Evans could be as "salty" as the next man when placed in a command position, insisting upon discipline and extolling the traditional virtues of seamanship. Upon assuming command of the 14th Coast Guard District in Hawaii, Evans had emphasized "that our purpose can be insured only if we run *tight stations* and *tight ships*. A tight ship or station is one where everyone knows exactly what he is supposed to do—and does it. A tight ship is naturally a happy ship."[34] Yet Evans also was a compassionate man, whose success as Superintendent would come largely from his effectiveness in interpersonal relations.

Building Character

The traditional Academy goal of character development, through a system of discipline and training that emphasized duty and honor, was one to which Evans assigned high priority. One of his first acts as Superintendent was to revise and update cadet regulations. Although the result was a considerable expansion of the rule book, the intent was not that of adding red tape. Instead, the objective was that of making the system more rational. Reforms in the swab system and definitions of cadet responsibilities that Evans had introduced when he was Tactics Officer before World War II were codified.

Special emphasis was placed on making a distinction that previously had been lacking between offenses that were merely breaches of discipline and those that called into question the personal integrity of the offender. Cadet officers themselves had made the suggestion to the Superintendent, in a meeting which he had called to elicit their views on a variety of cadet problems and activities, that a distinction was needed between the two types of offenses. Thus, within the category of Class I (major) offenses, a distinction was introduced between military offenses (designated M) and honor offenses (designated H). The latter—for example, cheating on an examination, fraud, theft—ordinarily resulted in a request for resignation or the dismissal of the offender. Evans chose to emphasize the positive approach to honor, however, devising the motto, "Who Lives Here Reveres Honor, Honors Duty," which was emblazoned on the deck in the entryway to Chase Hall, the cadet barracks.[35]

One of the first occasions that Evans had after assuming command to draw upon relevant experimentation elsewhere as the basis for a critical review of the Coast Guard Academy program was the March 1960 conference of service academy superintendents. By the time of the conference, the Air Force Academy under McDermott's leadership had moved to advocacy with Air Force Headquarters of a program that would enable them to award a master's degree to selected cadets in the fields of astronautics and public policy.

When Evans met the following month with the Academy Advisory Committee, he assured them that the Coast Guard Academy (like West Point and Annapolis) opposed the introduction of a graduate degree program for cadets. He noted further that he foresaw no major changes in the Academy curriculum. In particular, the essentially totally prescribed nature of the curriculum would be retained.

Yet Evans already had revealed some impatience with existing practices and performance; moreover, he had initiated some modest reforms. For example, he had called upon the Academic Council to take the entire curriculum under review once again with a view toward improving it. He had let it be known that he regarded the staffing limitations in the library and among the faculty a high priority problem, a view which the Dean echoed to the Academy Advisory Committee. The importance of academics had been emphasized to cadets by developing a program of special privileges for students whose performance was exceptional; moreover, the decision was made for the first time to award degrees with academic honors to deserving candidates.

The Advisory Committee, while indicating their pleasure with improvements they had observed since their last meeting, pointed out a number of observed deficiencies. First of all, the library was a facility that was not widely used by cadets or by faculty; the committee urged that it be open evenings and that its holdings be modernized. There was a critical need for a library cataloger. Secondly, the committee noted that some instructors appeared to "lack the necessary depth in the subject they have been assigned to teach." Third, the committee noted that English literature was the only humanities course, strictly speaking, in the Department of Humanities. They pointed out furthermore the failure of the cadets' preparation in history to extend back to the Greek and Roman eras. Fourth, the committee pointed out the need for modernization of equipment in the science and engineering laboratories.[36]

Modernizing the Physical Plant and Facilities

Evans responded with a plan for improving existing facilities and for increasing the size of the faculty, especially its permanent component. Rear Admiral Leamy had laid the groundwork, going brashly outside official channels in a successful effort to get action to replace the World War II "temporary" buildings that were still in use at the Academy. He had used the occasion of a parent's weekend to call the visitors' attention to the fire hazards to which their sons were exposed, suggesting that letters and other personal communications to members of Congress might be appropriate.[37] The scheme had resulted in congressional appropriations in fiscal years 1959 and 1960, with a phased program of construction authorized to begin almost immediately for completion by 1968. Supervision of the initial phase of construction fell to Evans.

Naturally, to Evans and his staff and faculty, modernization of the Academy required more than merely replacing wooden buildings with brick ones and providing additional space in classrooms, offices, and barracks. In the autumn of 1960, the Academy requested the assistance of Coast Guard Headquarters in the acquisition of a subcritical reactor simulator and a gamma ray spectrometer. By the following year these were installed. Early in 1962, Evans asked his Academic Council to consider the desirability of acquiring an electronic digital computer for educational use. With the council's support as well as that of the Academy Advisory Committee and that of the Board of Visitors, Evans submitted the request for the computer, which was delivered in 1963. A related development was the acquisition, at Evans's urging, of self-teaching equipment that cadets could use in academic areas such as chemistry, mathematics review, "rules of the road" in navigation, and for the improvement of spelling. An instructor in the Academy mathematics department had prepared a report drawing heavily from the experimental work that had been undertaken in recent years by B. F. Skinner of Harvard in the stimulus-response learning theory and its applicability to programmed learning. The Academic Council responded favorable to the report, and Evans directed the Academic Division to order a dozen of the machines on an experimental basis.[38]

Upgrading the Faculty

If the Coast Guard Academy was abreast of changes in educational technology, and indeed ahead of many institutions of higher education in the sophistication of the equipment it was acquiring, it still lagged in the qualifications of its faculty, and in the heavy reliance on faculty members who were assigned to the Academy for a tour of duty of only three or four years. This was a problem that commanded Evans's early attention. By August 1960, his staff had prepared for him to submit to Coast Guard Headquarters a plan for the increase of faculty in five phases. Within four years, there would be a shift of ten faculty positions from temporary to permanent status, with the addition of eight permanent faculty billets and two new temporary ones. In forwarding the plan to the Coast Guard Commandant for his approval, the Superintendent stressed "the necessity of having a teaching staff of highest possible caliber to insure achievement and maintenance of the highest possible educational standards in an age of fast, vast changes in educational needs and fields." He also urged that Coast Guard Headquarters support the implementation of the plan by ensuring "the assignment only of the most highly-qualified officers to the rotating faculty and the recruitment only of the most highly-qualified persons to the permanent faculty."[39]

The Academy had a long way to go in upgrading the academic qualifications of its faculty. Of sixty-six faculty members who were "on board" for the 1961-62 academic year, only four had doctorates—one in physical science, one in mathematics, and two in physical education. Nearly three-fourths of the faculty members possessed no academic degree beyond the baccalaureate, although most of the permanent cadre of faculty had had postgraduate study, many of them to the master's degree.[40] Concerned that those officers who came to the Academy on a rotating assignment basis began their teaching typically with no more than a summer refresher course as advanced preparation, Evans wrote to the Commandant again in February 1962, citing with approval an assertion made the previous year by the Board of Visitors to the Naval Academy that "it is imperative. . .that every officer ordered to the [Naval] Academy for instruction of midshipmen possess a master's degree as a minimum."[41] The Coast Guard Academy Advisory Committee lent their support with a unanimous motion in their meeting in March 1962 recommending that instructors assigned to duty on the academic faculty of the Academy should have completed a minimum of one year of graduate study before reporting to the Academy.

White House Pressures for Racial Integration

Evans was anxious not only to eliminate the assignment of poorly qualified or poorly prepared officers to the Academy faculty, but also to eliminate any vestiges of racial discrimination in hiring. The first black assigned to the Coast Guard Academy faculty, a chemistry instructor with a master's degree from Howard University, arrived in New London in the summer of 1961. Coast Guard Headquarters and the Treasury Department were quick to call the appointment to the attention of the White House.[42]

At the Inauguration parade in January, President Kennedy had noted with consternation the absence of blacks among the ranks of Coast Guard Academy cadets who passed the reviewing stand. The Commandant of the Coast Guard had been notified that the apparent exclusion of blacks was to be ended forthwith, and that the next class of cadets admitted to the Academy should include a number of blacks. Admiral Richmond, the Commandant, contended that there had been no policy of exclusion. On the contrary, candidates were admitted on the basis of a nationwide competitive examination. If the Academy learned the race of its candidates, typically it was only after the arrival of the successful ones at the Academy as new cadets. The Coast Guard Academy had taken great pride in the fact that, unlike the Military, Naval, and Air Force academies, its appointments were made *only* on a standardized competitive basis. At New London there were no congressional appointments, for instance, in which political "connections" might be decisive, or in which criteria other than performance on a competitive examination might become paramount.[43]

After weeks of presenting a strong defense of the existing cadet appointment system, the Commandant finally received the President's agreement that it be continued without change. The President continued to emphasize, however, that the service must do more than it had in the past to attract qualified applicants from racial and ethnic minorities. Coast Guard Headquarters and the Academy got the message. Special efforts to recruit qualified blacks and applicants from other racial and ethnic minorities were launched.[44]

Evans had resented the pressures from the White House, perceiving them as the intrusion of "politics" into Academy affairs.[45] Nonetheless, he himself felt strongly about the importance of freeing the Coast Guard from any racial or religious discrimination. Early in 1961 he had called a special meeting of cadets holding top leadership positions to express his dismay at a dittoed note that someone had placed under the doors of several cadet rooms, making scurrilous remarks about President Kennedy on the basis of his religion. Evans told the cadets that he hoped the climate of opinion in the corps "would be so clearly tolerant of race and creed that no one ever again will have the temerity to make so cowardly a gesture as this towards any individual or group of men." Noting that his own brother was one of thousands of Americans who had lost their lives in World War II for the principle of equality under the law and in opposition to Nazi intolerance and racism, Evans also reminded the cadets that the service academies exist through the will and the taxes of the people of the United States. "These 'people' are people of all the multitude of races and creeds that make up the body politic of the United States of America," he said. "For any individual or any group of individuals within one of these academies to raise the specter of intolerance based on race or creed, thus is no less illogical, from a political viewpoint, than it is immoral and unethical."[46]

The Continuing Problem of Attrition

Evans had become interested not only in the problem of attracting qualified applicants to the Academy from racial and ethnic minorities, but also in the more general need to improve the recruitment effort. He was concerned also about the other side of the coin—retention of cadets once they arrived at the Academy. Leamy had made an impact upon the problem of attrition, which was somewhat less acute than it had been in 1957, when only 31 percent of the class that had entered the Academy four years earlier had graduated. However, by the time of the Evans superintendency, attrition figures still were hovering around 50 percent of each class, in contrast to the other three academies, each of which was graduating roughly 75 percent of the classes that entered.

Evans saw the cadet retention problem as especially being one of motivation. Cadets entered the Coast Guard Academy with impressive academic qualifications. For example, one-third of those who entered the Academy in 1960 had graduated in the upper 5 percent of their high school classes, and over 90 percent of them in the upper quarter of their high school classes. In comparison with entering students at colleges and universities around the country on the College

Board Scholastic Aptitude Tests administered in 1960, cadets at the Coast Guard Academy ranked seventh nationally in the mean score attained on the SAT mathematical test. This placed them ahead of students from Haverford, Yale, Dartmouth, Lehigh, Brown, Chicago, Stanford, and Oberlin, to cite simply a handful of well-known institutions. On the SAT verbal test, the performance of Coast Guard Academy cadets was somewhat less impressive, placing them thirty-eighth nationally. However, this performance still placed them only slightly below students from academically prestigious schools such as Lawrence, Grinnell, and Colgate, and placed them ahead of students from institutions such as Knox, NYU, Beloit, and Vanderbilt.[47]

Curriculum Review

Topflight academic qualifications gave Coast Guard Academy cadets options in their pursuit of higher education to a somewhat greater degree than was true of cadets and midshipmen at the other academies, whose entering board scores and other indicators of academic potential, while impressive, were generally not on a par with those of Coast Guard Academy cadets. Unless the Coast Guard Academy could provide a program sufficiently challenging academically as well as attractive in other respects, cadets could be expected to resign. Thus a concern for retention on the part of Academy officials led naturally to a concern for the content of the academic program.

An added concern was provided by the widely publicized criticisms of the service academies by Vice Admiral Hyman Rickover. Rickover had fired his first volley of criticism at the U.S. service academies in testimony before the House Appropriations Committee in 1959, following his return from a trip to the Soviet Union to assess their educational system in the context of the Soviet success with *Sputnik*. In May 1961, appearing this time before the Defense Subcommittee of the Appropriations Committee, Rickover again voiced complaints about alleged inadequacies of the education provided by the service academies. Rickover's detailed criticisms were directed to particular facets of the program at his alma mater, the Naval Academy. (The role and views of Rickover are examined in detail in the following chapter.) Moreover, the House subcommittee was concerned only with the three Defense Department academies. Officials in the Treasury Department, in Coast Guard Headquarters, and at the Coast Guard Academy followed Rickover's criticisms closely, however, and circulated for internal comments correspondence between Representative Clarence Cannon, chairman of the House Committee on Appropriations, and Secretary of Defense Robert McNamara and his assistants.

Cannon had cited Rickover's testimony to the effect that the "Academies are not providing an education that is adequate to the present and future needs of our Armed Forces," and had urged McNamara "to undertake a searching appraisal of the Service Academies with respect to function, performance, and areas for improvement."[48]

Even before the additional spur to critical reappraisal provided by Rickover's comments, Evans had asked the Academic Council to undertake a comprehensive review of the curriculum. After several months of study, the Academic Council concluded that the curriculum was "sound, up-to-date, well-integrated, balanced, and satisfactory to present and foreseeable Coast Guard needs." In particular, the council precluded any major departure from a uniformly prescribed curriculum, noting that "The Academy does not exist to graduate specialists in specific fields." Nonetheless, as a result of the council's recommendations, the "enrichment" program was expanded slightly, making five elective courses available on an overload basis in the fall of 1961-62. Participation remained modest: eleven cadets were enrolled in conversational French in the fall semester, fourteen in the spring; fifteen cadets were enrolled in advanced mathematics for engineers; thirty-seven in modern abstract algebra; seventeen in nuclear physics; and thirty in classical foundations of Western thought.[49]

Because the practice of the Coast Guard during the postwar years remained that of assigning Academy Superintendents only for a short tour of duty, Evans had only two and a half years within which to review the needs of the Academy and initiate any required changes. But in the brief time allotted to him before he retired and was succeeded as Superintendent by Rear Admiral Willard Smith, Evans succeeded in effecting a number of changes that were badly overdue, and demonstrated a sensitivity to the rapidly changing times. Moreover, he maintained the atmosphere of receptivity to new ideas that had been created under Leamy. Although the senior department heads who constituted the Academic Council had been little disposed toward change, as Evans emphasized when he forwarded the report of the council's review of the curriculum to the Commandant of the Coast Guard,

> the Superintendent feels very strongly that the Academy's educational and training processes are in a dynamic rather than a static state; that the currents of change are being directed towards the Service's progress and improvement; and that the Academy, with the continued wise and generous counsel and support of the Commandant, the Advisory Committee, and the Board of Visitors, may be expected to maintain these evolutionary trends.[50]

The Willard Smith Superintendency

Willard Smith was likely indeed to maintain the dynamism of the approach to education and training that had been developed under Leamy and Evans. Smith had been Commandant of Cadets under Leamy, experiencing first-hand the dramatic transformation of mood and outlook at the Academy that had occurred with the energetic Leamy at the helm. The man who had appointed Smith as Superintendent, and the one in particular who would be assessing his performance, was the newly appointed Commandant of the Coast Guard, a man also deeply interested in the Academy. Admiral Edwin J. Roland had graduated from the Coast Guard Academy in 1929, and had served four years at the Academy in the 1930s as an instructor and assistant coach and another four there in the early 1950s as Commandant of Cadets. Upon learning of his concurrent appointments to flag rank and the superintendency, Smith had expressed his pleasure that the Commandant had been willing to depart from the precedent of appointing an officer who was close to retirement to the top position at the Academy. "That's exactly what we don't want any more," Roland had said; "We should be sending people up there who have a lot to contribute after they leave here. It will be better for the school and better for the Coast Guard."[51]

Smith and Roland would have been contemporaries as cadets had Smith gained admission to the Academy immediately after his graduation from high school in 1927 as he had intended to do. However, Smith's father, a warrant officer in the Coast Guard, had urged him to get a taste of a civilian engineering education first. Thus, Smith had studied engineering at the University of Michigan in Ann Arbor for two years before taking the competitive examination for entrance to the Academy. Entering the Academy in 1930, Smith found that the academic program offered little challenge; in fact, the instructors often were little more prepared to deal with the subject at hand than were cadets such as Smith who had had some previous college training. He was favorably impressed, however, by those who taught the professional subjects such as seamanship and navigation, and found that he liked the exposure to a life at sea that was provided by summer training cruises to the Mediterranean (in 1931) and to the South American coast (in 1932). Moreover, he did not object to the regimentation or the discipline, except for the "mickey mouse" demands that had come to characterize the swab system. He hated such petty requirements as "sitting on three inches of your chair at meals, and running errands for upperclassmen, and 'working out' with a rifle for no particular reason except [that] somebody thought you should. . .that kind of business."[52]

In his final year at the Academy, Smith had seized upon the opportunity provided by his designation as Battalion Commander (the top position in the cadet chain of command) to try to effect some reform of the swab system. In particular, he took pride in persuading his classmates to abandon the practice of requiring swabs to perform such chores for upperclassmen as shining their shoes, counting their laundry, cleaning their rooms, and generally serving as their personal servants. When Smith returned to the Academy as Commandant of Cadets, twenty-four years after his graduation, he would renew his efforts to minimize the petty harassment associated with the swab system. They were efforts he would continue as Superintendent, although he would find that the traditional attitudes among upperclassmen, that their role was to make life miserable for swabs, were highly resistant to change.[53]

As a junior officer, Smith had established a solid record, and was selected only three years after graduation to become aide to the newly appointed Commandant of the Coast Guard, Rear Admiral Russell R. Waesche. Waesche was an extraordinary man, whose stewardship of the Coast Guard through nine years of expansion (including integration with the Lighthouse Service) and wartime responsibilities left an indelible impression on the service in general and on Smith in particular. After three years of valuable experience as Waesche's aide, Smith volunteered for flight training, receiving his wings in June 1940. Shortly after the Japanese attack on Pearl Harbor brought the United States into the war, Smith was flying antisubmarine patrols over the Pacific, interspersed with assignments that took him to Alaska to conduct aerial surveys. In the final months of the war, he was called back to Washington to serve again as Admiral Waesche's aide. His tour as Commandant of Cadets at the Academy during the Leamy years was followed by an assignment in Seattle as Chief of the Operations Division of the 13th Coast Guard District, from which he received the assignment as Superintendent.

Continuation of the process of curriculum reform and the upgrading of the faculty were the subjects that Evans emphasized when he and Smith talked in the few days that the two had together in New London at the time of the change of command. Smith was not inclined to move precipitously to any further alteration of the curriculum. However, Captain Stanley Smith (no relation to Willard), who had succeeded Lawrence as Dean at the latter's retirement in 1961, was instructed to continue the review of the curriculum. The new Dean was aided in the review process by a group of civilian educators that made visits to the Academy (usually twice a year) as the Advisory Committee. This group had become especially energetic in its analysis and its proferring of advice since the assumption of the board chairmanship by Arthur S. Adams.

In their collaboration in the design of new curricular reforms, Adams and Stan Smith could draw upon approximately forty years and fifteen years, respectively, as educators. Curiously, each could draw also upon experience at another service academy. Adams had graduated from Annapolis in 1918, Smith from West Point in 1937.

His years as a midshipman and as a naval officer were long behind him when Adams became chairman of the Advisory Committee to the Coast Guard Academy. Nevertheless, as Admiral Willard Smith has observed, Adams retained a sensitivity to the distinctive needs of the service academies. As Superintendent, Smith discussed with Adams his growing conviction that the Academy curriculum ought to be broadened, especially in ways that would move it beyond "our overpreoccupation with the physical sciences and math and engineering." He found Adams in accord. Indeed, Adams proved to be "the greatest source of inspiration and help on this because he understood exactly what we were talking about, and he still had an appreciation of the fact that we just didn't dump everything and set up a liberal arts educational program."[54]

Adams had served recently as president of the American Council on Education, and before that as president of the University of New Hampshire. However, it was experience from his pre–World War II days with the faculty of the Colorado School of Mines that Adams remembers as having provided particularly useful vantage points from which to assess the program at the Coast Guard Academy. Most notably, this experience included membership in the Society for the Promotion of Engineering Education, and service on a curriculum committee that reviewed a widely cited critique of engineering education in the United States which had been prepared by the Dean of Engineering at Penn State University.[55]

The Dean at the Coast Guard Academy, Captain Stan Smith, had been a West Point cadet during the peak years of the Depression. In his final year, he became the top man in the cadet chain of command, the cadet First Captain, as Willard Smith had been at the Coast Guard Academy four years earlier. Stan Smith resigned from the Army three years after his graduation, joining the Coast Guard Academy faculty as an instructor in mathematics and tactics in 1941. However, with the American entry into World War II, he rejoined the Army, serving throughout the war, but then returned to the Coast Guard Academy in 1946. When Professor Dimick retired that year, Smith became head of the Department of Mathematics, a position he held until 1961, when he succeeded Lawrence as Dean.[56] Willard and Stan Smith had come to know one another well when both were serving at the Academy during the Leamy years. A solid basis of mutual respect had been established, providing the foundation for an effective relationship now that they were working together as Superintendent and Dean, respectively.

Stan Smith's ties to West Point provided the Coast Guard Academy with some useful informal channels of communication. Several of the ties were at high levels in the military command structure. For example, when Lieutenant General Davidson came to New London for the first superintendents conference, Smith still greeted him as "coach," as he had when he played Army football under Davidson at the Point in the mid-1930s. When Smith became Dean of the Coast Guard Academy, the West Point superintendency had passed to Major General William C. Westmoreland, whom Smith had known from cadet days, and who had been the cadet First Captain the year before Smith's selection for that coveted position.

The Opening of the Curriculum to Electives

When Adams and the other members of the Advisory Committee met with the two Smiths, Willard and Stan, and with other members of the Academy staff and faculty at a meeting in April 1963, ideas had begun to crystallize. A statement of goals for curriculum reform, incorporating suggestions of the Advisory Committee, was drafted by Academy personnel and then forwarded to the Commandant of the Coast Guard. Because the Advisory Committee formally reported to the Coast Guard and made periodic visits to Washington, Adams had the opportunity to add his personal recommendation that the Academy ought to strengthen its educational program especially in such areas as management, international relations, and the humanities. Admiral Roland lent legitimacy to the proposal to continue with more vigorous reform, by sending a directive that embodied the guidelines that had been drafted at the Academy. The Roland memorandum directed the Superintendent "to conduct a study as to the feasibility of introducing a modified curriculum that would offer some free choice of subject matter to cadets, in lieu of the traditional required and restricted course of study."[57]

Organizational folk-wisdom at the Academy long had sustained a totally prescribed curriculum with an argument (identical to that advanced at West Point and Annapolis) that any departure from total uniformity of experience would undermine group cohesion and would introduce the possibility of invidious comparisons and rivalries. Admiral Smith had countered this argument by noting that some diversity already had been introduced into summer training (with some cadets sent to participate in training with West Point cadets, for instance, and some to other types of training), without the adverse consequences that might have been predicted.[58] Moreover, even electives on an overload basis represented a departure from total uniformity of experience—again without dire consequences.

If the general idea of introducing electives could gain acceptance in principle at the Academy, however, there remained a more troublesome question: where would the cuts be made in the prescribed curriculum to make room for electives? Consideration of the introduction of electives *in lieu of* some required courses implied a reduction in the cadet hours over which one or more departments had maintained control, with a possible loss in status and a reduction in resources and personnel. In short, with the Commandant's directive to the Superintendent in the summer of 1963 authorizing the first inroads on the prescribed curriculum, a Pandora's box was being pried open.

Recognizing that future internal acceptance of whatever program of electives was developed would be conditioned by the procedures that were employed to shape the program, the Superintendent asked the Academic Council to review the curriculum. The council was able to reach early agreement on the proposition that a substantial core of prescribed courses would be retained, although electives would be introduced. The most heated discussions centered on the identification of the courses that were to constitute the "core curriculum." The so-called professional courses—the how-to-do-it courses—were especially vulnerable to elimination from the core. A thousand arguments could be (and were) advanced in terms of the proud traditions of a seagoing service for retaining each of the many courses in the prescribed curriculum in seamanship, navigation, communications, ordnance, gunnery, damage control, and the like. At least since the mid-1950s, however, when the withdrawal of ECPD accreditation had been predicated in part upon the heavy emphasis on apprenticeship-type "professional" courses, the Professional Studies Department had been on the defensive. Admiral Evans, although a believer in the value of practical training and experience, had made known his own skepticism about the necessity of retaining so many of these courses on a prescribed basis, and Admiral Smith had expressed similar skepticism.[59]

After months of discussion, the council produced a report which the Superintendent endorsed and sent to the Coast Guard Commandant for his approval. The report called for a common program for the cadet's first two years, followed by two years within which he could concentrate either in science and engineering or in management and the social sciences. All of the coursework during the second-class (junior) year would be prescribed, and roughly two-thirds of that during the first-class (senior) year would be prescribed. However, the composition of the prescribed and elective program during the final two years would depend upon which of the two concentrations the cadet had chosen to pursue.

As anticipated, the major cuts from the existing prescribed curriculum were to come in "professional" studies. Seamanship instruction was to be sharply reduced from the academic curriculum, with such skills to be imparted instead primarily during summer training or in waterfront instruction. Instead of five prescribed courses in communications there was to be but one. Instead of eight prescribed courses in ordnance and gunnery there were to be but three. There were to be only three prescribed courses in navigation instead of six. Overall, whereas slightly more than a third of the prescribed curriculum had been devoted to professional subjects, slightly less than a fourth of the core curriculum was to be devoted to such subjects.[60]

Each cadet was to be permitted to take at least five electives (or more, if he chose an overload program) during his final two years, three of which were to be within his area of concentration. Proposed electives included such courses as solid state physics, nuclear physics, matrix theory, advanced calculus, advanced electrical engineering, heat transfer, organic chemistry, Shakespeare, American literature, modern European literature, classical foundations of Western thought, personnel and labor relations, and law of the sea. Since most of these were of a more specialized or advanced nature than courses that had been offered heretofore, the proposal implied an active continuation of the start that had been made under Evans toward upgrading the qualifications of the faculty. Such an implication was fully warranted, because the Dean, the Superintendent, the Commandant of the Coast Guard, and the civilian Advisory Committee all strongly supported a policy of upgrading the faculty. In the first two years of the Smith superintendency, the number of civilians on the faculty had been more than doubled, bringing the total to nineteen (all of whom except a couple of faculty members in physical education had graduate degrees). The permanent commissioned faculty had been increased only slightly, to eight, but provisions had been made to send most of those who lacked the Ph.D. back to graduate school to attain the degree. Moreover, the qualifications of the forty-six commissioned officers who were on the faculty on a short-term basis (typically four years) reflected the increased priority that was being given in Coast Guard Headquarters to the assignment of officers with postgraduate education. Captain Joseph McClelland, who became Admiral Roland's Chief of Personnel in 1964 (and in 1973 would become Superintendent of the Coast Guard Academy), fought successfully to increase the number of billets for which a Ph.D. was authorized.[61]

The proposed new curriculum was given approval by the Commandant for implementation beginning in the 1965-66 academic year. In addition, other reforms of the academic program were being introduced. A computer center was opened in 1963, organized around an IBM Model 1620 computer and auxiliary equipment. Rather than augment or reconstitute the core curriculum with courses focusing narrowly on computer usage, the decision was made to incorporate a few hours of such instruction into each of several existing courses in a variety of disciplines, ranging from English to chemistry to seamanship. Concurrent with the opening of the curriculum to electives and "areas of concentration" (not yet termed "majors"), an "Academy Scholars Program" was launched. The program provided the opportunity to a small number of cadets each year, who were selected for their potential ability to carry out independent research, to design and execute special research projects in lieu of approximately two-thirds of the coursework which otherwise they would have taken during their first-class year.[62]

In July 1965, Willard Smith left the Academy to assume his new post as Commander of the 9th Coast Guard District (the Great Lakes region). A year later, Smith was selected for Commandant, the top position in the Coast Guard. Interestingly, the pattern of career success that he thereby established would be followed by his successor as Academy Superintendent, Rear Admiral Chester R. Bender.

Conclusions

In a sense, change at the Coast Guard Academy was inescapable. Leaving questions of engineering accreditation aside, it is highly doubtful that the Academy could have continued even as a four-year accredited institution of higher education without making some major departures from its totally prescribed, "nuts-and-bolts" curriculum. Nor is it likely that the Academy could have hired and retained capable faculty members without updating their classroom and laboratory facilities, introducing more modern educational technologies, and providing opportunities for faculty members to advance their own education and engage in research.

Yet if change had been inescapable, it was far from inevitable that the Academy would move in the direction that it did during the Leamy-Evans-Smith years. Under less aggressive, open-minded leadership, the Academy might have foundered to a point where the Congress or the Treasury Department would have found it necessary to close the Academy down, or to impose drastic alterations from without.

What Leamy had done, most notably, was to unfetter the Academy, to break the shackles from the thinking of those who had been in positions of key responsibility at the Academy, and thereby to encourage those in lesser positions also to generate ideas that otherwise might have been stifled. The morass in which the Academy had found itself in the pre-Leamy years reflected not only the limitations of the top leadership of the time, but also traditional perspectives which had become ingrained in most Coast Guard officers. The organizational subculture of the Coast Guard had encouraged a "make do" philosophy, in which its members took grim satisfaction in struggling along on a shoestring budget, shying away from bold experimentation, eschewing publicity, and taking solace in the informal motto "in obscurity lies security."[63] Leamy, in contrast, thought big. Thus, Otto Graham was hired as football coach. The Coast Guard Academy gained recognition that it previously lacked among the other service academies with the initiation of the superintendents conferences. The flimsy wooden buildings from World War II were razed, and the basis was established for a major new building program. The faculty was organized into an academic division, headed by a dean, and told to update the curriculum. Evans, and then Smith, sustained the Leamy philosophy of updating and upgrading the Academy.

Morale at the Academy improved "110 percent" (as Al Lawrence, the Academic Dean under Leamy and Evans, put it).[64] Paradoxically, however, the changes that occurred served to accentuate the Spartan-Athenian duality of organizational mission, thereby generating tensions which, while mostly latent during the period of organizational "renaissance," would rise to the surface in more troublesome form in the future. While Leamy, Evans, and Smith each in his own way strove to modernize and reform the academic program of the Academy, none of them wavered from his insistence that this was a *military* school. Indeed, the swab system was altered only slightly. Disciplinary rules for cadets of all four classes remained extensive and strict, although privileges were expanded somewhat (as in granting limited drinking privileges to those otherwise eligible to drink under Connecticut law, or permitting cadets to wear civilian clothing when on liberty).

The idea of a firm but positive approach to discipline accorded with the view expressed by Smith, but doubtless shared by Leamy and Evans, that, "If you treat these young people like prep school kids right up to the point where you hand them a commission, you're doing them a disservice." Still, if the goal was the cultivation of maturity, lingering prep-school morés and customs did not disappear readily.[65] For example, a letter that a cadet wrote to his father in the summer of 1959 (when Leamy was Superintendent and Smith the Commandant of Cadets) is revealing. The cadet was experiencing his second month of swab year, and observed:

> It's starting to be fun—Collision Drill, Relay Races, Rowing to Norwich, and Toothbrush drill are some of the games played by the 2nd class. We're having a mock funeral tomorrow. I'm giving the eulogy.[66]

Three years later, after returning from summer leave to begin his final year at the Academy, the same cadet wrote to his father:

> I'm afraid you don't understand what I meant when I said I did not especially feel "Rah-Rah" when I returned here, and called it a "zoo." To dislike at times a routine of doing things, a conforming system of living, does not mean that one does not love the ideals of the environment and what it produces. . . . Since I've been here, I've seen one cadet killed and four go insane. The last one was sent to the hospital just a week ago in Staten Island. Two of these people I've had command of (and been responsible for), at one time or another, and I've had to crap on them for one reason or another. No matter how justified a person may be, he would naturally wonder if he was responsible in some small way for this. No system made by man is perfect, and this one certainly has its faults too. In short, though the finished product may be good, it's still a "zoo."[67]

The monitoring of cadet behavior had become more ubiquitous with the introduction of full-time company officers (on the West Point model). Although ostensibly the company officer served as counsellor to the cadets under his supervision as well as disciplinarian, invariably he subordinated the former role to the latter, as the Academy staff psychologist had warned at the outset.[68] Moreover, it was the enthusiasm of the company officers with the West Point model of discipline (which they observed on a visit early in the Leamy administration) that had led to the imposition of punishment tours for those cadets who exceeded the monthly quota of demerits. Disciplinary offenders previously faced only confinement (or ultimately expulsion, which continued to be a threat for chronic offenders or for extremely serious offenses). Now cadets marched one hour, rifle on the shoulder, for each demerit they compiled in excess of the quota.

With faculty members no longer assuming the company officer role as a collateral duty, and with greater emphasis placed upon their academic qualifications and scholarly responsibilities, the distinction between academic roles and military ones had become heightened. The basis was thereby established for rivalry and mutual mistrust of the sort that often had characterized relations between faculty members and company officers (or at a higher level, between the Dean and the Commandant of Cadets or Midshipmen) at the other service academies.

In short, under the leadership of Admirals Leamy, Evans, and Smith, the Coast Guard Academy entered a new era. It was an era which the vast majority of those associated with the Academy greeted with enthusiasm and pride. Yet it was also an era that highlighted the dilemma heretofore largely obscured, of reconciling the commitment to Sparta with the commitment to Athens, a dilemma that future generations of Academy personnel would experience in more acute form.

7

Listing to Starboard: The Quest for Reform at Annapolis

The Naval Academy was subject to many of the same reform stimuli in the late 1950s and early 1960s that had triggered change at West Point and New London, most notably including the innovative example of the Air Force Academy. However, departures from the seminary-academy tradition were made with painful hesitancy at Annapolis, reaching fruition only during the superintendency of Admiral James Calvert (1968-1972). The reasons for this pattern of change are explored in this chapter.

The Annapolis case holds special interest for the role played by one of its graduates, Admiral Hyman Rickover, as a persistent critic. The reforms that were begun in 1959 and continued into the early 1960s were far too tame to suit Rickover. In 1962, the new Secretary of the Navy, Fred Korth, endorsed some of Rickover's proposals with a mandate that required the Naval Academy to appoint a civilian as Dean and move toward the replacement of military faculty members with civilians. But Korth's tenure was short-lived, and it is fascinating to see how quickly and effectively uniformed Navy officials were able to move to undercut much of the thrust of the Korth directive.

Adverse publicity in 1966 which highlighted the Naval Academy practice of assigning academic grades according to prescribed quotas led to a revival of high-level concern for reform. Civilian faculty members, who had become more effectively organized to express their concerns during the debate over grade quotas, played an increasingly active role on behalf of further reform. However, the tendency of Superintendent Draper Kauffman, and of the Navy captains upon whom he called to study the existing situation, was to push for a reemphasis of the Spartan dimension of the Academy mission, on the grounds that the Academy already was becoming too Athenian in orientation.

The Calvert era, which followed, began by continuing the emphasis on "naval professionalism." However, not only did Calvert soon prove to be fundamentally in sympathy with the pleas of faculty members for reinvigorating the academic program, but also he had a number of ideas of his own for change. The dynamics of interaction among Calvert, his Dean, an internal objectives review board, the Faculty Council, an academic advisory board of external consultants, and the Board of Visitors to the Academy provide important insights into the process of organizational change.

Rickover and the Naval Academy: The Early Confrontations

For officials at the Naval Academy in the 1960s, the appearance of spring each year meant, among other things, that it was time to catch hell from Rickover again. It had started in 1959 with his report to the House Appropriations Committee on his trip to assess Soviet scientific and technological education in the wake of the successful launching of *Sputnik*. In describing what he saw as glaring deficiencies in American education in general, he had managed to include some scathing commentary on the alleged cultivation of intellectual mediocrity that was occurring at Annapolis.[1] Thereafter, it was an annual affair. Members of the House Subcommittee on Department of Defense Appropriations would call on Vice Admiral Hyman Rickover to testify. Short, wiry and sometimes wiley, articulate but abrasive, the admiral would launch a scathing attack on the apparent failure of the Naval Academy to move into the twentieth century. Here was one alumnus-made-good whose continuing interest in his alma mater Academy officials found frequent reason to regret.

Despite his reputation for brilliance as "father of the nuclear submarine," and despite glowing evaluations in his file from those who had supervised his work, Rickover had been passed over twice by Navy selection boards for promotion to rear admiral. Indignation expressed in the mass media and in Congress, coupled with a change of administration, led a new Secretary of the Navy to ask the selection board to make an exception to rules and consider Rickover one more time for promotion. The guidance which was provided to the board (one of those selected must be an "Engineering Duty Captain experienced and qualified in the field of atomic-propulsion machinery for ships") made his promotion to rear admiral in 1953 inevitable.[2] In 1958, with further prodding from Congress, Rickover was promoted to vice admiral.

Thus, when Rickover in 1959 began what would be an annual lecture to the Navy and the Naval Academy on their abysmal failure to adapt to the needs of the twentieth century, he did so from a formidable power base. First of all, he now was a three-star admiral. Although characteristically he wore civilian clothes rather than a uniform, those whose views and practices he challenged in the Navy were as conscious of the three stars as if he had had them imprinted on his forehead. Secondly, he was in control of the Navy nuclear power program, with a dual position established in the Navy Bureau of Ships and on the Atomic Energy Commission. As Calvert has observed,

> Only a student of Washington bureaucracy can fully appreciate what an enormous coup this [dual assignment] was. It was as if Casey Stengel, while remaining manager of the Yankees, had gotten himself installed as president of the American League—with, incidentally, control of all the umpires.[3]

Rickover personally interviewed every individual considered for admission to the nuclear power program. Just as Admiral Elmo Zumwalt, Chief of Naval Operations from 1970 to 1974, still could recall Rickover's interrogation of him in 1959 so vividly that he could recite it verbatim, so a whole generation of applicants for positions in nuclear power were alternately captivated and terrorized by Rickover's interviews. As Zumwalt noted, Rickover was fast becoming an "independent baron in the Navy."[4] Moreover, a block of Rickover supporters had developed in Congress, quick to champion policies that Rickover espoused and seemingly eager to torment those among the Navy brass who differed with Rickover.

Navy officials grew upset with Rickover and his congressional devotees, not only because the barrage of criticisms directed at the Navy seemed ceaseless, but also because they felt that much of it was ill-founded. In the case of the criticism of the program of education and training at the Naval Academy, for instance, several members of Congress had echoed Rickover's criticisms at precisely the point in 1959 when officials at Navy Headquarters and at the Academy believed that important reforms were under way. For example, four senators had inserted a supplementary statement into the report of the Board of Visitors to the Naval Academy in 1959, urging the kinds of changes that Rickover said he desired. From the view of Navy and Academy officials, however, the senators were oblivious to changes that were already being implemented.[5]

The Melson Reforms

At the urging of the Board of Visitors to the Academy in the spring of 1957, and with additional impetus provided that October by the national soul-searching in the wake of *Sputnik*, the Naval Academy began a major review of the curriculum.[6] Rear Admiral Charles L. Melson, who succeeded Rear Admiral William R. Smedberg III as Superintendent in June 1958, continued the review and translated its findings into academic reforms.

His curiosity peaked by discussion in Navy circles about the many bold departures from traditional service academy practice that were being instituted at the Air Force Academy, Melson paid a visit to Colorado Springs before assuming command at Annapolis, receiving a briefing from Brigadier General McDermott.[7] Upon his arrival at Annapolis, Melson urged the committee reviewing the Naval Academy academic program to come up with a framework for needed reform.

By the end of the fall semester, 1958, the report was ready. Basically, it called for some updating of the curriculum, supplemented by opportunities for validation and overload electives of the sort that had been adopted by the Air Force Academy in the first wave of reform under McDermott (and that were being studied concurrently at West Point and at New London).

Modernization of the curriculum meant, especially, a shift away from the "trade school" orientation that had been dominant at Annapolis (as it had been in New London), in favor of an emphasis in each academic field on helping students to understand basic principles and to acquire analytical skills.[8] Much of the applied work in seamanship and navigation that had been taught during the academic year was to be taught instead only during the summer training cruises. Courses such as "naval boilers" and "naval machinery" were to be dropped entirely, with more theoretical courses such as "thermodynamics" and "fluid mechanics" increased in classroom hours.

The validation proposal provided that midshipmen who entered the Naval Academy with previous college experience (one midshipman in three), as well as those who had had exceptional secondary school preparation, would be permitted to take qualifying examinations to exempt them from various introductory courses. More advanced or specialized courses were to be developed which could be taken in lieu of the introductory courses for those who validated the latter. In addition, some electives were to be available in each department on an overload basis (that is, available in addition to the normal load of prescribed courses).

In January 1959, the Navy Department approved the changes. At the recommendation of Admiral Melson, however, the Chief of Navy Personnel (who had primary responsibility in Navy Headquarters for Academy affairs) appointed an external curriculum review board to assess the new "space age curriculum" (as it was being described to the mass media). The board was chaired by Dr. Richard G. Folsom, the president of Rensselaer Polytechnic Institute. The Folsom Board gave its early endorsement to the proposed curriculum changes, and to the provisions that were to be made for validation and overload electives. Encouraged, Academy officials put the reforms into effect beginning with the 1959-60 academic year.

The board cautioned, however, that in order "to do justice to an improved academic program," there would have to be an upgrading of the faculty and an improvement in the conditions under which they served. Commissioned officers who were assigned to the faculty should receive relatively long tours of duty at the Academy, rather than the two- or three-year stints that had been the practice. More civilians should be added to the faculty, each qualified in his field at least to the master's degree and preferably to the doctoral level.

Melson was sympathetic to the board's recommendations. The number of officers in the Navy or in the Marine Corps with advanced academic degrees was relatively small.[9] Moreover, the most promising career officers in the Navy did not seek assignments to Annapolis; on the contrary, they sought to avoid such assignments on the grounds that they were not career-enhancing. The officer who had served as Commandant of Midshipmen from 1956 to 1958, for example, has acknowledged that it is only that position, which carried high status professionally, that could have brought him back to the Academy.[10]

However, Melson found Navy Headquarters unresponsive to the plea for adding more civilians. Lack of funds was given as the excuse; but Melson sensed "the feeling (which I didn't subscribe to) that we ought not let the number of civilians at the Academy become too large."[11]

The Rickover Rebuttal

Melson had not consulted Rickover in advance on any of the changes that were being initiated. Indeed, it had been with considerable reluctance that he had complied eventually with a request from the Navy Chief of Personnel that Rickover be invited to the Academy for a visit. Moreover, Rickover's behavior had not encouraged repeat invitations. According to Melson, upon his arrival at the Academy Rickover had announced, "I'm not here for a briefing; I'm here to tell you what to think." With that, the meeting had been terminated.[12]

When Melson left the Academy in June 1960 to assume command of the First Fleet, he was relieved as Superintendent by Rear Admiral John F. Davidson. Rickover was not long in making his presence felt by the new Superintendent. "He called when I first arrived," Davidson recalls, "saying, 'I'm sorry you have the job since you have no qualifications for it.'"[13]

It was true that Davidson lacked impressive academic qualifications of the sort to which Rickover probably was alluding. In common with every Academy Superintendent to that time other than Admiral C. Turner Joy (who had attained a master's degree from the University of Michigan), Davidson had no postgraduate degree from a civilian educational institution. Moreover, unlike his four immediate predecessors as Superintendent, Davidson had not attended the Naval Postgraduate School. However, Davidson had had a career of command and staff assignments in the Navy in which he had compiled a record of outstanding performance comparable to that of most of his predecessors. For example, he had won the Silver Star and the Legion of Merit for his leadership in command of the submarine *Blackfish* during World War II. More recently, he had headed the prestigious Politico-Military Policy Division in the Office of the Chief of Naval Operations, followed by an assignment as chief of the Navy Group in the joint U.S. military mission in Turkey. Moreover, he had had previous experience with military educational institutions. He was a 1951 graduate of the Canadian National Defense College, and had served at the Naval Academy as head of the Department of English, History, and Government from 1951 to 1954.

After hearing these credentials challenged for their adequacy to the job of Academy Superintendent, Davidson learned to anticipate calls frequently from Rickover on various matters of Academy policy and practice. Even though the Rickover barrage of advice and criticism could be expected to continue, it appeared that soon he would have to forsake his base of power within the government. In January 1962, Rickover would reach the compulsory retirement age (sixty-two). However, prompted by Rickover supporters in Congress, and ultimately by a directive from President Kennedy, Secretary of the Navy John B. Connally announced early in 1961 that Rickover had been asked to extend his active duty for two more years.[14] Thereafter, every two years Congress would "recall" Rickover from retirement for an additional two years. Thus, from a position of influence, Rickover continued to voice his consternation at the deficiencies which he saw in Naval Academy education.

In contrast to a generally favorable image which he had developed of NROTC graduates on the basis of his interviews of applicants for the nuclear power program and his observations of officers once in the program, Rickover presented a dismal picture of the average Academy graduate.

> A more serious defect even than this deficiency in scholarship [which he had noted] is the attitude the midshipmen acquire at the Naval Academy. Once he graduates he will in 9 cases out of 10 stop thinking and studying. I see this clearly when I talk with young officers; and I have talked to thousands of them.
>
> Once they get out their mental growth comes to an end. They putter around the house, repair cars, and build furniture; they do little to improve themselves. They become "nest-builders" and "bird-hatchers." This is the thing we have to change.[15]

Rickover had a number of specific changes to propose: lower the maximum age for admission to the Academy; tighten the scholastic entrance requirements; make physical admissions criteria more flexible; introduce more theoretical and liberal arts courses, while reducing the emphasis on practical training; expand the program of electives; rely more heavily on qualified civilian faculty members and less heavily on naval officers; drastically cut down on the demands that were being made on midshipmen's study time by the administrative routine and by the plebe system; reduce the emphasis on extracurricular activities in general and on varsity sports in particular; subordinate the function of the Executive Department, under the Commandant of Midshipmen, to the academic function of the Academy. "If drastic steps are not taken immediately to improve the service academies," Rickover advised, "I would advocate that you consider abolishing them."[16]

The Korth Directive

Shortly after Fred Korth succeeded his fellow-Texan John Connally as Secretary of the Navy early in 1962, he ordered drastic steps taken at the Naval Academy of the sort Rickover had recommended. (Rickover's testimony had been followed by a letter to the Pentagon from Representative Clarence Cannon, the powerful chairman of the House Appropriations Committee, asking for a statement as to changes that would be made at Annapolis.) On May 22, two weeks before a scheduled appearance at Annapolis to deliver the commencement address, Korth announced major policy changes. Academy scholastic entrance requirements were to be raised "to the maximum practicable extent." *All* commissioned officers on the Academy faculty were to be replaced, except in the Division of Naval Science, by civilian instructors, with the replacement being implemented "gradually." Finally, a civilian was to be appointed as Academic Dean at the Academy.[17]

Heretofore, the Superintendent had supervised the academic sphere through the Academic Board, with the assistance of a secretary to the Academic Board. Davidson had appointed a committee to study the desirability of altering the structure by creating the position of Dean, but the idea had been rejected. A phone call from Korth in May 1962, however, alerted Davidson that he would "read in the papers" about the decision to appoint a civilian academic Dean.[18] It was not Davidson, however, to whom fell the task of implementing Korth's directive. Davidson retired from the service in the summer of 1962 and was succeeded as Superintendent by a Texan who had spent much of his career with submarines, Rear Admiral Charles Cochran Kirkpatrick. Kirkpatrick had most recently served as commander of the Training Command of the Pacific Fleet. His recent service also included valuable Pentagon duty, with the chairman's Staff Group of the Joint Chiefs of Staff in the early 1950s, then in the late 1950s as Navy Chief of Information.

Kirkpatrick quickly became caught in the middle of the furor which Korth's order created. Naturally, Rickover and his congressional supporters were pleased with Korth's order. Within the Academy, civilian faculty members generally were hopeful that their status and influence in the formulation of academic policies would be enhanced by the implementation of the order. However, commissioned officers on the staff and faculty tended to resent the fact that a civilian secretary was imposing policy upon them. Many (probably most) senior alumni were outraged at the prospect of having the Academy transformed into an institution that was to have a major portion of its operations run primarily by civilian professors and a civilian dean.

Korth had been able to cite the valuable support of Vice Admiral William R. Smedberg III, the former Naval Academy Superintendent (1956-1958) who, as Chief of the Bureau of Naval Personnel, was now the key official in Navy Headquarters responsible for Academy affairs. But it was by no means unqualified support. In the course of translating Korth's directive into action, Smedberg at the bureau and Kirkpatrick at the Academy pursued a successful strategy of cooperating energetically with Korth in the search for a civilian dean and in moving actively to increase the number of civilians on the faculty, while resisting the proposed total elimination of officers from academic teaching positions by taking advantage of a loophole in Korth's May 22 directive.

In that directive, Korth had indicated that the policy of replacing officers on the faculty with civilians did not "preclude the future assignment of Naval Officers to the Naval Academy faculty in those instances where such officers possess the educational and teaching qualifications to be required of civilian professors."[19] Even before receiving the directive, the Academy had recommended to the Bureau of Naval Personnel that its civilian faculty be increased by eighty-five, with a reduction of six officers assigned to the faculty. This shift upward in the civilian-to-military ratio was maintained as a goal after receipt of the Korth directive. However, a feverish new effort was made in mid-1962 to identify officers in the Navy with advanced academic degrees. By late 1962, Smedberg was able to inform Kirkpatrick that fifty such officers had been located and would be assigned to the Academy the following June.[20]

If Korth had found it impossible to obtain the whole loaf in his dealings with the Navy brass on Academy matters, he also found that his efforts to represent the Navy in the larger arena of Defense Department politics were yielding less return than he had expected. Thus, in October 1963, Korth abruptly resigned from government.[21]

Continuing Ferment

Four months before his resignation, Korth had announced the successful conclusion of the search for a civilian dean for the Naval Academy. A. Bernard Drought, who had been serving as dean of the College of Engineering at Marquette University, was hired initially on a one-year pro-tem basis, subsequently on a regular appointment. Drought had graduated from Milwaukee State Teachers College in the mid-1930s, and obtained a master's degree from Northwestern University in 1942. Immediately thereafter he had gone to Annapolis for the five-months' accelerated program for Naval Reserve midshipmen, spending the remainder of the war at Harvard and at MIT as a radar instructor. He remained at Harvard after the war long enough to complete an M.S. and a D.S. degree in electrical engineering, returning then to Milwaukee to join the faculty of Marquette University as an assistant professor in electrical engineering. Drought had become assistant dean of the College of Engineering in 1956, and dean the following year. In the summer of 1957, Drought had made a brief visit to the Naval Academy as a consultant, broadening ties that would lead to his becoming a prime contender once Korth had made the decision that the Academy should have a civilian dean.

One of Drought's first assignments was to work with the Superintendent and his staff to consider the kinds of academic reorganization that might be appropriate, given the creation of a new position in the authority structure. In the spring of 1964 a reorganization plan was ready and was routed through channels at the Academy and at Navy

Headquarters without opposition. The three divisions into which academic departments had been grouped under Melson were abolished. Thus the division directorships were disestablished, with each department head reporting directly to the Dean, who in turn reported to the Superintendent. Furthermore, the position of Secretary of the Academic Board, which had assumed some importance because of the role which the incumbent military officer assumed for coordinating policy among the various departments on behalf of the Superintendent, was abolished. The Academic Board, which had been reduced in size under Melson, was kept small (partly because of an awareness of difficulties which General McDermott was beginning to experience with a larger Academy Board at the Air Force Academy, and which superintendents at West Point had experienced with the relatively large Academic Board for decades). The Dean was to be the only civilian on the board. Its other members were the Superintendent, the Commandant of Midshipmen, and three Navy captains selected on a rotating basis from among the heads of the various departments. Even in his own office, the civilian Dean was surrounded by military personnel (although Drought assured Academy officials that the Jesuit character of Marquette had accustomed him "to working with a uniformed service").[22] The new organization created billets for two assistant deans, each of whom was to be a naval or Marine Corps officer.

Meanwhile, even before Drought arrived to assume his new position at the Academy, some additional modifications of academic procedures and programs were being made. A two-year internal study that had been made of the system that had been used in assigning numerical grades (in tenths ranging from 0 to 4.0) resulted in a change to the more conventional use of letter grades (A to F) beginning in the 1963-1964 academic year.[23] Faculty members had come to rely less than in the past on daily quizzes and demands for rote memorization, and progressively more on understanding a subject in depth. Less frequent grading reduced somewhat the obsession with comparing cumulative grade averages calculated to the third decimal point, and thereby facilitated a shift to letter grades.

It was in the autumn of 1963 also that the Naval Academy designated its first "Trident Scholars," particularly outstanding midshipmen who would be able to devote a substantial portion of their final academic year to independent research in lieu of the coursework that otherwise would have been required of them. During the 1963-64 academic year, plans were developed for making the first significant inroads on the core curriculum. Beginning in the fall of 1964, all midshipmen were to take at least six electives—15 percent of their program, with the remainder still to consist of required courses. Since the majority of midshipmen had been unable or unwilling to validate required courses or to take electives on an overload basis, the change would provide them with the first real opportunity to take electives.

Admiral Kirkpatrick, who was succeeded as Superintendent early in 1964 by the man who had been his Commandant of Midshipmen, Rear Admiral Charles S. Minter, Jr., described the academic reforms that had been effected or developed at the Academy over the past several years as "revolutionary innovations." Hanson Baldwin, the noted military analyst for the *New York Times* (who had graduated from the Naval Academy in 1924), picked up the theme in a feature article in the *Times* in December 1963.[24] These glowing assessments of new programs and practices at Annapolis were sharply contested, however, by Rickover (who had been a contemporary of Baldwin as a Naval Academy midshipman) in his annual appearance before the House Defense Appropriations Subcommittee in March 1964. "The appearance of education is there," Rickover acknowledged, "but not the reality."

Rickover had more support among Academy faculty members themselves for this harsh appraisal than perhaps he realized. The effect of the response of Academy officials to pressures to hire increasing numbers of civilians, all of them reasonably well qualified academically, had been to bring a new "generation" to the faculty. They were joined, beginning in the summer of 1963, by a new cluster of academically qualified officers (with a turnover every two or three years), most of them reservists and thus less wedded to the traditional institutional mores and practices than were most of the Academy graduates on the faculty. While the influx of relatively well educated newcomers, civilian and military, was emotionally threatening to many of the civilian professorial "old hands" (most of whom had joined the Academy faculty during or just after World War II), to some of them it provided the support and encouragement that long had been lacking for their views regarding needed increased emphasis upon academics at Annapolis and a greater voice for civilian faculty members in Academy governance.

Also, faculty members remained subjected to regimentation and restrictions to a far greater degree than many thought appropriate or necessary. Newly hired faculty members typically came fresh from civilian graduate schools, harboring strong feelings about the respect that ought to be accorded to their professional competence in the development and teaching of courses. They often found, instead, that the textbook for a course which they were assigned to teach had been selected for them (to preserve uniformity if there were several sections of the same course). Moreover, they would be expected to provide course outlines to the department head for approval, and to route draft examinations through the senior civilian professor in the department as well as through the Navy captain who headed the department.[25] They would find also that there was little opportunity to pursue one's individual research interests, and that the lip service that was paid to "academic freedom" did not provide wide latitude for the expression of views that differed from those that had received at least implicit official sanction.

David Boroff, in a widely read article on Annapolis in *Harper's* magazine in January 1963 (one of a series by Boroff on the U.S. service academies and a number of other American institutions of higher education), had sensed "subdued elation" among civilian faculty members about the Korth directive that was to have created an all-civilian faculty (except for naval science) headed by a civilian dean.[26] However, it had not taken the civilian faculty long to recognize that the all-civilian faculty idea was being squelched. Some found even the appointment of a civilian dean to be largely "eyewash."[27] Not only did the new Dean fail to become the champion of the civilian faculty that some of the latter briefly had hoped he would; but soon after assuming his position he became the instrument of a policy on the assignment of grades that put him at odds with many of his faculty.

The Grade-Quota Issue

Following the transition to the use of letter grades in the 1963-64 academic year, there had been an increase in the number of midshipmen who were receiving grades below the passing mark (C). Upon taking over as Dean in 1964 (having been on hand during the year on a pro-tem basis), Drought had instructed all department heads (each of whom was a Navy captain) to impose limits on the number of Ds and Fs to be assigned in their departments. Grades of D and F could be given to as many as 12 percent of the plebes in a given course, for instance, but no more than 4 or 5 percent of the first-classmen in a course could be given such grades.[28]

There had been some pressures on faculty members for years to keep the number of academic failures at "acceptable" levels. However, the new, explicit, quota system infuriated many of the faculty. In the Department of English, History, and Government (still referred to colloquially as the "bull" department), twenty-seven civilian faculty members signed a petition demanding that the quota system for grades be dropped. The Navy captain who headed the department responded by calling a department meeting at which he denounced the petitioners as "mutineers."[29]

Although the captain had angrily refused to forward the petition to the Dean and the Superintendent, one of the petitioners managed to get a copy to the Superintendent through unofficial channels. Moreover, an effort was made to develop new organs through which the collective voice of the faculty might be expressed more effectively. A chapter of the American Association of University Professors (AAUP) was established. The idea that faculty members should organize to promote academic freedom and to represent their interests was viewed by many officers and by some civilian faculty members as a dangerous departure from precedent, divisive and essentially incompatible with adherence to military order. The conventional argument was that it was the task of military leadership to keep *official* channels of communication open to inputs from subordinates. If existing channels should prove to be inadequate, a farsighted leader would create new official forums. Thus, early in 1966, a long dormant faculty council was revived at Annapolis, ostensibly to provide a forum which would satisfy the clamor among some faculty members for a greater voice in governance. However, its top-heavy representation of six department heads (all military officers) and six civilian senior professors led many faculty members to regard the revival of the council as but a feeble gesture at a time when a positive act of responsiveness was sorely needed.

By the spring of 1966, the AAUP chapter had grown from its handful of founders to a group of about seventy members who were committed to the general goal of gaining broader representation for faculty in Academy governance, and to the particular goal of eliminating grading quotas. In a strong statement of protest which was released to the press in April 1966, the chapter members warned that "Manipulating grades and exerting pressures on individual instructors to conform to a quota strike at the heart of academic and professional integrity."[30]

Two days previously, an assistant professor had resigned from the Academy faculty, denouncing the system of grading quotas and stating in disgust his feeling that "The academy views its civilian faculty as a commodity which it has bought like provisions for the mess hall."[31] Nine days earlier, another assistant professor had informed the *Washington Post* that his refusal to comply with the grade-quota system had led to the termination of his teaching contract.

Distressed as he was by adverse publicity in the mass media, the Superintendent, Rear Admiral Draper Kauffman, had reason to be even more anxious about what might appear in the forthcoming report of an accreditation team of the Middle States Association of Colleges and Secondary Schools, which recently had made a week-long visit to the Academy. When Kauffman received the team's confidential report the end of April, he took the unusual step of releasing it to the press, hoping that it would "clear the air." The Academy had had its accreditation renewed. Moreover, as Under Secretary of the Navy Robert Baldwin pointed out to the press, the only portion of the report that the evaluation team had underscored was a complimentary statement:

> The strongest and firmest impression we have is one of rapid academic growth and a well-conceived effort to strengthen the academy's academic program.[32]

This "strongest and firmest impression" was supplemented by other less laudatory ones, however, in the team's twenty-nine page report. The practices of imposing grading quotas and of giving reexaminations to students who failed were described in the report as ones that "no good college" would permit. The demands on midshipmen's time that were made by the various military, academic, athletic, and extracurricular aspects of the Academy program were seen as excessive. Formally, the Academy had provided study time in the schedule for each midshipman. Realistically, however, the time for contemplation that is essential to intellectual growth was lacking. The inadequacy of the library also hindered intellectual development. Moreover, pervading the Academy milieu, the evaluation team found, was the problem of reconciling the commitment to Sparta with the commitment to Athens.

> To fail to observe that this [dual responsibility] has led to a certain tension between the "academic" academy and the "military" academy would be to fail to comment on a major aspect of the Naval Academy in 1966.[33]

The tension was not entirely destructive, the report added; moreover, Academy officials seemed to be resolute in their search for the proper balance. Yet there was "a danger that the faculty may develop into two camps and that the cohesive quality so necessary if a faculty is to function as an effective force will not be achieved."[34]

No action was taken until 1970, under a new Superintendent, to open departmental chairmanships to civilians, as the evaluation team recommended. Moreover, no action was taken on a recommendation that the number of educators serving on the Board of Visitors be increased and that the board schedule visits more frequently. The latter change awaited congressional action. However, the Secretary of the Navy, Paul Nitze, through his Under Secretary, Robert Baldwin, moved rapidly to appoint a twelve-member academic advisory board, as recommended by the evaluation team.[35] And in the autumn of 1966, Kauffman authorized the establishment of a faculty forum consisting of fifty-eight members, including civilians and military officers.

Although members of the Academic Advisory Board came to the Academy only two or three times a year, the Faculty Forum represented a constant potential challenge to existing academic policies and procedures. Kauffman confessed to the members of the forum at its first meeting in 1966 that "he was convening it with 'trepidation' because it was 'not the Navy custom' to summon junior officers and civilians to offer complaints and suggestions."[36] Many faculty members, in turn, found the creation of the forum an empty gesture, since the group was denied the status of a standing organization and denied the right to adopt resolutions. Some openly called it "a fraud," others "a bitter joke."[37]

Sparta Resurgent: Restoring the "Balance"

Kauffman recognized, however, that in the view of many of his key naval-officer subordinates at the Academy, he already had gone too far in the direction of satisfying the demands of faculty members for upgrading the academic program and for increasing their representation in policy making. Kauffman himself was inclined to agree. "Making the Naval Academy the same as a darn good university is not really the point of having a Naval Academy," he asserted.[38] In June 1966, he had reported to Secretary of the Navy Nitze that "our greatest challenge at the Academy for the next couple of years would be to markedly improve our professional training and education in order to bring it into balance with the academic improvements."[39]

A starting point was a reexamination of the process of indoctrination of plebes to Academy life. In December 1966, Kauffman convened a "Midshipman Time and Effort Board" (a modern managerial approach that the accreditation team had recommended).[40] Special attention was paid to the plebe indoctrination system, with comparative data provided by the Military Academy on the procedures used there to indoctrinate West Point plebes. The board's recommendations led to modest reforms designed to prevent undue harassment of plebes by upperclassmen, especially during study hours. The fundamental rationale for the plebe system, however, remained unchallenged. For the plebe, life was characterized by regimentation and discipline that were far more severe than those experienced by upperclassmen, with very few "rates" (privileges). Prompt compliance was the order of the day, with incessant demands for high levels of performance under physical and emotional stress. Nonetheless, the attention being devoted to the improvement of the plebe system, especially by Captain Sheldon Kinney, the Commandant of Midshipmen, seemed to be reaping dividends at least on a short-term basis—perhaps because of the Hawthorne effect.[41] During Kinney's second year as Commandant (1965-66, plebe-year attrition fell below 20 percent for the first time in five years. Kauffman had high praise for Kinney's accomplishment. The hope that such results could be sustained was short-lived, however. The following year attrition rose again to 22 percent.

In February 1967, Kauffman assembled an internal committee to examine professional training and education at the Academy more broadly, and to make whatever recommendations for change they deemed appropriate. The formal mission statement had been revised under Melson to commit the Academy to developing "graduates who are dedicated to a career of naval service and have potential for future development," in contrast to the earlier trade-school emphasis on graduating "capable junior officers." By 1967, however, as Kauffman explained to his Professional Training and Education (PT&E) Committee and later to a congressional special subcommittee on the service academies, he felt that the time had come for a reemphasis on the production of "a very good, immediately employable, professional, junior officer."[42]

Kauffman's PT&E Committee strongly shared the view which congressional subcommittee chairman F. Edward Hébert expressed, that the Academy had shifted "too far to the academic left."[43] In a preliminary report, the committee of five Navy captains warned that opportunities for valuable learning during summer cruises were being squandered, and midshipmen were not demonstrating the proper motivation for at-sea training. The findings persuaded Kauffman to put several changes into effect in the summer of 1967, including an assessment of midshipman performance during at-sea training to be included in the aggregate multiple rating that was used to determine class standing. An interim report approved by the Academic Board also increased the weight to be accorded to physical education in the aggregate multiple rating. The result of these actions was a considerable de-emphasis on academics, and an increase in emphasis on the applied training provided during the summer cruise, and on physical training.[44]

In its final report in November 1967, the PT&E Committee reemphasized the primacy of the military mission of the Academy. Kauffman strongly agreed with the committee that the trend toward increasing the number of civilians and reservists on the faculty posed a threat to the maintenance of an aura of military professionalism. He acknowledged that some of the civilians who had been "on board" for many years had become "so Blue and Gold, that I think sometimes they do a better job of motivating midshipmen towards a naval career than some officers do."[45] But the recent influx of civilians and reservists disturbed him, as it disturbed the committee. Kauffman obtained the backing of Navy Headquarters to reverse the trend, reducing the number of civilians on the faculty from the all-time high of 303 during 1967-68, to 287 for the 1968-69 academic year. The number of commissioned officers, in turn, was increased from 241 to 274.[46] A large number of the officers assigned to the Academy continued to be reservists, because the war in Southeast Asia continued to impose a heavy demand for the assignment of career military professionals to that combat area. However, the Chief of Naval Personnel assured Kauffman that the Naval Academy would have first priority after Vietnam in the assignment of officers.[47]

The Changing of the Watch

The assignment of Rear Admiral James. F. Calvert as Superintendent revealed that at its top level, at least, the Naval Academy was receiving priority in assignments. Calvert was the very model of modern naval professionalism. At age forty-seven, Calvert was the youngest admiral ever to be selected as Superintendent of the Naval Academy. It was the culmination of a naval career marked repeatedly by "firsts."

Calvert had attended Oberlin College for two years during the Depression before entering the Naval Academy in 1939. Upon graduation in June 1942 (with the Class of 1943, which had completed an accelerated wartime program), he began what would prove to be extensive service with submarines. Wartime service in the Pacific brought him two Silver Stars and two Bronze Stars. Assignments with submarines continued in the early postwar years, including his first two years of command.

In 1955 Calvert was informed that Admiral Rickover wanted to interview him. If the interview were successful, orders that had been cut for a staff assignment to Hawaii would be cancelled and Calvert would report instead to the Naval Reactors Branch, which Rickover headed, of the Atomic Energy Commission. Calvert was "thunderstruck" at the prospect of an interview with Rickover, whose reputation for intimidating interviewees already had spread throughout the Navy. After a two-hour wait in the anteroom outside Rickover's office, Calvert was admitted for "interrogation." Why had he graduated so low in his class at Annapolis (114 in a class of 616), Rickover asked Calvert. "You're either dumb or lazy. Which is it?" Before Calvert could choose between these no-win options, Rickover continued with a merciless probing of other suspected shortcomings. He chided Calvert for playing an occasional game of golf, noting that Calvert had admitted that he was always pressed for time. At the conclusion of the interview, Calvert "walked out of his office a beaten man." But the following morning he received the news that his orders to Hawaii were cancelled and that he was to proceed to Washington, D.C., to the Naval Reactors Branch of the AEC.[48]

Nearly two years of intensive study of nuclear energy and related topics followed. The effort was rewarded in November 1956 when Calvert was placed in command of the nation's third nuclear-powered submarine, the U.S.S. *Skate*. In March 1959, almost exactly fifty years after Robert Peary's expedition had reached the North Pole, the U.S.S. *Skate* became the first ship to surface at the Pole. Calvert would relate the exciting account of the voyage for the *National Geographic* and, with the encouragement of the magazine's editors, develop the account into a book.[49]

Calvert had held the grade of commander at the time of the historic voyage. Subsequently he had been promoted to captain, two years ahead of the time that his graduation class from the Naval Academy normally would come due for promotion (and a year ahead of another fast-rising classmate, Elmo Zumwalt). Another feather was added to Calvert's cap when he was selected to attend the National War College. While on the staff of the Armed Forces Staff College a few years later, Calvert completed another book that would draw wide attention in military circles, *The Naval Profession*.[50] By the time the book appeared in print in the summer of 1965, Calvert had learned of his promotion to rear admiral. His classmate and friendly rival, Bud Zumwalt, had made it also. They were the youngest officers in U.S. Navy history to attain flag rank.[51]

Calvert's promotion to flag rank was accompanied by return assignment to the Office of the Chief of Naval Operations with the Politico-Military Division, this time as director. Two years later he was placed in command of Cruiser Destroyer Flotilla Eight in the Atlantic Fleet, an assignment which gave him an advance opportunity to evaluate the behavior and attitudes of Annapolis midshipmen who were assigned to the flotilla for summer training. Admiral Charles Duncan, who as commander of the Second Fleet was Calvert's immediate superior, was tremendously impressed with Calvert's performance, as most of Calvert's previous superiors had been. It probably was Duncan's recommendation, Calvert would later surmise, more than any other that led to his selection as Academy Superintendent. The extensive and distinguished experience that he had had with nuclear submarines "certainly wasn't held against me," Calvert acknowledges. But as to whether Admiral Rickover expressed his own views regarding the selection, Calvert remains uncertain.[52]

The Continuing Emphasis on Naval Professionalism

When Calvert arrived at Annapolis in the summer of 1968, Kauffman briefed him on current Academy affairs. The "serious loss of professional emphasis" in the program that had occurred during the years of upgrading the academic program was a key problem that merited continuing concern, Kauffman said.[53] He gave Calvert the three-inch thick report of the Professional Training and Education Committee as a guide for further action. Calvert was predisposed to agree with Kauffman on the need for further emphasis on professionalism; reading the PT&E report strengthened this conviction.

For Calvert, a logical starting place for the continuation of the process of reform of professional training was the summer cruise. Each summer, midshipmen of the third and the first classes were assigned to ships at sea. This provided them not only with valuable experience in learning various shipboard duties, but also with interesting travel to far-flung ports (such as Hong Kong and Tokyo for a Pacific cruise, London and Oslo for an Atlantic cruise). The quality of training for the past several years, however, had been dependent upon the organizational ability and disposition of the captains of each of the dozens of ships to which various clusters of midshipmen were assigned. Calvert decided to continue the practice of assigning small numbers of first-classmen to each of a wide variety of ships, so that each first-classman could be given the opportunity to perform the duties of a junior officer. However, the at-sea training for the third class would be concentrated on two LPDs (attack transports) in the First Fleet in the Pacific, and on two in the Second Fleet in the Atlantic, thereby insuring more centralized control over the quality and uniformity of the training that was provided.[54]

The cultivation of professional attitudes in midshipmen started earlier than the third-class cruise, however. Calvert directed his attention also to two concepts that he deemed to be fundamental: the plebe indoctrination system and the honor system. Reform efforts were concentrated on the summer indoctrination phase of plebe year. Beginning in 1969, the second-classmen who were to conduct the indoctrination were given more intensive preparation (50 hours of instruction) than in the past. Moreover, doubling the number of second-classmen assigned to these duties rather than to other training assignments meant that plebes could be provided with more individualized supervision. Calvert emphasized that the purpose of the change in approach was not to make life easy for the plebes. On the contrary;

> It will remain tough but the goal is to remove humiliation and embarrassment, to emphasize the positive and professional approach, and to more nearly approach the leadership situations that young officers will find themselves in when they reach the Fleet.[55]

The goals were virtually unassailable. It was translating them into reality that was difficult. Once having endured plebe year, upperclassmen—or indeed alumni—typically were resistant to the imposition of restrictions that seemed to them to be "softening" the system, however distasteful they might have found the plebe experience themselves. Consequently, like the efforts of Sheldon Kinney as Commandant of Midshipmen and of Draper Kauffman as Superintendent a few years earlier, the changes that Calvert was able to effect in the plebe system were more evident in principles enunciated than in the behavior of upperclassmen.

Reform of the honor system, Calvert thought, also required a more careful indoctrination than in the past. The key, in his view, lay in tapping latent idealism from the midshipman's earliest days at the Academy, and then progressively cultivating his sense of discipline, responsibility, and willingness to be accountable for his actions. Concepts such as discipline and honor were "old fashioned," Calvert often acknowledged; but he offered no apologies for emphasizing them. On the contrary, these were the core values that made the service academies distinctive. However, the facts that such values no longer were nurtured as they once had been in homes throughout the country, and that unrest and permissiveness on civilian college campuses reflected contrary values, made the Academy's task difficult.[56]

A rash of incidents of cheating and plagiarism among midshipmen raised the spectre of an honor scandal such as ones that had wracked the Military and Air Force Academies. Thus, Calvert ordered a comprehensive review of the honor system during the 1970-71 academic year, and urged officers and midshipmen to renew their commitment to the traditional ideals of duty, loyalty, and honor.[57]

The Recruitment Effort

Maintaining traditional ideals and reforming the program in other respects would enhance the Academy's appeal. However, more energetic efforts than in the past were also needed, Calvert was convinced, to seek out prospective applicants. Since 1962, the number of young men admitted annually had varied between approximately 1,300 and 1,400. However, the number who had satisfied the minimum academic requirements for admission typically had been between 2,000 and 2,100; the number who also were fully qualified physically and medically had been less than that. In the summer of 1968, for example, there were only some 1,400 fully qualified applicants, of which the Academy had admitted 1,368.[58] In short, the Academy had had little selectivity among applicants—none at all in a number of instances where only one applicant from a particular congressional district was fully qualified.[59]

During his final months as Superintendent, Admiral Kauffman had expressed his deep concern about what seemed to be a growing recruitment problem.[60] When Calvert took over, he gave the matter high priority, allocating resources to a major effort to reach a broader pool of potential applicants than in the past. For the most part, the various components of the recruitment effort had been in operation when Calvert arrived as Superintendent. However, he not only demonstrated great personal concern in seeing to it that the efforts bore fruit, but also increased the staff that was assigned to recruitment. Whereas previously responsibility had lain with two Navy lieutenants, Calvert appointed a Navy captain as Director of Recruitment, reporting directly to the Superintendent, assisted by a lieutenant commander, and by four lieutenants to coordinate recruiting in the various geographic regions. In addition, another officer was assigned to handle liaison with members of Congress, and still another, a Navy black lieutenant, to promote and coordinate the recruitment of young men from racial and ethnic minorities.[61]

Calvert was pleased to be able to tell the alumni at his final Homecoming with them as Superintendent in the fall of 1971 that "with 7350 men nominated by the Congress and the President, we had not only the largest number of men that had been nominated to the Naval Academy but also the largest number that had ever been nominated to any service academy."[62]

Calvert's figures were more optimistic than those released by his Dean of Admissions. The latter showed that the summer to which Calvert had been referring had witnessed an upturn in the number of qualified applicants although not on the scale Calvert had indicated. The following summer, however (representing the final recruitment year of the Calvert superintendency), the number of qualified applicants dropped again. Over Calvert's four years, the ratio of qualified applicants to applicants admitted was not much different from what it had been in 1968 when he assumed the superintendency (the year that he described as especially bleak). Moreover, high school performance and test scores suggested a remarkable continuity over the Calvert years from 1968 to 1972 (of high quality, it should be noted). In the realm of minority recruitment, however, there was clear and dramatic success, with a record-high 73 blacks entering with the Class of 1976, together with 21 midshipmen from other ethnic minorities. (The number of blacks entering the Academy peaked the year after Calvert's departure, although over 10 percent of each of recent classes entering the Academy have been from racial minorities.)

Fostering the Will to Win

The general and minority recruitment efforts were supplemented, as they had been for years, by a more specialized program of athletic recruitment coordinated by the director of athletics. Calvert's predecessor had given renewed emphasis to the athletic program, including recruiting. Kauffman's special concern for athletics had been provoked by the discovery, when he assumed command at the Academy, that Navy teams had gone eleven consecutive sports seasons without winning the majority of their games against West Point.[63] When Calvert took over the reins, some members of the press had interpreted remarks by him as suggesting an intention to downgrade the importance of football, especially, in the Academy program. The response of Academy alumni to such rumors was immediate and indignant. Calvert, in turn, was quick to provide reassurance that "Football is not going to be de-emphasized in any way at Annapolis. There will be no change in our scheduling policy. We will play the best because we belong with the best."[64]

An emphasis on athletics in general and football in particular was perceived by Calvert and his close associates as being entirely compatible with—indeed, mandated by—the mission of the academies. As Navy Captain John Coppedge, the director of athletics, explained to a gathering of parents of plebes in the fall of 1971, the athletic program contributed to the midshipman's pride, his poise, and his confidence. "Today there are many environments in which competition is not in fashion," he observed. "It's always in fashion here. We believe in competition, and we play to win!"[65]

If you learned to play to win, you would fight to win in combat. That reasoning was accepted almost universally at Annapolis, as it was at the other academies. The only philosophy acceptable to the real naval professional was that of James Lawrence ("Don't give up the ship"). When this will to persevere despite overwhelming odds was lost, one had to expect incidents such as that of the *Pueblo* in 1968—surely the darkest day in American naval history, Calvert pointed out to the midshipmen and to Academy faculty and staff members on numerous occasions.

Annapolis (like the other academies) recruited young men who had shown that they thrived on competition. Of an average entering class in the 1960s and 1970s, three-fourths would have participated in varsity athletics in high school. At Annapolis, this spirit of competition was nurtured and the keen competitor rewarded, not only in athletics but in all facets of midshipman life. An unintended consequence of the ubiquitous emphasis on competition, and on scoring or grading the midshipman's performance at every turn, was widespread cynicism among midshipmen regarding their own commitment to institutional goals. One learned to "play the game," many midshipmen decided, in order to "beat the system" rather than be overcome by it.[66] For example, as Calvert warned incoming faculty members in the summer of 1971, midshipmen had developed some unfortunate attitudes and behaviors which he hoped to correct. What he termed "union rules" governed their approach to academics. If a midshipman devoted too much of his free time to studying, in the eyes of his fellow-mids he had violated the "union rules."[67]

Midshipmen remained intensely competitive despite their professed nonchalance, however. Even their efforts to outdo one another at being "cool" in response to the demands of "the system" were revealing. Moreover, oftentimes their behavior in unguarded moments of leisure provided evidence of a desire to prove their mettle or their masculinity which they managed to conceal or suppress in the performance of prescribed routines. Off duty, for example, midshipmen frequently enjoyed comparing the informal ratings which they made of one another's dates, and of the automobiles which midshipmen were permitted to have (displaying a strong preference for sports cars) during their final year at the Academy.

Of course, competitive behavior in all-male subcultures in America frequently manifests itself in what some observers have termed the "hairy chest syndrome."[68] Thus, what is notable about Annapolis during the Calvert years (and before and since) is not the mere presence of attitudes of *machismo* among midshipmen, but rather the extent to which these attitudes were fostered by nearly all facets of the Academy environment.

The Revival of Concern for Academics

In his support for athletics, in his expression of concern for enhancing military professionalism, and in the frequency of his references to concepts such as honor and personal accountability, Calvert associated himself with the traditional values of the institution. His commitment to retaining the requirement that all midshipmen attend chapel service on Sunday was further evidence of the importance he attached to traditional practices.

But respect for tradition was not to be confused with complacency, in Calvert's view. Once convinced that reform of a particular policy or program was needed, Calvert was not hesitant to act. The most impressive illustration of the development of such conviction, and action pursuant to it, came in academics. Despite Calvert's agreement with Kauffman on the primacy of professionalism among Academy concerns, and despite his feeling after arrival that "academically I think we are doing just fine,"[69] the academic realm quickly became one of paramount concern.

Before becoming Superintendent, Calvert had been aware that "the relationship between the office of the Superintendent and the civilian faculty had deteriorated. . .I think largely over those grading things."[70] When he became Superintendent, the faculty took little time before raising through the Faculty Forum and the AAUP their mounting concern about what they perceived as lost opportunities to build upon the academic reform that had been started in 1959 and the early 1960s. Moreover, concern was expressed lest the recognition that had begun to be accorded to the important contribution which civilian faculty members made to the Academy be thwarted by the quest for a restoration of military professionalism.

Although expressing his general skepticism regarding the need for radical academic reforms, in acknowledging the desirability of enabling midshipmen to pursue an academic concentration sufficient to constitute a major, Calvert lent his support to a change which the Faculty Forum had been recommending since its inception.[71] His ideas regarding the need for further academic reform were being shaped not only by persistent communications from his faculty, but also by the more detached appraisal that was being provided by the

Academic Advisory Board. Suggestions that the Academic Advisory Board made were discussed with the Board of Visitors to the Academy, including most notably Edwin Harrison, who had graduated from Annapolis the year Calvert had entered (1939) and who was serving as president of Georgia Tech. Calvert also placed the ideas that he was developing regarding an expansion of electives and opportunities for majors before the Objectives Review Board (ORB). The ORB was one of four boards that Calvert had created internally to provide the Superintendent with a forum for discussion of proposed policy changes and to provide "some degree of formality to structure his decisions."[72] In addition, Calvert made it a point to talk individually with as many faculty members as possible about the proposed reforms, to midshipmen, to alumni groups, and to the Chief of Naval Operations, the Vice Chief, and the Chief of Naval Personnel.

"Let a Hundred Flowers Bloom"

By spring 1969, Calvert had touched all bases, and had secured approval of plans for a major revision of the curriculum to go into effect that fall. The core curriculum would be greatly reduced, with electives expanded proportionately. Calvert's continuing strong commitment to the cultivation of professionalism among midshipmen was reflected, however, in the nature of the core curriculum. The number of professional courses was to be doubled, approximately, and only these courses were to be required by name and number. Each midshipman was to be required to select a certain minimum number of other courses from among those offered in mathematics, the sciences, and the social sciences and humanities, respectively. He would be required to select an academic major from among twenty-four available initially (the number expanded slightly the following year). In pursuit of his major, the midshipman would select at least a required minimum number of electives in his major field, completing his program with electives selected from among any of the dozens available.[73]

The nucleus of Ph.D.'s in each department bore the prime responsibility for the development and teaching of the specialized and advanced courses that were being added. The faculty in mathematics, for example (who were equal in number to the conglomerate of English, history, and government), had been offering such electives as "differential equations" and "statistics." To these were added new and relatively specialized topics, such as "mathematical logic" and "introduction to mathematical economics," bringing the total number of electives in mathematics to twenty—the number of Ph.D.'s on their faculty.

The science faculty, numerically the largest at the Academy, provided midshipmen with opportunities to explore various scientific frontiers through electives in astrophysics, quantum physics, biochemistry, environmental dynamics, and synoptic meteorology. Fortuitously, the master plan for the modernization and expansion of Academy facilities, which had been started under Admiral Davidson, was beginning to reach fruition when Admiral Calvert took over. Thus during his first year, he was able to dedicate a handsome new glass and granite mathematics building, named for William Chauvenet, father of the Naval Academy, and its architectural twin, a new science building, named for Albert A. Michelson, whose Nobel Prize-winning measurements on the speed of light had been completed while he was serving at Annapolis as a faculty member in the 1870s. The facilities provided within the new structures were dazzling examples of modern technology—closed circuit television; remote, computer-linked terminals in every classroom; laboratories for complex experiments in chemistry, physics, electronics.

Paul Quinn, of the science faculty, became the director of the Academic Computing Center when it was first established early in 1966. The dozen or so participants in a workshop that Quinn conducted in the summer of 1966 returned to their respective departments as disciples of a new technology that had a multitude of potentially exciting applications in research and teaching. Physicists, electrical engineers, mechanical engineers, mathematicians began utilizing the computer in their courses (although many mathematicians were apprehensive initially that the introduction of computers might lead midshipmen to downgrade the importance of their discipline). A groundswell of interest among faculty members and midshipmen was given significant support by Admiral Calvert when he became Superintendent. The number of remote terminals was increased to approximately a hundred, allowing access from points throughout "the yard" (as the area occupied by the Naval Academy is termed) to a Honeywell Model 635 Digital Computer on a time-sharing basis. The involvement of faculty and students was expanded to include the social sciences and humanities.[74] Basic instruction in the use of computers was required of all midshipmen.

Faculty in engineering had shown their commitment to academic experimentation not only by being among the most active in the development of computer usage, but also in other ways. As late as 1963, the Department of Engineering had been alone among the academic departments in being without a single Ph.D. However, in subsequent years, there was a dramatic resurgence within the department. First under Captain Wayne Hoof, who hired the first dozen Ph.D.'s, and later under the departmental leadership of Captain Randolph King, an Annapolis graduate who had done postgraduate work at MIT, a new spirit of innovation was injected.[75] With the shift to the majors program, a wide array of engineering electives was developed, ranging from aerospace engineering, orbital mechanics, and computer methods in structural mechanics, to oceanography, underwater acoustics, and reactor physics.

The combined math-science-engineering developments were designed in part to meet the standards of engineering accreditation. Success in this regard was attained when majors in aerospace, electrical, mechanical, and systems engineering received accreditation by the Engineers Council for Professional Development.[76]

Prompted both by Admiral Kauffman's complaints about the Western bias of course offerings and by the mounting interest of Americans in the military involvement in Southeast Asia, the Department of English, History, and Government had been developing new courses focusing on Asia, as well as new ones focusing on Africa, Latin America, and the developing nations in general. Courses such as contemporary non-Western civilization, political and military development of Southeast Asia, development of Luso-Brazilian civilization, and Afro-Asian culture were added. The Department of Foreign Languages added Chinese to a list that already included French, German, Spanish, Italian, Portuguese, and Russian. Other electives focused on subjects such as cultural anthropology, philosophy of religion, Chaucer, American black literature, civil-military relations, economics of labor relations, econometrics, imperialism, modern poetry, and science, technology, and international relations. The justification for such courses had to be derived from the general rationale for the usefulness of a broad, liberal education, rather than resting on the argument that immediate relevance to a naval ensign or to a Marine Corps lieutenant could be demonstrated.

That Admiral Calvert accepted, and even encouraged, the liberal education justification gave hope to faculty members in these areas that at last the social sciences and humanities were coming into their own at the Naval Academy. As expressed by Stuart Pitt, a senior professor of English with a Ph.D. from Yale, Admiral Calvert was "a glorious example of the new faith."[77] Or, as some midshipmen wryly observed, the slogan of the Calvert era had become, "Let a hundred flowers bloom."[78]

However, as some faculty members reminded their students, this slogan, which had gained prominence in the 1950s as expressing Mao Tse-tung's concept of the spirit of Chinese society, had been replaced subsequently by others placing the emphasis instead on the virtues of loyalty and conformity. Drawing upon more directly applicable experience, faculty members who had served at the Academy for many years recalled numerous swings of the pendulum between relative openness on the part of a Superintendent and relatively heavy-handed command.

Whatever apprehension skeptical faculty members or midshipmen might feel about the future, however, few denied that Calvert's early performance was impressive. He had given a powerful boost to sagging faculty morale, while simultaneously reassuring commissioned officers on his staff and among the alumni of his strong commitment to the development of professional skills and attitudes. He had revived the momentum of reform that had begun under Melson and then gradually slowed. He had broadened participation in discussions of various policy issues, down to and including the assignment of midshipmen to committees that were reviewing policies. Yet there never was any doubt who was in command. "Lord Jim," he was termed by many faculty members and most midshipmen (although never to his face). Indeed, Calvert commanded with a regal air—masterful but with a common touch. Moreover, he was articulate, and had developed a philosophy of education which the "golf and cocktails" Superintendents of earlier years at Annapolis sorely lacked.

Perhaps most important to the success that Calvert was enjoying in translating ideas for reform into action, however, was the fact that he had a keen sense of organizational politics. From the beginning of his reform effort, he had managed to isolate those military officers on the staff who opposed making any concessions to further academic reform—the "real Neanderthals," as they were termed by civilian faculty members.[79] He had eased the suspicion of his intentions that civilian faculty members had had by attending meetings of the Faculty Forum regularly and letting the faculty members "have at me."[80]

Although Calvert encouraged faculty members to express whatever ideas they might have for reform of the academic program, he had maintained the Objectives Review Board as the definitive policy-making body in academic matters; and he made sure that he "had the votes" before any crucial issue was to be decided. The ORB, which had about a dozen members, half of whom were senior civilian faculty members or administrators and half military officers, met monthly. As Calvert recalls,

> Ben Drought [the Academic Dean] and I were the two senior members of that board, and we sort of made an informal agreement among ourselves to begin with that we would not have differences of opinion before the board. . . . And when these fairly pivotal changes in the academic structure of the school—namely the move to the majors program—were coming up at the ORB . . . I was able always, I think, to present a situation where Ben Drought and I were in agreement. And in most cases . . . we were [able] to introduce a resolution of the Academic Advisory Board stating that they were in agreement with these proposals.[81]

Calvert's own views of what reforms were needed were close enough to those which faculty members also deemed desirable that even in instances where Calvert had rammed particular changes through the ORB, most faculty members found little reason to complain. To some faculty members, however, Calvert's style and procedures revealed a commitment to faculty participation that they deemed to be largely superficial.

A confrontation that brought criticism of Calvert's style into the open was that over academic reorganization. Most civilian faculty members long had thought that a reorganization was badly needed in order to accord proper recognition to the various academic disciplines. Instead, conglomerates existed such as "science" (which included chemistry, physics, and electronics, neglecting the biological sciences), "engineering" (which ranged from metallurgy to aerodynamics), and "English, history, and government." Calvert consulted widely on the matter and endorsed the creation of departments according to academic discipline. However, the final reorganization decision was presented to the faculty as a *fait accompli* by the ORB, and ignored some of the other recommendations which had had broad support in the faculty. Especially irksome to some faculty members was the grouping of the now-separate Departments of English and History in a division that separated them administratively from the other social sciences and humanities in the Division of U.S. and International Studies.[82]

In a letter to the Superintendent, the president of the AAUP chapter, who also had been one of the founders of the Faculty Forum, expressed his deep concern that the reorganization decision had been made by procedures which

> cast serious doubt upon the future of a reorganization plan purported to initiate a fundamental transfer of authority and responsibility to faculty

hands, undermines [*sic*] faculty morale and confidence and respect for their representative organizations, and contradicts [*sic*] practices generally accepted throughout the academic community.[83]

Calvert himself realized that the reorganization plan was not fully defensible on the grounds of grouping departments according to the relationship of academic disciplines to one another. However, there were additional factors that had to be considered, realistically, such as the personalities of those who would occupy various key positions following the reorganization. Calvert thought that it was important, for example, to get the political scientists out from under the thumb of some of the senior historians. Thus, the former were placed within the Division of U.S. and International Studies; the latter were within the Division of English and History.[84]

The reorganization was put into effect in July 1970. Despite the annoyance of some faculty members at the procedures that had been utilized to effect the change, most realized that the change substantially embodied many of the recommendations which they had been advocating for years. Indeed, the "emancipation" of individual departments from preexisting configurations (such as "English, history, and government") was, some thought, the "most significant single act of the renaissance" that the Calvert era represented.[85] Moreover, for the first time in recent history, some (more than half) of the departments would be headed by civilians.

Curricular reform had been a major substantive achievement of the Calvert years; departmental reorganization was an essential structural component. The reform process had been agonizingly slow, those members of the Academy faculty who had been working for reform for over a decade realized. But at last, and at least, Jim Calvert had brought the Naval Academy "kicking and screaming into the twentieth century."[86]

However, it was neither the major academic reforms nor the impressive modernistic changes in physical plant that were completed or under way that Calvert chose to emphasize in June 1972, upon turning over his duties to his successor, Vice Admiral William Mack. Rather, Calvert spoke of professionalism, and its roots in what he termed the "non-rational nature" of nationalism. "We are professional nationalists," he observed, "and we must pick up the tab for all the shortcomings of that idea, while we are privileged to participate in much [*sic*] and many of its privileges." In commenting on the relationship of the Naval Academy to the profession as a whole, Calvert offered a commentary on the views of those who would proceed with modernization more completely than he had done. "This profession needs a touchstone, a shrine," he observed,

> and it is that function more than any other which is furnished by Annapolis to the naval service. There are always those who point out that we can train our officers much more cheaply elsewhere than we can at Annapolis or West Point. And this is true. And there are those who point out that we could give a better education here if we would knock off all this military stuff and stop participating in intercollegiate athletics. And that is also true. But *what these people fail to understand is that magic non-rational ingredient which is the catalyst for excellence in our profession, and without which, indeed, we have nothing.*[87]

The "magic ingredient," Calvert observed, could be sensed in parades, when the band began to play; it could be felt in the "tearstained exultation which comes from a great football victory." Knowledge of machines and of people was essential to modern naval professionalism; ". . . but, above all," Calvert emphasized,

> we must have that last final spark that I've tried to talk about this morning. And if we have it here at Annapolis it is because many fine men in the past have determined . . . that it will be preserved against all the onslaughts of logical [*sic*], cost effectiveness and keen analysis.[88]

It was an eloquent statement by a reformer who at heart was a traditionalist. But the debate about the adequacy of the service academies and the relevance of their traditions and their current programs to the needs of the 1970s and 1980s was far from over.

PART III

Analysis

The Determinants and Consequences of
Organizational Change

8

Internal Leadership and Organizational Change

The case studies in the preceding chapters made evident the impact that service academy executives can have on the structure and programs of their institutions. The changes that followed Robert McDermott's assumption of the role of dean of the faculty in the critical early years of the Air Force Academy are a dramatic example of such impact. The enormous stimulus to organizational activity and morale provided by the leadership of Frank Leamy as Superintendent of the Coast Guard Academy is another example. Although the role of the commandants of cadets or midshipmen received less sustained attention in the case studies than did that of the superintendent, reforms such as those initiated by John Throckmorton at West Point or by Sheldon Kinney at Annapolis suggest that the role of the academy commandant can be important too.

Highlights from the careers of key leaders at the academies were presented in the case studies not merely for the human interest such details may add, but to provide clues that might be relevant to explaining their actions as superintendents, commandants, or deans. The discussion of Garrison Davidson's early coaching experience at West Point, which included association with Earl Blaik, was designed to provide some insight into Davidson's decision as Superintendent to alter priorities in athletics, which had become dominated by the Blaik-directed football program. The discussion of Willard Smith's role as Commandant of Cadets at New London under Leamy contributes to an understanding not only of the policies of the Leamy years but also of the policies which Smith initiated when he became Superintendent. Ruminations of former superintendents, commandants, or deans on their own days as cadets or midshipmen were included when doing so helped to explain the individual's commitment to institutional tradition or, in some cases (for example, Robert McDermott and James Briggs), to breaking from past practices.

Illuminating though it may be to view the actions of an individual within the context of his previous experience, to what extent do the examples that have been described provide a basis for generalization? Particularly, what generalizations, if any, are we able to develop about the relationship between career patterns and the proclivity of academy executives for innovation on the one hand, or for resistance to change on the other? To put the question in different terms, to the extent that one can identify leaders who have presided over periods of change at the academies, and distinguish these leaders from others whose tenure has been characterized by the maintenance of the status quo, can the differences in policy be explained by contrasting patterns of previous career experience and preparation? Evidence from the case studies is supplemented in this chapter with data on the career patterns of all post–World War II superintendents, commandants of cadets or midshipmen, and academic deans, in an effort to answer these questions.

Interpretation of the evidence is complicated, in part because we know from the case studies that even leaders who are reform-minded in some respects often prove to be traditionalists in other respects; James Calvert, who presided over a vitally important period of change at Annapolis, is a notable example. Moreover, one event after another in the recent history of the academies has demonstrated that no leader operates with complete automony. Quite to the contrary. Superintendents, commandants, and deans at various times were influenced significantly by the actions or desires of faculty members, boards of visitors, accreditation-review committees, members of Congress, chiefs of the parent arms of service, distinguished alumni, the press. Furthermore, whatever desire executives may have had to launch needed reforms, the opportunities that were afforded them for experimentation in most cases were severely constrained by their limited time in office. The concluding portion of this chapter provides a detailed consideration of the constraints upon academy leaders. These include not only the duration of their appointment but also the policy climate externally and the milieu within which they must operate internally.

The Superintendents

All academy superintendents in the modern era have been flag-rank officers. Their careers, therefore, have been exemplary of the credentials regarded in the military service as the *sine qua non* of success, and virtually all superintendents have possessed such impressive credentials upon assumption of the academy command.

Command

The watchword of success in the military profession is "leadership," a quality which is most readily demonstrated in command assignments. Only six of the forty-seven superintendents in the years 1945-1978 (Table 4) assumed their academy responsibilities without previously having had what would be considered a "major" command: a division or larger unit in the Army or Air Force; a battleship, carrier, or cruiser division in the Navy (deep-draft command, as it is known); one of the seventeen regional districts for which the Coast Guard has responsibility; or, in any of the services, command of an important military installation such as a service school. However, even the careers of the exceptions to the rule of major command included assignments that might well have been viewed by promotion and selection boards as meritorious alternatives to a major command.

The credentials of the forty-one superintendents who had had a previous "major" command reflected three types of such experience in particular: command in combat; command of a unit the prime technology of which is of growing importance to the service; or command of a military school or training installation.

The selection of Maxwell Taylor as Superintendent of the Military Academy and that of Aubrey Fitch as Superintendent of the Naval Academy at the end of World War II are illustrative of the appointment of leaders who had gained fame in wartime commands. Taylor had jumped into Normandy at the head of the 101st Airborne Division, and had led the division through the fighting in France and into Germany. Fitch had commanded the Air Task Force in the crucial battle of the Coral Sea in 1942, and subsequently had been in command of aircraft for the South Pacific Fleet. Similar appointments to the academies were made at the end of the Korean War and in the final years of the U.S. involvement in Southeast Asia.

Although Taylor at West Point and Fitch at Annapolis represented the assignment of prestigious combat leaders as academy superintendents, each also exemplified the recognition by his respective service of the growing importance of a particular technology—airborne in the case of Taylor, naval aviation in the case of Fitch. Frank Leamy became the first Coast Guard aviator to become Academy Superintendent when he assumed command at New London in 1957, having served as Chief of the Aviation Division at Coast Guard Headquarters during World War II. In 1960, John Davidson became the first submariner to be appointed to the top post at Annapolis.

Table 4 Previous Career Experiences of Academy Superintendents Who Served in the Years 1945–1978

Academy		Career Experience		
		Major Command[a]	Top Staff or Aide Assignment[b]	War College
USMA	yes	11	12	8
	no	1	0	4
USNA	yes	14	14	9[c]
	no	0	0	5
USCGA	yes	8	13	3
	no	5	0	10
USAFA	yes	8	8	7
	no	0	0	1
total	yes	41	47	27
	no	6	0	20

[a]For Army: Brigade, division, or higher. For Navy: Battleship, carrier, cruiser division, or higher. For Coast Guard: Coast Guard district or higher. For Air Force: Air division or major installation.

[b]For Army: Department of the Army Headquarters, Office of the Secretary of Defense, JCS staff, or Chief of Staff of a field army. For Navy: Chief of Naval Operations or Deputy CNO staff or aide, aide to the Secretary of the Navy, or key staff or aide position with U.S. Atlantic or Pacific Fleet. For Coast Guard: Aide to the Coast Guard Commandant, bureau chief or Chief of Staff in Coast Guard Headquarters, Chief of Staff in Coast Guard district. For Air Force: staff of Air Force Headquarters or Strategic Air Command Headquarters.

[c]Includes attendance by Rear Admiral John Davidson at the Canadian War College, and service by Rear Admiral Draper Kauffman on the staff of the Navy War College without attendance as a student.

If Taylor, Fitch, Leamy, and Davidson can be described as the "first wave" of successful officers associated with a relatively new military technology, there are numerous examples of superintendents with similar qualifications appointed subsequently, at a time when these specialized skills had become more prevalent among flag-rank officers. The "second wave" appointments signal the rise or ascendancy of these specialized skills within the respective arms of service.[1] A "third wave" is by now making its appearance, in the form of officers trained in even newer military technologies. The appointments in the summer of 1974 of Sidney Berry as Superintendent at West Point and of James R. Allen as Superintendent at Colorado Springs are notable recent examples. Berry was qualified both in airborne training and as a helicopter pilot. He reported to the Military Academy directly from command of the 101st Airborne Division, which by now was equipped with 422 helicopters as America's only air-assault division. Allen reported to his command of the Air Force Academy from Washington, where he had served for two years as Special Assistant to the Air Force Chief of Staff for the B-1 bomber.

That combat command or command of a unit characterized by an advanced military technology should be deemed important in the selection of an academy superintendent attests to the continuing salience in the minds of the service secretaries and chiefs who make the selection, of the Spartan component of the academy mission. It does not follow, however, that the latter deem the Athenian component to be unimportant. Rather, the selection process for superintendents reflects an outlook analogous to that which prevailed before the Civil War in the selection of civilian college presidents, when it was widely believed that training in the clergy was sufficient preparation for the role.[2]

Although previous experience in an educational role by no means is a prerequisite for selection as superintendent of a service academy, under some circumstances it has given a candidate for the position a competitive edge. For example, Hubert Harmon was an obvious choice to become the first Superintendent of the Air Force Academy, having been involved in the planning for several years as Special Assistant to the Air Force Chief of Staff for Air Academy Matters. Following a massive cadet cheating scandal, which called into question not only the success of the Military Academy in maintaining traditional values but also the integrity of the academic program, the Army in 1977 called from retirement an officer to become Superintendent at West Point who combined a reputation as "a real honor-duty-country man"[3] with academic credentials that exceeded those of any previous academy superintendent; General Andrew Jackson Goodpaster, who had graduated from West Point in 1939, had received both a master's and a doctor's degree from Princeton shortly after World War II. Moreover, after serving as the presidential staff secretary throughout the Eisenhower years, Goodpaster had had experience as Commandant of the National War College.

Staff and Aide Assignments

Only three superintendents in the years 1945-1978 previously had held the key subordinate command position of Commandant of Cadets or Midshipmen: Frederick Irving at West Point; Willard Smith at New London; and Charles Minter, Jr., at Annapolis. However, all superintendents have been academy graduates, and previous service on an academy staff or faculty has been common, although by no means universal, among superintendents (Table 5). The extent to which such service is perceived to be career-enhancing varies among the services.

Table 5 Previous Service Academy Assignments of Officers Who Served as Academy Superintendents in the Years 1945–1978.

Academy	Nature of Assignment				
	Academic	Military	Academic and Military	Other (Athletics, Aide)	None
USMA	Taylor Knowlton Berry	Bennett Koster	Irving[a]	Bryan Davidson	Moore Westmoreland Lampert Goodpaster
USNA	Joy[b] Smedberg Davidson[b]	Hill Kirkpatrick Minter[a]		Fitch Melson	Holloway Boone Kauffman Calvert Mack McKee
USCGA	Hall Leamy Evans Engel[b] McClelland Clark[c]	Pine Smith[a]	Derby Jenkins[b]		Mauerman Bender Thompson
USAFA[c]	Harmon Briggs Stone	Allen Tallman			Warren Moorman Clark

[a]Commandant of Cadets or Midshipmen
[b]Academic department head
[c]The earlier academy experience of Air Force Academy superintendents was at West Point, the alma mater of each of them.

For decades, assignment to West Point has been regarded as beneficial—although not essential—to the career of regular Army officers.

It is not surprising that eight of twelve Military Academy superintendents in the years since 1945 have had earlier Academy experience in one capacity or another. The career rewards attached to Military Academy service became generally applicable in the Air Force also, when it became independent of the Army. In the Coast Guard, assignment to the Academy has been a career "plus" rather than a "minus"; thus, ten of the thirteen Coast Guard Academy superintendents since 1945 had earlier Academy experience.

Only in the Navy does one find a sharply different pattern. Relatively few billets at the Naval Academy have been regarded as career-enhancing: Superintendent, his aides, Commandant of Midshipmen, academic department head. Assignment to the Academy in another capacity, such as company officer or academic instructor, was regarded as detrimental rather than beneficial to one's career, although the Navy has been assigning more highly regarded officers to such billets in recent years in an effort to change this image. Six of the fourteen Naval Academy superintendents who served in the years 1945-1977 had had no previous Academy duty. Of the eight with such earlier service, three (Joy, Smedberg, Davidson) had served as department heads, one (Minter) as Commandant of Midshipmen, and one (Melson) as aide to the Superintendent. Only three of the superintendents had served in lower-status positions, and none of these in a strictly academic capacity.

Whereas approximately two-thirds of academy superintendents in the years since World War II had had previous assignments at the service academy from which they graduated, all of them—without exception—had had other high-level staff assignments before assuming the superintendency. "High-level" staffs are defined here to include those of the service headquarters in Washington, those of the headquarters of key functional components of each service (such as the Strategic Air Command, or the Corps of Engineers), and those of major joint or allied commands (such as NATO). Behind the apparent uniformity of emphasis on high-level staff experience, however, one finds some interesting differences among the services in the types of staff or aide assignments that are most prevalent in the backgrounds of academy superintendents.

Important previous staff experiences of Air Force Academy superintendents include varied areas of specialization, but have as a common denominator at least one important tour in the Washington, D.C. area. Each of the superintendents of the Coast Guard Academy in the years since World War II, like the Air Force Academy superintendents, had had previous high-level staff experience in service headquarters in Washington. Three of the four superintendents who were aviators had been called upon relatively early in their careers to serve as aide to the Commandant of the Coast Guard: Willard Smith from 1936-39, and again from 1943-45; Chester Bender from 1947-50; and William Jenkins from 1950-54. Frank Leamy, the fourth superintendent-aviator, had been Chief of the Aviation Division in Coast Guard Headquarters from 1940-43, and then served as Chief of the Floating Units Division from 1946-49.

The staff backgrounds of Military Academy superintendents tend to differ in emphasis from those of their Air Force and Coast Guard counterparts. Experience in Washington is prevalent among superintendents in all the services. However, the Army in particular has tended to attach increasing importance to staff assignments which provide exposure to military policy issues within a broad social, political, and economic context, and which involve interaction with civilian officials. The earlier staff experience of Andrew Goodpaster, who assumed command at West Point in June 1977, is especially impressive. However, it should be noted that Goodpaster differs from all of his predecessors in having had the advantage of a full career before being called from retirement to become the Military Academy Superintendent. Goodpaster had been President Eisenhower's staff secretary from 1954 to 1961. From 1962 to 1966, he had served as assistant to the chairman, Joint Chiefs of Staff, and the following year had served as director of the Joint Staffs.

The Navy, like the Army, has accorded increasing emphasis to the acquisition of broad policy perspectives and a capacity to deal effectively with civilian officials. The career patterns of Naval Academy superintendents also reveal the desirability of having served as aide to a high-level official in the defense establishment. Especially notable were the appointments as aide to the Secretary of the Navy of William Smedberg and Draper Kauffman, each in mid-career. Smedberg was aide to James Forrestal during the crucial period of the struggle over the unification issue in late 1946 and early 1947. Kauffman was aide to Thomas Gates in the mid-1950s when the latter served first as Under Secretary of the Navy and then as Secretary of the Navy.

The Commandants

In March 1978, Brigadier General Thomas C. Richards, a graduate of the Virginia Military Institute, assumed the position of Commandant of Cadets at the Air Force Academy. The appointment marks the first departure from the practice of placing West Pointers in this key role. The other academies have consistently selected the commandants of cadets or midshipmen from among their own graduates, and with graduates of the earliest Air Force Academy classes (1959, 1960) soon becoming eligible for promotion to flag rank, the newest of the academies may adopt that appointment tradition also.

The career patterns of commandants of cadets or midshipmen are similar to those of superintendents in respects other than reliance on "old-school" credentials. Of course, appointees to the more junior position are likely to be officers with less extensive experience than superintendent appointees have acquired.[4] However, academy commandants characteristically are officers who already have established reputations as outstanding leaders. Their academy appointment typically is a stepping-stone to promotion and more prestigious assignments. Well over half have gone on to three- or four-star rank. The pattern of success is most evident at West Point, six of whose postwar commandants have become four-star generals (Paul Harkins, John Waters, John Michaelis, John Throckmorton, Richard Stilwell, and Bernard Rogers). Two commandants of cadets at New London ultimately became Commandant of the Coast Guard, the only four-star billet in that service (Willard Smith, who also served as Coast Guard Academy Superintendent, and Edwin Roland). Louis Seith, Commandant of Cadets at the Air Force Academy from 1963 to 1965, was a four-star general as Chief of Staff for SHAPE less than ten years later. William Bringle attained four-star rank thirteen years after having served as Commandant of Midshipmen at Annapolis.

With the exception of a few officers appointed during or immediately after World War II, those selected for the position of academy commandant have had significant previous command experience. As with the selection of superintendents, a familiar pattern is the appointment of an officer who recently has gained fame as a combat commander. During the years of American involvement in Vietnam, each of the four services assigned veterans of that war to its academy. The warrior image is promoted not only through emphasis on combat credentials, but also through a bias in favor of the more "heroic" career specializations. For example, officers of the infantry and cavalry (or armor, its modern equivalent), the branches of the Army most closely associated with the warrior tradition, tend more so than officers of other branches to be assigned as West Point commandants of cadets.[5] The Navy tends to assign naval aviators as commandants of midshipmen. At the Air Force Academy, during the years of American involvement in Vietnam, four fighter pilots in succession were selected as commandants of cadets. Otherwise, however, Strategic Air Command bomber pilots have been selected.

Most academy commandants are war college graduates (although the pattern is evident at New London only in recent appointments). Since World War II, roughly half of those appointed had had a previous tour of duty at the academy, as instructors or as members of the Commandant's or Superintendent's staff. Other notable previous staff experience invariably has included assignments in service headquarters in Washington or with one of the major commands (such as SAC or one of the operational fleets). In addition, nearly half of Military and Naval Academy commandants had served as high-level aides (for example, to the President of the United States, to a service secretary, or to the service chief).

The Issue of Patronage

That an assignment as aide to a prominent official or a high-level staff position should serve to enhance one's career prospects raises the question of patronage. To what extent has the career advancement of those who become academy superintendents or commandants—and indeed of flag-rank officers generally—been contingent upon sponsorship by a higher-ranking patron? The military personnel system frequently prescribes short tours of duty, but nonetheless demands annual evaluation of performance by one's superior officer (whom one may have served only for a few months). With the ratio of those selected for promotion to those who are eligible sharply declining as one moves to the higher grades, it is inevitable that choices are made by promotion and selection boards among individuals whose records appear virtually indistinguishable. With the inflation of performance ratings that has occurred, with a "superior" rating the norm rather than the exception, the problem of identifying and rewarding truly outstanding performance has become acute.[6] In such circumstances, it is not surprising that among several dozen "superior" officers, the one whose file includes glowing comments, say, from a service secretary or from a service chief, might receive special attention from a promotion board.

To read into such behavior organizational infirmities or perversity peculiar to the military would be to miss the ubiquity of ascriptive criteria in job assignment and promotion throughout organizational as well as political life. It is important to emphasize that the tendency in modern organizations and governments is not one of applying ascriptive criteria in lieu of achievement criteria, but rather that of supplementing the latter with the former—especially when the former fail to provide clear evidence that one contender for a position or a promotion has qualifications that are markedly superior to those of his competitors. Thus, in the case of academy superintendents and commandants, one finds that many—perhaps most—benefited at some point in their careers not merely from aide assignments but also from other close associations with high-ranking persons whose praise carried weight in the minds of members of promotion and selection boards. Yet, given the competitive nature of a military career, and the rapid reassignment from one job to the next, none of the officers who later would become academy superintendent or commandant could rest on his laurels. Indeed, the opposite is more nearly accurate. The career patterns of academy superintendents and of commandants are those of ambitious "hard chargers" (to use the idiom of the military), many of them propelled by a what-makes-Sammy-run drive to succeed.

Such an orientation might suggest that the typical superintendent or commandant would approach his responsibility with a zeal for innovation. Such a supposition, however, fails to take into account several important disincentives to innovation that often have been associated with the top academy positions.

Academic Deans

In the years since World War II, approximately one-fourth of the men who have been appointed as academy superintendents have had neither postgraduate education nor previous responsibilities in military educational institutions. Consequently, the extent to which effective academic reform could be achieved at the academies has been contingent upon the ability of the Academic Dean and permanent members of the faculty to fill in the void of expertise, and upon the ability of the Dean to work harmoniously with the Superintendent.

Before World War II, the most senior faculty member at each academy often was known as "the Dean." In 1945, when the role was formalized for the first time at West Point, the individual selected as acting dean of the Academic Board (Dean on a permanent basis a year later) was the senior member of the board. Roger Alexander, promoted to brigadier general upon becoming Dean, had graduated from the Academy in 1907. His early service with the Corps of Engineers had included a four-year tour as an instructor of surveying and topography at West Point. He had served with General Pershing's staff in France during World War II, but then had returned to West Point as professor of drawing during the MacArthur superintendency in 1920, remaining on the faculty ever since.

Alexander retired in 1947, succeeded as Dean of the Academic Board by Harris Jones. In the appointment of Jones as Dean, the

Military Academy departed slightly from the principle of selecting the senior member of the board. Herman Beukema had graduated from West Point in 1915, whereas Jones had graduated in April 1917; moreover, Beukema had joined the faculty in 1930, a year before Jones. However, Beukema had been a professor in the Department of English and History (later English, Government, and History; then Social Sciences). Jones, in contrast, had been with the Department of Mathematics, regarded traditionally as the core discipline at the Academy.

To the mid-1960s, the pattern of appointment to the top academic position at West Point was maintained. Selection was made from among the most senior members of the Academic Board, with the tacit proviso that eligibility was limited to those associated with mathematics, engineering, or science, in contrast to the social sciences and humanities. The role of the Dean, who typically only had a few years of service remaining before retirement, was simply that of *primus inter pares* relative to the other permanent professors on the Academic Board.

The appointment in 1965 of John Jannarone as Dean represented no departure from the science-engineering bias in selection criteria. However, Jannarone was somewhat younger (fifty-one) with fewer years of service on the faculty (eight) than had been true of his predecessors in the position. During his term, the authority of the Dean was increased, with academic department heads required to report through the Dean rather than directly to the Superintendent as they had previously.

Even with the modified structure, the Academic Board has remained powerful at West Point, limiting the independent influence that the Dean is able to wield. The structure that contrasts most sharply with this has emerged at the Air Force Academy. A position of Academic Dean had been in existence from the founding of the Air Force Academy. However, the first incumbent in the position, Brigadier General Don Zimmerman, found himself besieged by enormous difficulties (described in Chapter 4) and was reassigned to a staff position in the Pentagon midway through the first academic year. It was under his successor, Robert McDermott, that the potential of the Dean for significant influence became manifest. McDermott's enormous impact on the shaping of policy at the Air Force Academy was attributable not only to the force of his personality, but also to his ability to limit the composition and authority of the Academy Board (the analog of the West Point Academic Board).

Shortly after McDermott became Academic Dean at the Air Force Academy (on an acting basis initially) the position of Dean at the Coast

Guard Academy was formalized. Albert Lawrence, who had been teaching humanities at the Coast Guard Academy since 1938, and before that had taught at MIT, was appointed to the position in 1957 by reform-minded Superintendent Frank Leamy. The appointment of a professor of humanities as Dean, like that of the selection of a social scientist (McDermott) for the position at the Air Force Academy, stood in contrast to the selection pattern at the Military Academy. Neither at New London nor at Colorado Springs did the lack of engineering credentials of the Dean hamper his ability to wield influence. (However, the most recent incumbents have been an engineer at New London and a physicist at Colorado Springs.)

At New London, the Dean's influence has derived primarily from his ability to work harmoniously with the Superintendent. Formal structures and rules of governance that allocate policy-making authority have been less important at New London than at the larger academies. Lawrence and his successors, Stanley Smith, Paul Foye, and Roderick White, in varying degrees have developed close working relationships with a succession of superintendents and thereby have had an important impact on curriculum development and program reform. Similarly, various permanent professors have been influential, with their influence derived largely from their informal access to top decision makers rather than from formally prescribed authority.

The Academic Dean of the Naval Academy has been more totally reliant on formal authority as the basis of his influence on policy than have his counterparts at the other three academies. Whereas all of those who have been appointed to the position of Dean of the other academies were selected from among permanent members of the academy faculty, each of the two men who have held regular appointments as Academic Dean of the Naval Academy was recruited for the position from a civilian academic institution, and had had no previous service on the Naval Academy faculty. Indeed, the creation of the position of Academic Dean and the decision to fill the position with a civilian were imposed upon the Naval Academy by the Secretary of the Navy, Fred Korth, in 1963. Under those circumstances, it is not surprising that the first incumbent to the deanship, A. Bernard Drought, enjoyed the respect but not the full confidence of the Academy community.

Nevertheless, Drought possessed a doctor's degree, which made him unique among service academy deans to that time. Such credentials were still quite exceptional among Naval Academy faculty members, also. Moreover, Drought had gained recognition from previous service as an academic consultant to the Academy. Thus, the series of superintendents under whom Drought served tended to accord weight to his views regarding curriculum development. Yet many civilian faculty members had hoped for more from the Dean. It has taken several years for the current Dean, Bruce Davidson, to become sufficiently accepted by members of the Annapolis faculty to give him a base of support from which he can wield effective influence.

Career Experience and Academy Leadership: A Summary

The evidence discussed above reveals that academy superintendents, commandants, and deans in the years since World War II have been, above all, individuals whose careers are testimony to dedication and professional talent. In their academy roles, typically they have displayed, and demanded from their subordinates, a "can-do" spirit in response to assigned tasks.

As in most military organizations, however, the milieu at the service academies has been one in which there has been a recurrent demand for action, but only an infrequent change of structures or procedures other than changes of a cosmetic nature. The incentives which the military profession provides for dedication to duty should not be confused with incentives for experimentation. On the contrary, the successful executive in the military as in much of the federal bureaucracy often is one who has learned to contain whatever talents he might have for creating sharp departures from precedent.[7]

Yet one must note also Morris Janowitz's more optimistic finding of innovating perspectives among the "elite nucleus" within the military establishment. Briefly, from a survey of the career patterns and behavior of hundreds of officers who had attained flag rank within the American military establishment, Janowitz found that most had reached that level through competent performance which adhered closely to convention. Many of those who had become the real "prime movers" at the top of the career ladder, however, had been innovators or mavericks at important stages of their careers.[8] The question of how unconventional members of the "elite nucleus" are likely to be once they have attained flag rank is not fully answered by Janowitz, and Maureen Mylander's more recent data provide a basis for skepticism regarding expectations of nonconformity.[9] In any event, among those whose careers lead them back to a service academy as a top executive, other disincentives to innovation must be confronted.

Probably the most important of these is the brevity of the time span that characteristically had been provided for the tour of duty. Of the forty-three academy superintendents since 1945 who had completed their terms by mid-1978, twenty-one had served for less than three years and only seven had served as long as four years. (Strictly speaking, the number serving less than a full thirty-six months was higher than twenty-one. However, if the tour included three full academic years, it is treated as a full three years.) The tours of academy commandants typically have been even more abbreviated, except at the Coast Guard Academy, where the tendency has been for the tour of a Commandant of Cadets to be dovetailed with that of the Superintendent he serves. Thirty-two of the thirty-nine post–World War II commandants at the Defense Department academies who had completed their tours by mid-1978 had served for less than three years—many of them for less than two years.

Lieutenant General Garrison Davidson was one of the rare Military Academy superintendents in the postwar years to enjoy a full four years of the job (although his request for an extension of the tour in order to see through the implementation of reforms which he had set in motion was denied). As Davidson noted subsequently, in an effort to alert the Army to the organizational costs of short command assignments:

> The short length of time some senior officers stay in command . . . provides a pro forma type of assignment that neither develops them further, utilizes them fully, nor affords an adequate test of their abilities. Further . . . the frequent changes in command have a disturbing effect on the operations and morale. . . . Policies are continually changing; nothing is stable, and progress is limited.[10]

The paradox of frequent turnover of command, Davidson might have added, is that although it serves as a stimulus to surface change and instability, it tends to impede enduring change and sustained reform. For the incumbent to a short-tour command, a premium is placed on the "quick fix," on obtaining quick though superficial results, rather than on systematic analysis of organizational problems and long-range planning. The rediscovery of the wheel takes the place of actual innovation. Numerous examples of such behavior are found in the actions of academy commandants of cadets and midshipmen, who periodically over the years since World War II have proudly announced that the approach to the indoctrination of plebes was being reformed through the introduction of an emphasis on "positive leadership." In almost every instance, the successor to a "positive approach" commandant has announced with equal pride that he was introducing measures designed to "restore discipline to appropriate standards," thereby leaving the stamp of his command on the student body in the brief time allotted to him.

Periods of important reform at the academies invariably have occurred when top executives have had a tenure in office of at least two and a half years and have received cues from the task environment that supported rather than discouraged change.

Phases of Change

Fluctuations in the reform pattern can be suggested through a synopsis of phases in the evolution of the academies since World War II. In Figure 2, we have plotted the term of each academy superintendent who served for at least two and a half years, with the name of the superintendent listed above the arrow that spans his tenure in office. If a commandant of cadets or midshipmen served for at least two years concurrently with a superintendent, the name of the commandant is listed below the arrow, to the left of the diagonal. If an academic dean served for at least two years concurrently with a superintendent, the name of the dean is listed below the arrow to the right of the diagonal.

Periods of major reform have been ones in which a superintendent not only had at least two and a half years in his command, but also enjoyed the support of a commandant or dean, or both, with a similarly long tenure. Furthermore, invariably the major periods of change have been ones in which the task environment provided support and encouragement of, if not the demand for, change. At a broad level of generality, one can identify shifts that have occurred over time in the dominant emphasis of cues from the task environment. Of special interest are the fluctuations that have occurred between support or demand for the improvement or modification of academic programs or structures (the "Athenian climate") and support or demand for increased attention to the cultivation of discipline and the traditional military virtues (the "Spartan climate"). The time-lines that distinguish one change in climate from the next are necessarily imprecise, and are not necessarily identical from one service to the next. (Shifts in concern that affect the Defense Department do not necessarily affect the Coast Guard, for example.) However, the lines serve to demarcate shifts that are explicable not merely as random shifts of a pendulum but rather as understandable fluctuations in the response of national leaders to changing policy needs. The pattern of change at the service academies, in turn, becomes comprehensible within a context of broad shifts in the pattern of cues provided by the task environment.

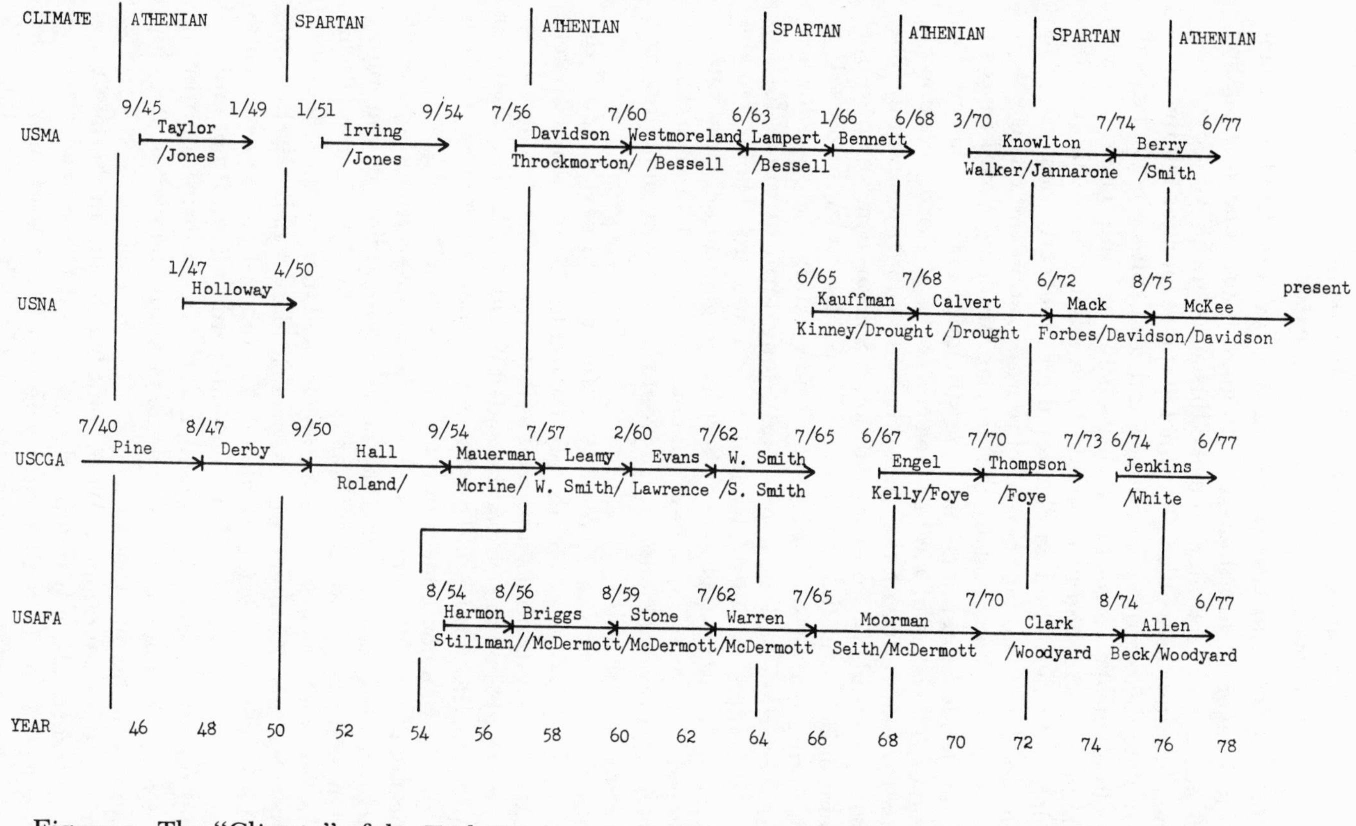

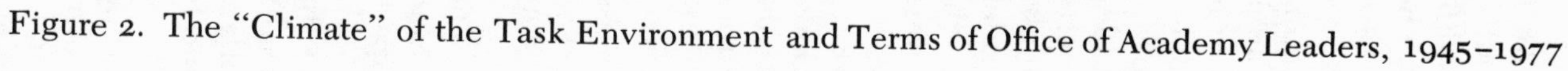

Figure 2. The "Climate" of the Task Environment and Terms of Office of Academy Leaders, 1945–1977

At the end of World War II, for example, each of the three extant academies reverted to a four-year program. Curricular and staffing modifications were required to make this adjustment. Moreover, service headquarters as well as personnel at the academies felt the need to assimilate the wartime experience and resultant strategic, tactical, and technological changes into education and training. The result was a climate that was highly favorable for academic innovation at the academies. As described in chapter three, the early postwar years were indeed an important period of reorganization and change.

The issuance of a report by the Service Academy Board in 1950, however, tended to endorse existing structures and programs at the academies. The onset of war in Korea and of the McCarthy era domestically further reduced the incentive for experimentation. The emphasis, instead, was on the traditional Spartan component of the academy mission, in the production of capable combat leaders. It is worth noting that in such a climate of low external support for academic experimentation, a relatively long term of office for top academy leaders in and of itself is no stimulus to change. On the contrary, if the external climate is hostile to organizational innovation, a long term for a Superintendent well may be better assurance that the status quo will be maintained than would be true with more frequent turnover of command.

With the end of the Korean War, the pressures for using the academies to develop immediately utilizable combat skills abated. However, the incentives for academic reform were not nearly as marked as had been the case at the end of World War II. The Cold War remained intense in the mid-1950s, with a continuing preoccupation with the Spartan component of the academy mission.

The founders of the new Air Force Academy were encouraged to indulge in some experimentation, however, regardless of continuing Spartan emphases and Cold War concerns in Washington. By no means was such encouragement unrestrained, nor were the founders of the Air Force Academy predisposed to depart radically from the traditional seminary-academy model. Yet innovations which eventually were introduced under McDermott were difficult if not impossible for the other services to ignore, and became a major contributing factor to the change in climate of the task environment at the older academies in the late 1950s.

Organizational ferment at West Point had begun in the late 1950s under Davidson largely independent of the Air Force Academy experience. At the Coast Guard Academy, the Air Force Academy provided less the stimulus for change than the model of what was possible. The major stimulus came from a combination of abysmally low internal morale and intense concern from Coast Guard Headquarters that the situation be rectified. Reforms were initiated under Leamy and were continued under Evans and Smith.

Many of the factors that contributed to an external climate conducive to academic reform at West Point and New London were present in the task environment of Annapolis in the late 1950s as well. Some modest changes were introduced during the period beginning with the Melson superintendency and extending through the Kauffman superintendency. However, throughout this period, in which the external climate not only encouraged but sometimes insisted on change, no Naval Academy Superintendent served for more than two years. Moreover, by the mid-1960s, when Draper Kauffman assumed command at the Naval Academy for a three-year tour, external support for academic reform had waned. The other three academies had moved from major academic change (such as the introduction of electives) to a period of consolidating and evaluating the new programs that had been introduced. A major honor scandal at the Air Force Academy in 1965 had led some persons to charge publicly that academics had been overemphasized during the era of early experimentation. The major escalation of American involvement in Vietnam had begun. This was a time when a majority of the American public still were supportive of, if bewildered by, the war with which military officials were increasingly preoccupied. Thus, for a variety of reasons, the climate of the task environment for all four academies tended to shift toward support for traditional Spartan values in the mid-1960s.

By the late 1960s, support for the American involvement in Southeast Asia had begun to wane, however, and vocal opposition to the war, especially among college-age youth, increased. American military institutions in general and the academies in particular were under pressure to become less authoritarian and more in tune with the rest of society. Academy officials, finding it difficult to attract enough qualified applicants for existing vacancies, were under pressure to make their academies more attractive. These external pressures were being generated at a time when internal links to the civilian sector—especially to civilian higher education—had been increased. The decision in the late 1950s to the mid-1960s at each academy to upgrade the qualifications of faculty had resulted by the late 1960s in a far higher percentage of faculty members than ever before who had spent an extended period of study at civilian graduate schools. Their numbers were augmented at the Defense Department academies by an influx of reservists to replace officers with regular commissions who were being assigned to Southeast Asia.

It was in this period that Calvert assumed command at the Naval

Academy and began instituting major reforms. The external climate was especially conducive to change at Annapolis, which experienced not only the factors described above but also the brunt of Rickover's criticisms of service academy education. However, the accumulated aspirations of civilian faculty members at the Naval Academy for reform provided an especially compelling incentive for change.

The severe problem which the DOD academies especially had been having in attracting sufficient numbers of qualified applicants seemed to have been surmounted by the early 1970s. Moreover, the "winding down" of the war in Vietnam and the end of conscription helped to relieve the pressures on the services to "civilianize," and in turn alleviated the pressures on the academies for modelling their programs more closely to those of civilian institutions of higher education.[11] Changes continued to be made during the period of Spartan emphasis in the early 1970s; but many of the changes (for example, reductions in numbers of electives, the termination of programs for immediate civilian graduate schooling, a reemphasis on engineering and military-professional studies) had the effect of reversing earlier initiatives.

By 1976, however, a variety of pressures had been building for further reform at the service academies. The admission of women to the academies, a dramatic step in itself, in turn required academy officials to reexamine their various programs, requirements, and facilities to determine what further modifications might be needed. Less readily diagnosed because they had been developing over a period of years were problems stemming from the growth and attendant bureaucratization of the academies. Growth, in the form of increasing enrollments, had been the product primarily of decisions made in the parent arms of service and in Congress. Bureaucratization, in contrast, primarily reflected decisions that had been made by internal leadership of the academies. As we shall demonstrate in chapter ten, to a large extent the growth-bureaucratization process had been one with consequences which neither internal leaders nor external monitors anticipated or desired. However, by the middle to late 1970s, a variety of critical investigations of the academies had identified symptoms of the problem in the form of student alienation and malaise. These investigations included one by the Borman Commission into a major honor scandal at West Point, with extensive follow-up studies by study commissions appointed by the Army; massive studies of the problems of high student attrition at all the academies and of programs of academics and military training, by the General Accounting Office; and detailed reviews of academy education and training by the Clements Committee in the Pentagon.

The thrust of the various investigations was a call for further change—possibly of greater proportions than any that had occurred for many years. The extent to which the academies should or will experience radical changes in the foreseeable future is an issue which we address in the concluding chapter.

9

The Academies and Their Task Environment

In very general terms, it is useful to describe changing cues to the academies from their task environment as fluctuations between a Spartan and an Athenian "climate." However, the influence which particular elements in the task environment have been able to wield over academy affairs can be described with greater specificity. Key elements have been alluded to in earlier chapters and are identified in Figure 3. The relatively clear and stable channels of formal authority and communications are highlighted, whereas the often-important informal channels, such as those through which the mass media and academy alumni make their influence felt, can only be hinted at through graphical representation.

Drawing upon findings from Parts 1 and 2, supplemented by other pertinent evidence from recent experience at the academies, this chapter is designed to clarify (1) the domain of activity over which particular components of the task environment are most likely to exert influence, and (2) the circumstances under which such influence is most likely to be felt. The following environmental influences are discussed: linkages among the four academies; civilian higher education; the parent arms of service (in relation to which the roles of the media and of the courts also are discussed); academy alumni; the Congress; boards of visitors; and the General Accounting Office. In a concluding section, limits of the responsiveness of the academies to their task environment are described.

The Set of Service Academies

The infancy of an organization typically is a time of borrowing of ideas from older organizations with similar goals and purposes, although creation of a new organization also provides opportunities for experimentation. As we have seen, West Point in the formative Thayer years drew heavily upon the experience of the École Polytechnique; the Naval Academy in turn drew heavily upon the West Point model,

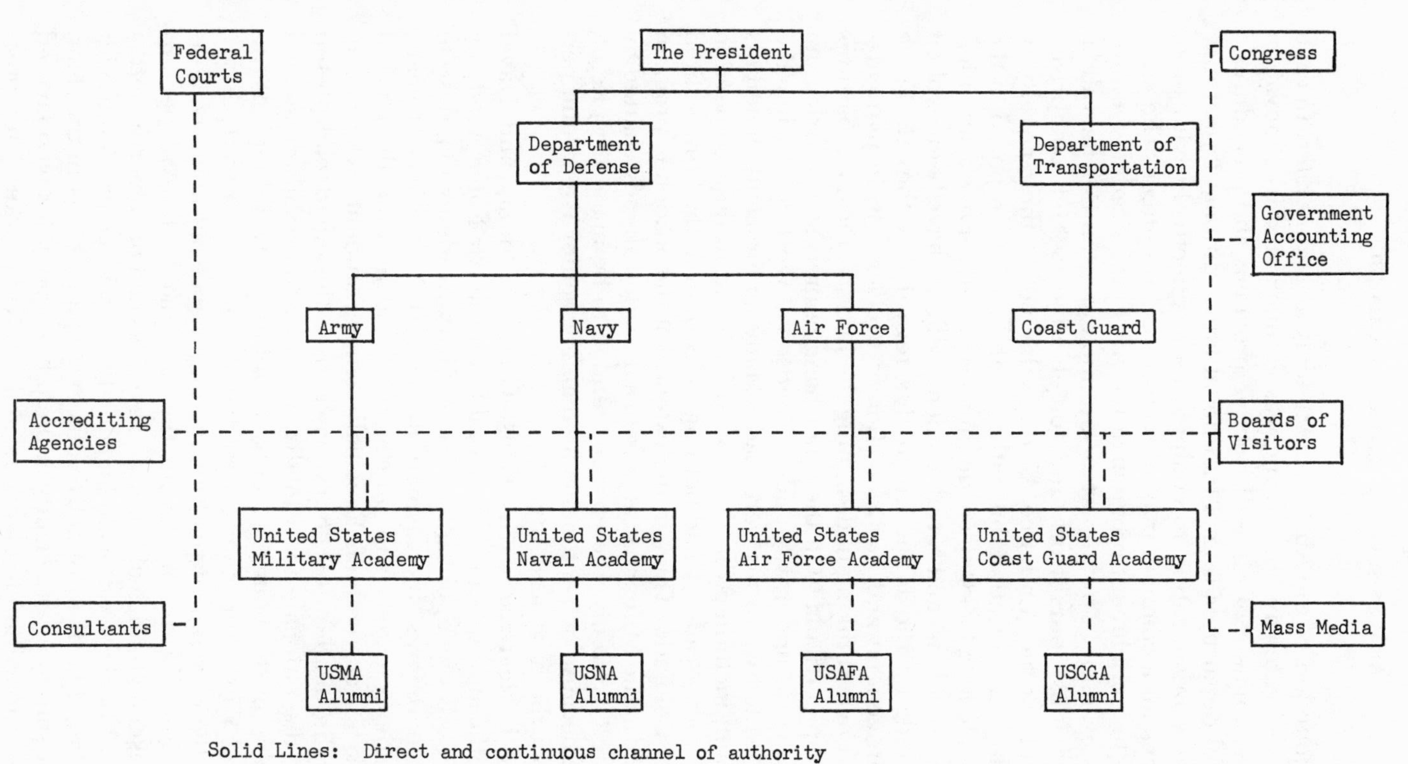

Figure 3. The U.S. Service Academies and Their Task Environment

as did the founders of the Air Force Academy a century later. In each case, however, extensive borrowing was accompanied by innovation and the development of some distinctive programs and procedures.

Successful departures from a traditional organizational format may enable a new organization to compete more effectively with older organizations for a common pool of recruits and resources. The early success of the enrichment program at the Air Force Academy affords an obvious example. The importance of the opportunities for validation of previous coursework and for electives which comprised "enrichment" at the Air Force Academy lay not in their novelty in American higher education, but rather in the challenge posed to the hoary rationale for a lockstep curriculum for armed service academies. Coast Guard Academy and Naval Academy officials have been explicit in acknowledging the influence in this regard that the Air Force Academy in general and its Dean, Robert McDermott, in particular had on their own thinking in designing the first significant departures from lockstep curricula.[1] On the other hand, a process of curriculum study already was under way under Davidson at West Point when the Air Force Academy reforms were made. Many senior faculty members at West Point, including the Dean, were adamant in their conviction that the Military Academy had nothing to learn from the "blue-suited upstarts" in Colorado. Thus, to the extent that the academic program was changed at the Military Academy in a direction already charted by the Air Force Academy, the changes tended to be made in spite of, rather than because of, sentiments held among senior West Point personnel about the new academy.

Still, the basic reason why Coast Guard Academy and Naval Academy officials came to look favorably on the Air Force Academy innovations was one applicable to the Military Academy too, despite prevailing sentiments. Increasingly in the post–World War II years, service academy cadets and midshipmen had become dissatisfied with a lockstep curriculum. Especially among students who had to repeat at the academies coursework that they had completed elsewhere, the rationale for a totally prescribed curriculum seemed absurd. Many of the younger faculty members at New London, Annapolis, and West Point thought so, too. To those who were dissatisfied with the lockstep curriculum, the importance of the innovations at the Air Force Academy lay in the fact that high-level personnel in at least one branch of the armed services had been persuaded of the untenability of the rationale for a totally prescribed curriculum for the education of their prospective officers. The foot was in the door. Cadets, midshipmen, and faculty at the other academies who favored opening the curriculum to electives could utilize the Air Force precedent to help them overcome the resistance that heretofore had kept the door slammed shut in their own services. A process of diffusion of curricular innovation, beginning generally with the Air Force Academy, soon became evident (Table 6).

Table 6 The Abandonment of the "Lockstep" Curriculum

Curriculum Change	Academic Year in Which the Change Was Put into Effect			
	USAFA	USCGA	USNA	USMA
Validation, substitution, or overload electives	1957-58	1958-59	1959-60	1948-49
Electives for all students	1957-58	1965-66	1964-65	1960-61
"Concentration" equivalent to academic "minor"	1957-58	1965-66	1963-64	1969-70
Academic "majors" for all students	1964-65	1970-71	1969-70	no

The annual superintendents conference (which includes other top officials and staff members also), launched at the initiative of the Coast Guard Academy Superintendent, Admiral Leamy, beginning in 1958, has institutionalized communications among the academies.[2] Although the Coast Guard Academy still is not linked as closely to the other three academies as they are to one another, all four are sufficiently aware of one another's structure, problems, and operations that any important change at one becomes quickly noted at the others. Moreover, parent organizations, the Congress, the General Accounting Office, and others in the task environment increasingly have been inclined to demand justifications of deviations that exist at a given academy from successful structures or programs that are found at the others. On the one hand (as we shall stress in the concluding chapter), this tendency puts a premium on standardization and thus on cautious incrementalism rather than audacity in the approach to organizational change. On the other hand, when important changes do occur at one, the multitude of linkages among the academies makes it probable that such changes will be diffused throughout the set of academies, unless the change is clearly service-specific as with particular types of training.

Formal Links to Civilian Higher Education

If diffusion of innovation is central to an explanation of the first significant departures that were made from the lockstep curriculum at New London, Annapolis, and West Point, the question remains: why did the Air Force Adademy adopt its "enrichment" program in the first place? The answer lies mainly in the fertile imaginations of McDermott and his colleagues, although it should be noted that in many respects they were simply carrying the early validation experiments at West Point (with which they had had direct experience) to their logical conclusion. However, one should note also the powerful stimulus to academic reform that was provided by the desire to attain accreditation in time for the graduation of the first class (1959).

A concern for meeting accreditation standards or for being able to respond to specific criticisms by accrediting bodies has served in varying degrees as a stimulus for change at all the academies. The most obvious, and in some respects the most important, example of such change is provided by the upgrading of faculty that has occurred at each academy. An influx of relatively well-qualified civilian faculty members occurred at each of the three academies during World War II, to release regular officers for combat duty. After the war, some of those who had come "on board" during the war stayed on at the academies. In addition, the Naval and Coast Guard Academies began insisting upon at least a master's degree for their civilian appointees to the faculty. West Point, in turn, began sending nearly all officers who were to be assigned to the faculty to civilian graduate schools for at least a year of study, often until completion of the master's degree. Little further upgrading of the faculty at any of the academies occurred, however, until the periods of ferment and academic reform in the late 1950s and early 1960s, with increases in the percentages of faculty members with advanced degrees continuing into the 1970s.

The Air Force Academy had insisted upon academic qualifications for its faculty at least to the master's degree from its first academic year (1955-56). Roughly a third of the initial cadre of faculty possessed the Ph.D. Although no absolute reduction in the number Ph.D.'s occurred subsequently, with a rapid expansion of the faculty, consisting mainly of officer-instructors with the master's degree, there was a decline to just over 10 percent of faculty members with the Ph.D. in 1958-59. By the 1970s, virtually all permanent faculty members and a fair number of the officers who were assigned on a temporary basis possessed the Ph.D., bringing the figure back up to approximately a third of the faculty. At Annapolis, New London, and West Point, policies also had been instituted to make the Ph.D. a prerequisite for the permanent, professorial appointments. Thus, the number of Ph.D.'s was on the increase, although the percentage at these academies still lagged somewhat behind the figure at the Air Force Academy.

A comparison of the academic qualifications of faculty members at the service academies with those at selected civilian institutions of higher education is instructive (Table 7). Institutions included in the comparison are the Virginia Military Institute (VMI), Rensselaer Polytechnic Institute (RPI), the Georgia Institute of Technology (Georgia Tech, formerly the Georgia School of Technology), Purdue University, the Massachusetts Institute of Technology (MIT), Notre Dame, Dartmouth College, and the California Institute of Technology. VMI, founded in 1839, has a long and proud tradition of training and educating prospective military officers. RPI, Georgia Tech, Purdue, MIT, and Cal Tech are engineering schools, each of which has had ties with one or more of the service academies. RPI, in particular, has had long historical ties to West Point and more recent ones to Annapolis. Each of the others has provided graduate training to many prospective service academy faculty members. Moreover, until relatively recently, each of these engineering schools except Purdue was a relatively small, basically if not exclusively all-male institution, and in these respects resembled the service academies (and VMI). Dartmouth and Notre Dame also were roughly comparable in size to West Point and Annapolis; moreover, each had long had an all-male student body, with substantial support into the 1960s for an ROTC program on campus.

Although the academic qualifications of faculty members at VMI and at each of the civilian institutions listed in Table 7 were superior to those of faculty members at the service academies on the eve of World War II, a "trade school" outlook remained prominent at VMI and at civilian engineering colleges much as it did at the academies. Consequently, practical experience in the application of engineering skills which one had acquired as an undergraduate still was widely believed to be sufficient preparation for taking on the responsibilities of a faculty member. This situation persisted into the 1950s. A widespread trend in the 1950s toward upgrading the academic qualifications of faculty is evident in a comparison of the figures for 1950-51 with those for 1958-59. By the latter academic year, the vast majority of faculty members at VMI and at the civilian engineering schools were qualified at least to the master's level, and increasing numbers possessed the Ph.D. At Notre Dame, Dartmouth, and Cal

Table 7 Percentage of Faculty Members with Academic Degree(s) beyond the Baccalaureate: Service Academies and Selected Civilian Institutions of Higher Education Compared, 1938–1970 (Percentage of Faculty with Ph.D. indicated in parentheses)[a]

	1938-39	1946-47	1950-51	1954-55	1958-59	1962-63	1966-67	1970-71
USNA	15 (6)	38 (10)	38 (11)	37 (10)	41 (12)	49 (12)	77 (22)	74 (24)
USCGA	b	37 (5)	33 (4)	37 (8)	40 (10)	45 (8)	56 (6)	71 (17)
USMA	b	b	b	b	b	79 (5)	93 (8)	97 (11)
USAFA	c	c	c	c	91 (12)	97 (20)	98 (27)	99 (32)
VMI	63 (20)	61 (20)	59 (18)	68 (19)	90 (33)	81 (33)	84 (41)	91 (56)
RPI	71 (33)	43 (17)	44 (16)	68 (31)	89 (53)	94 (54)	90 (67)	92 (71)
GaTch	76 (18)	73 (17)	71 (20)	56 (22)	72 (33)	90 (49)	92 (58)	d
Purdue	71 (30)	75 (29)	74 (35)	88 (41)	85 (39)	d	90 (65)	93 (62)
MIT	66 (28)	b	b	68 (46)	68 (51)	b	97 (82)	94 (83)
Notre Dame	88 (33)	87 (37)	88 (39)	81 (45)	88 (57)	95 (60)	90 (63)	96 (73)
Dart	87 (55)	85 (55)	83 (51)	91 (54)	84 (63)	92 (71)	90 (68)	95 (78)
CalTch	85 (73)	91 (74)	95 (84)	97 (92)	96 (90)	99 (93)	97 (91)	99 (98)

[a]Percentages have been computed on the basis of data from the following editions of *American Universities and Colleges* (Washington, D.C.: American Council on Education): C. S. Marsh, ed., 4th ed. (1940); A. J. Brumbaugh, ed., 5th ed. (1948); 6th ed. (1952); Mary Irwin, ed., 7th ed. (1956); Mary Irwin, ed., 8th ed. (1960); A. M. Cartter, ed., 9th ed. (1964); O. A. Singletary, ed., 10th ed. (1968); W. T. Furniss, ed., 11th ed. (1973). Percentages are based only on those classified as holding the baccalaureate, master's, or Ph.D. degrees. Occasionally, but erratically, institutions report persons as holding "professional" or "other" degrees. These have been excluded from the present analysis.

[b]Data on the academic qualifications of faculty were not supplied by the institution to the *AU&C*.

[c]The USAFA was established in 1955.

[d]Data in the *AU&C* entry for this year are so inconsistent with those reported in preceding or succeeding editions of the *AU&C* that they were deemed by the present author to be spurious.

Tech, where possession of at least a master's degree had been characteristic even in the pre–World War II years, the upgrading that occurred in the 1950s is evident in increasing percentages of faculty members with the Ph.D.

From the turn of the century up until World War II, and to a lesser extent even into the early 1950s, the service academies had managed to remain insulated from many of the important trends that were occurring in civilian higher education. However, as degree-granting institutions in the postwar years, they were subject to inspection at least once every ten years by civilian accreditation teams. These same teams were also making visits to civilian institutions of higher education. Likewise, the academies began to turn to academic advisory teams and consultants (in several instances including representatives from institutions listed in Table 7). Inexorably, as those civilian institutions which academy officials themselves had come to view as being at least partly comparable to the academies in terms of mission and structure upgraded their faculties, academy officials felt compelled (or inspired) to follow suit. Cues to do so were provided by accreditation or advisory teams.

As increasing numbers of academy faculty members in the past two decades have come to their assignments after having spent two or more years in civilian graduate schools, the influence of civilian higher education on the service academies has thereby become cumulative. Not only have service academy faculty members in growing numbers brought with them perspectives acquired in the atmosphere of civilian campuses, but also many of them have maintained contacts with civilian academicians, through correspondence and professional meetings, to a degree unprecedented for the academies. The growing reliance of the academies on faculty members who have received at least a crucial part of their higher education at civilian institutions, coupled with the growing dependence on civilian accreditation and advisory teams, thus has made the academies far more sensitive to currents of change in civilian education than they were in an earlier era.

The Parent Arms of Service

Although civilian academic accrediting agencies or advisory committees have played an increasing role in the affairs of the academies in recent years, their influence has been somewhat spasmodic. The external bodies that tend to exert influence on the academies on the most continuous basis are, of course, the parent arms of service:

the Departments of Army, Navy, and Air Force over the affairs of the Military, Naval, and Air Force Academy, respectively, and the Coast Guard over the affairs of the Coast Guard Academy.

The parent organization typically is consulted at an early stage regarding impending major policy decisions, such as a reorganization of the departmental structure or a change in the type of training that is to be provided during the summer months. When such decisions have been made at the academy, they are forwarded to the parent headquarters for final approval. It is to the parent service that the academy must appeal for annual funding and for authorization of new billets or facilities (although such appeal also might involve joining representatives from the service headquarters in an appearance before the appropriate congressional committee). Moreover, the Superintendent and other key academy personnel are selected by top officials at the parent service.

In each of the services, the style of command that prevailed until relatively recently was one in which commanders of major installations, such as the academies, were given minimal initial guidance from headquarters, and could expect to carry out their responsibilities free of dictates from above, unless a blatant command failure or organizational crisis necessitated intervention by the parent service. Especially in the period before World War II, the academies had much more autonomy than they have enjoyed in recent years.

The concern at the end of World War II over possible unification of the armed forces had the effect of generating pressure on West Point and Annapolis to exchange personnel and ideas with one another, and to familiarize cadets and midshipmen with the organization and mission of each of the armed services. The unification debate had the additional and potentially far-reaching effect of calling into question the rationale for independent academies, each of which indoctrinated its students in the perspectives of a particular arm of service. However, the recommendations in 1950 of the Service Academy Board served to legitimize the continued existence of West Point and Annapolis and to pave the way for the founding of a separate Air Force Academy. For the next several years, the concern with the academies at the Pentagon level was considerably diminished.

The Trend Since the 1960s

Especially beginning with the McNamara era of the early 1960s, a trend toward increasing centralization of control has been evident not only in the three DOD arms of service but to a lesser extent in the Coast Guard as well. Various factors have served as stimulants of centralization of control. In addition to McNamara's personal predilection for insisting that all defense-related programs be justified according to standardized criteria of cost effectiveness, three other contributing factors particularly merit discussion.

First, the growth of the electronic media, coupled with the use by most newspapers in the country of national wire services, has transformed problems which in an earlier era would have remained localized into objects of nationwide attention. For example, the practice among West Point cadets of "silencing" fellow cadets who had committed honor offenses but who had not resigned from the Academy had a long, though dour, history. The offenders were effectively excommunicated from the community of the faithful. When the *New York Times* and the leading television networks learned in 1973 that Cadet James Pelosi was being subjected to such treatment, however, their critical revelations of the practice subjected the Military Academy to a barrage of criticism. In September of that year, the Cadet Honor Committee formally discontinued the practice of silencing, ostensibly because the corps of cadets in a referendum had decided to abandon this particular tradition. It was apparent, however, that the cadets were aware of the sentiments that had been expressed in the mass media, and of the embarrassment that the corps as well as Academy officials had experienced from the intense adverse publicity.[3]

Second, officials at service headquarters have been inclined to assert control directly over academy policies in proportion to the increasing tendency of cadets and midshipmen to define their relationship to the academies in legal terms, in which contractual obligations are reciprocal rather than unilateral (student to academy). A related trend was evident throughout American society in the 1960s. At the academies, grievances which earlier generations of cadets and midshipmen had borne with grudging acceptance (or had resigned in order to escape) became the object of court cases in the late 1960s and early 1970s. For example, evidence going back at least to the early 1950s suggests that the majority of cadets and midshipmen objected to being compelled to attend Sunday chapel service, despite the fact (or in some cases because of it) that most of them viewed religion as an important part of their lives. However, in 1969, nine Annapolis midshipmen and two West Point cadets enlisted the support of the American Civil Liberties Union in an ultimately successful challenge to compulsory chapel attendance.

Likewise, cadets and midshipmen were seeking legal help to protest other facets of the academies' programs. One of the distinctions that Lieutenant General William Knowlton had attained by the end of his tour as West Point Superintendent in mid-1974 was that he had become "the most sued Superintendent in the history of the Military Academy."[4] Knowlton's record probably also surpassed that held by superintendents at any of the other academies, although his contemporaries at Annapolis and Colorado Springs were not far behind. In every case, the service headquarters found its concern about academy affairs heightened.

Third, antimilitary sentiment in American society, which became widespread beginning in 1968 as the war in Southeast Asia dragged on with mounting casualties and no end in sight, led American military officials to become increasingly defensive about virtually all facets of military operations, training, and discipline. The defensiveness manifested itself in part in a tendency of the top military command to intrude into the decision-making process at lower echelons. The insistence by the Army Chief of Staff, General Westmoreland, on prescribing sideburn length for troops through the Army is but one example of such a tendency. One result of Westmoreland's concern was the issuance of a memo from headquarters which described the hair length that had been displayed in a *previously* issued poster as the maximum permitted the day *before* a haircut, not the day after.[5] The so-called Z-grams that flowed incessantly from the Chief of Naval Operations to the lowest echelons while Admiral Zumwalt held the top post (1970-1974), providing paternalistic guidance on matters ranging from "Lockers and Wash Facilities for Personnel who have to Work in Dungarees Ashore" (Z-20) to "Good Order and Discipline" (Z-117), are another example, although one that received more sympathetic treatment in the press, where Zumwalt characteristically was described as a young maverick pushing reform despite the resistance of "old guard" admirals.[6]

There have been at least two recent illustrations of the tendency of officials in service headquarters and in the Department of Defense to prescribe policies that in an earlier era might well have been left to academy officials. One is the creation in 1973 of a DOD committee on excellence in education. The second is the formulation by the Secretary of the Army, Martin Hoffmann, of policies for reinstating West Point cadets who had been found guilty in 1976 of academic cheating by the Cadet Honor Committee. The framework for such policy formulation was provided by a commission, appointed by Hoffman and headed by former astronaut (and West Point graduate) Frank Borman.

The Committee on Excellence in Education (Clements Committee)

As revealed by Deputy Secretary of Defense William P. Clements, Jr., the idea for creating a committee at DOD level that would investigate the programs of the various service schools originated in conversations that Clements had early in 1973 with then Secretary of Defense Elliott Richardson.[7] Sharp criticisms of the academies had appeared in the mass media, and were followed by a congressional request to the General Accounting Office in the spring of 1973 for an investigation of the academies (discussed below). These acts prompted Clements to convene a meeting of the secretaries of the Army, Navy, and Air Force, and the Assistant Secretary of Defense for Manpower and Reserve Affairs, with Clements chairing the meeting. The focus of discussion was precommissioning and officer education at all levels. However, priority was attached to the service academies, not only because of the urgency of some of the problems they were experiencing but also because of the high visibility of the academies throughout the armed services.

Organized formally in December 1973 as the Committee on Excellence in Education, this high-level group made periodic visits to the academies and to other service schools each year thereafter. The strategy was to make their presence and their critical scrutiny felt quickly, and to follow their visits with a barrage of questions which would provoke internal study.[8]

Early in 1977, the initial three-year charter of the committee was extended indefinitely. A new mechanism for monitoring the programs of the DOD academies on a regular basis thereby became institutionalized. However, the committee's early zeal for standardization had become greatly dissipated. The Naval Academy, in particular, had fought with considerable success for retention of their distinctive curriculum and for the distinctive civilian-military mix for faculty. The Clements Committee's argument that military officers were to be preferred to civilians as role models in constituting the faculties of the academies was further undercut both by the General Accounting Office and by the Congress. The former revealed that the reliance by the Military and Air Force academies on all-military faculties was considerably more costly than the Naval Academy's utilization of a civilian-military mixed faculty.[9] The latter included an amendment to the Defense Appropriations Authorization Act, introduced in May 1976 by Senator John Glenn of Ohio, calling on the Secretary of Defense to

> conduct a study as to how greater utilization of civilian faculty may be accomplished in the service academies, intermediate and senior war colleges. This study *shall recommend an equitable ratio between civilian and military faculty* in general academic subjects.[10]

Thus, the prospects were that the committee would continue to serve as a supplement to existing channels of communication among the DOD academies, but would not serve as a powerful force for change.

The Borman Commission

Independent of his concern with the service academies as a member of the Clements Committee, the Secretary of the Army, Martin Hoffmann, found it necessary to spend much of his time during the spring and summer of 1976 seeking to diagnose and remedy blatant organizational deficiencies that had come to light at the Military Academy. Since early April, when a phone call from a West Point cadet to the *New York Times* resulted in a story that brought into public view the news that well over a hundred cadets had been implicated in possible honor violations, Hoffmann had been under increasing pressure to act.[11]

Hoffmann had denied the request of defense lawyers for creation of an independent panel to review the honor system, supporting instead the plea of the Academy Superintendent, Lieutenant General Sidney Berry, that the Academy be permitted to complete its own investigation by an internal review panel. Despite the protestations of Academy officials to the contrary, however, allegations flourished in ensuing weeks that the internal review was a mere cover-up for serious institutional deficiencies. The Senate Armed Services Committee launched an investigation into the West Point honor system in early summer, and by August the House Armed Services Committee had established a subcommittee to launch a similar investigation. At least two members of the House had indicated that they planned to introduce legislation calling for the establishment of an outside investigation of honor codes at the service academies. On August 11, the Speaker of the House, Carl Albert, announced that he and 172 other congressmen had asked Army Secretary Hoffmann to intervene personally in the cheating scandal, because "The academy has failed to handle the current honor code problems properly."[12] In addition, by now the Academy's own board of officers had recommended to the Superintendent that an impartial outside inquiry be conducted into the cheating incident, and Superintendent Berry had concurred.

It was in this context that Hoffman, in an appearance before the Manpower and Personnel Subcommittee of the Senate Armed Services Committee, announced the creation of an investigative panel chaired by Frank Borman. In addition to Borman, the commission included a former Army Chief of Staff, a retired major general who headed an industrial firm in Indianapolis and was serving as chairman of the Board of Visitors to the Military Academy, a former university president, a law school dean, and an Episcopal bishop. With the assistance of a small staff, the Borman Commission moved rapidly to initiate its investigation, interviewing cadets, ex-cadets, Academy staff and faculty members, and numerous other persons who had served at the Academy or otherwise were in a position to provide commentary. Approximately three months after they held their first meeting, the commission members released their final report.[13]

In early January, 1977, Hoffmann announced that cadets who had been ousted in the 1976 cheating scandal would be permitted to reenter the Academy in the summer of 1977 (but not "as soon as possible," as the commission had recommended). He rejected the recommendation of the commission that continuing internal investigations into cheating be halted. However, Hoffmann announced his support for a change of regulations that would provide for penalties other than expulsion under some circumstances for violations of the honor code. He also announced that the Commandant of Cadets was being reassigned almost immediately, and that a new Superintendent would take command in June, upon General Berry's completion of his third year as Superintendent.[14]

By the end of January, Hoffmann had turned the reins of the Army over to his successor in the Carter administration, Clifford Alexander. However, detailed scrutiny of Military Academy operations continued under three study groups that had been appointed by the Chief of Staff to formulate recommendations of specific policy changes which might be needed to remedy deficiencies which the Borman Commission had identified. Because General Andrew Goodpaster worked with the study groups before his assumption of command at the Academy in June, his appointment provided considerable assurance that the Academy would move to directions consonant with the wishes of the parent organization. Yet the appointment of an officer with the stature and experience of Goodpaster also was a signal that the Pentagon wished to restore some of the autonomy for day-to-day operations which the Academy had lost in the months of crisis.

Academy Alumni

A crisis, such as the 1976 honor incident at West Point, which raises doubts about the ability of organizational leaders to maintain appropriate standards of performance, leaves the academy especially vulnerable to external influence. Service academy alumni, as well as officials in the Pentagon, members of Congress, and the mass media, were aroused to action by the 1976 incident. There was some overlapping of roles. Colonel Frank Borman, who chaired the commission that investigated the incident on behalf of the Secretary of the Army, had graduated from West Point in 1950, and in 1975 was elected to the board of the Association of Graduates USMA. One of the lawyers who brought suit against the Military Academy on behalf of young men who had been dismissed for alleged violations of the honor code was Michael Rose, a 1969 graduate of the Air Force Academy. Rose's participation in efforts to effect change in service academy honor systems extended back to the early 1970s. After graduation, he had availed himself of the opportunity provided by the Air Force for pursuit of legal studies at a civilian law school on a leave of absence from the military. At New York University Law School, Rose had made an extensive study of the conduct, honor, and ethics systems at the U.S. service academies. The investigation which he and a team of coworkers launched, published in 1973, received attention in the *New York Times* and other media, and contributed to pressure upon West Point, especially, to end the practice of "silencing" and to expand the provision of legal guarantees to cadets accused of honor violations.[15]

In May 1973, five West Point cadets, who were among twenty-one who had been found guilty of cheating on an examination given by the physics department in April and who therefore were to be dismissed from the Academy, drew upon Rose's analysis of deficiencies in the honor system to bring a lawsuit against the Academy Superintendent. The suit alleged that the cadet honor code was unconstitutionally vague in its use of the terms "lie, steal, or cheat" and that the honor system violated constitutional guarantees of due process of law. Both the U.S. District Court and the Court of Appeals upheld the West Point honor system against these claims, and in 1975 the Supreme Court denied a petition to review the appellate court decision.[16] However, in the 1976 incident, with over a hundred cadets formally accused of honor violations, and hundreds of others under suspicion, Rose and other lawyers went to court again on behalf of accused cadets.

Borman and Rose each represented expressions of concern that service academy officials could ignore only at their own peril (as the early reassignment from the Military Academy of the Commandant of Cadets after allegedly heavy-handed dealings with Army lawyers who were defending accused cadets would seem to substantiate). However, the influence base of the two service academy alumni differed considerably. Borman, who had graduated near the top of the West Point Class of 1950, had been a cadet concurrently with Lieutenant General Berry (USMA 1948), the Superintendent, and with Brigadier General Walter F. Ulmer, Jr. (USMA 1952), the Commandant of Cadets. Borman had gained fame as an astronaut, the commander of the first spacecraft in circumlunar orbit (in 1968). He had retired from the military in 1970 to join with his contemporary, Annapolis graduate H. Ross Perot, a Texas billionaire, to found the American Horizons Foundation to promote public discussion of national policy issues. Later the same year he had become president of Eastern Airlines, keeping close ties to the Military Academy, and becoming a member of the board of the Association of Graduates in 1975.

Rose, in contrast, had been a cadet at the Air Force Academy during the turbulent years when the Academy was experiencing the second major honor scandal in two years (the first in 1965, the second in 1967), and when academy officials were attempting to shore up academy morale against the mounting tide of antimilitary sentiment in the country. His credentials as a recent graduate of the Air Force Academy gave him no significant access to the Military Academy. During the years of his efforts to effect reform in the honor systems of the various academies, especially that at West Point, he has lacked the support of some highly placed military officials which Borman has enjoyed. Thus, to the extent that Rose has been able to exert leverage upon the academies, it has come from his credentials in and knowledge of the law, and from persons in the mass media and in Congress who have been sympathetic to the criticisms of the academies that he and others have been expressing.

Alumni Associations as Channels of Influence

The tendency of most service academy alumni is to direct any efforts to exert influence on the affairs of their alma mater into private or semiprivate channels, in contrast to public ones such as those which Borman and Rose each were able to utilize in varying degrees. A "Dear Sid" letter to the academy superintendent from one of his classmates or from one of his predecessors in the position, or a suggestion passed along to an academy official over cocktails during a reception on Homecoming weekend, is a form of expression that most alumni deem appropriate. Or, an alumnus might take advantage of one of the semiprivate forums provided by his alumni association to share his ideas, complaints, or suggestions with fellow academy graduates. Annual meetings of the associations provide such an opportunity.

In addition, the *Alumni Bulletin* of the Coast Guard Academy runs essays as well as letters to the editor which debate existing policies or programs at the Academy (and in the service at large). In recent years, under the editorship of retired Coast Guard Captain W. K. Earle (USCGA 1940), the *Alumni Bulletin* has encouraged a lively exchange of views among its readers. For example, an essay in mid-1974 by a graduate of the Class of 1970 which aroused much discussion called upon the Academy to admit women as cadets.[17] An article the following year by a Coast Guard lieutenant who was serving on the faculty at the Academy elicited even more impassioned commentary. The author of the article proposed that the existing program at the Academy be abolished, and replaced with a one-year graduate-level program.[18] Earle himself has written articles expressing concern about the apparent diminution of training in practical seamanship skills at the Academy; the articles seem to have served as a catalyst for the expression of similar concern by other graduates, leading to increased emphasis upon seamanship in cadet training.[19]

In its magazine, *Shipmate*, the Naval Academy Alumni Association annually publishes questions that graduates attending the Homecoming meeting of the association have posed to the Superintendent or to other Academy officials regarding the rationale for various programs or policies, along with answers that the officials have provided to the questions. The format lends itself less to structured debate than that provided by the USCGA *Alumni Bulletin*. However, frequently in posing a question, an alumnus is able to make it obvious what his own preferences are. Moreover, the editors of *Shipmate*, like those of the *Alumni Bulletin*, publish letters which contribute to dialogue. Furthermore, a note of iconoclasm often is provided by retired Navy Captain Paul Schratz (USNA 1939; Ph.D. Ohio State 1972) in his regular column, "Sea Breezes." In his April 1972 column, Schratz scolded his fellow alumni for discussing the admission of women to the Academy as if it were simply a joke, pointing out that "every objection raised also was raised by chauvinists of the previous generation when people first mentioned getting WAVES into the Navy. We got the Waves."[20] The same issue of *Shipmate* contained a longer article by Schratz speculating on the needs of the Naval Academy in the future, which provoked much subsequent discussion in letters to *Shipmate*. Schratz proposed a merger of Naval Academy and ROTC programs, with all prospective Naval officers beginning their undergraduate education at civilian institutions but completing it at the Naval Academy, which would provide also a year of graduate study.[21]

The *Association of Graduates Magazine* of the Air Force Academy publishes about five to ten letters per year from its graduates, selecting ones that offer relatively lengthy and thoughtful commentary on concepts such as "honor" and "leadership." Although this editorial policy provides the opportunity only to a tiny segment of the population of alumni of the institution, the Association of Graduates in 1974 did undertake a survey of all persons who had graduated from the Academy from 1959 to 1973, asking for their opinions on virtually all facets of the Academy program. Summary reports of the results of the survey have been published in the magazine's Spring, Summer, and Fall 1975 issues.

Among academy alumni magazines, that of the Military Academy has the most consistent record over the years of providing thorough documentation of policy and program developments at the alma mater. However, *Assembly* does not print letters to the editor, nor do its editors encourage debate in published essays.

"Loyalists" and "Turncoats"

In short, there is a considerable diversity among the four academies in the encouragement or opportunity provided for debate by alumni of academy policies and programs within the semiprivate confines of alumni journals or forums. There is much greater consensus regarding the expression of criticism in public forums such as the mass media by alumni: it is taboo. In the socialization process at the academies, the concept of institutional loyalty is closely intertwined with that of loyalty to one's country. Academy graduates (especially those who, having returned to the academies in top leadership positions, must cope directly with criticism) thus are inclined to perceive fellow alumni as "loyalists" or as "turncoats," depending upon the public posture that the alumni assume toward their alma mater.

A further distinction is made in terms of the status of the alumnus. Douglas MacArthur is the archtypical high-status West Point loyalist. From his memorable "old soldiers never die" speech to Congress in 1951 after being relieved of command in Korea by President Truman, to his farewell speech at West Point in 1962, two years before his death, MacArthur repeatedly invoked the memory of his long association with the Military Academy.[22] It is especially the 1962 speech that Academy officials—and indeed senior American military commanders at various installations throughout the world—continue to cite as exemplary of an alumnus remaining loyal to his school and its ideals.

If MacArthur exemplifies the high-status loyalist among service academy alumni, Vice Admiral Hyman G. Rickover is the archtypical high-status turncoat. Rickover's persistent needling of the service academies in general and of his alma mater, Annapolis, in particular was described in detail in chapter seven. The point to be made here is that although any alumnus who becomes a public critic of his alma mater may be perceived by many of his fellow alumni as a turncoat, one who enjoys high status is likely to be dealt with cautiously by academy officials. In contrast, relatively low-status turncoats are likely to receive much more peremptory treatment from academy officials.

Status is determined primarily according to one's seniority and record of accomplishment in a military career, although prestigious credentials attained in the civilian sector can partially offset their absence in the military sector. Thus, Robert Bowie Johnson, Jr., who had graduated from West Point in 1968, was basically a low-status (and thus highly vulnerable) turncoat when he and fellow author K. Bruce Galloway wrote a scathing indictment of Johnson's alma mater, *West Point: America's Power Fraternity.*[23] The fact that the book was published by a major New York publishing house, that it carried a glowing introduction by disgruntled Vietnam hero Colonel Anthony Herbert, and that it received a favorable review by Gore Vidal in the *New York Review of Books* in October 1973 served to enhance the status of the authors (and also to make their criticisms more threatening). The increase in status meant that Military Academy officials had to tread more gingerly in their public response to the criticisms. Ironically, however, the caustic tone of hyperbolic assault which Herbert and Vidal evidently found refreshing (for example, reference to West Pointers as "the gray hogs") reduced to zero the probability that the critique by Galloway and Johnson would receive a careful hearing at West Point.[24]

Ex-staff and Ex-faculty Critics

Those who have served on the staff or faculty of a service academy, even though they may have done their own undergraduate work elsewhere, often develop ties to the institution which they have served analogous to those developed by alumni. Likewise, if they choose to go public with criticisms of the academy, they can expect a reaction from academy officials similar to that which publicly dissident alumni receive.

In recent years, books and articles critical in varying degrees of the academies have appeared by a former member of the library staff at the Air Force Academy; by a former history instructor and a former English instructor at West Point; by the former head of the political science department at the Air Force Academy; by a former psychiatrist at West Point; and by a former instructor in history at West Point.[25] Such authors doubtless entertain some hope that their friends and contacts at the academies will recognize many of their criticisms as persuasive and justified, and thus will become internal agents for translating the authors' proposals for reform into action. However, ex-faculty members who become public critics, like turncoat alumni, are likely to enjoy greater credibility outside the academy than within it (at least within the academy power structure). Thus, by and large, it has been only when other elements of the task environment have echoed the criticisms that turncoat alumni or ex-faculty authors have voiced that such criticisms have served as stimuli for change at the academies. The most significant of such other elements of the task environment has been the United States Congress.

Congressional Oversight of the Academies

Academy officials have become well acquainted with the old Washington adage that "Congress giveth, and Congress may taketh." The service academies exist by act of Congress, and it is within the power of Congress to close them down. Although threats to utilize the latter power have been invoked only rarely over the years, members of Congress have been much less hesitant about reminding academy officials where the authorization for annual appropriations for the academies originates. Such reminders are largely superfluous. Academy officials are keenly sensitive to the control that Congress exercises, or may choose to exert, over the academies. Consequently, members of Congress are accorded great deference by all academy personnel, from the Superintendent on down to the lowliest plebe.

To inspire deference, however, is not synonymous with inspiring innovation. On the contrary, deference toward Congress typically has been accompanied by assurances from academy officials that traditions remain intact and that no bold departures from past practices are planned. With the notable exceptions of the founding of the Air Force Academy (legislation for which passed Congress in 1954, seven years after a bill for that purpose had been introduced), subsequent actions by Congress to seek standardization of structure and programs among the three DOD academies, and the opening of academy admissions to women, few important changes at the service academies in the period since World War II reflect direct congressional influence.

Four primary modes of access have served members of Congress in their efforts to monitor the activities of the academies and to influence academy policies and programs: (1) inquiries or expressions of concern by individual members of Congress on behalf of their constituents; (2) congressional representation on boards of visitors to the academies; (3) investigations by congressional committees; (4) investigations by the General Accounting Office undertaken at the behest of Congress. Any of these modes is sufficient to strike fear in the hearts of academy officials; nonetheless, each mode has its distinct limitations.

Members of Congress and Their Constituents

All applicants to the Coast Guard Academy compete on a nationwide basis for the vacancies that exist in any given year. Approximately two of every three recruits to each of the three DOD academies, however, are appointed by members of Congress. Many members of Congress have delegated the selection to the academies according to competitive criteria, although some members continue to use academy appoinments as patronage to dispense. However, even when there is no element of patronage, the fact that a young man or woman has been appointed as the nominee of a particular member of Congress usually assures the nominee of continuing concern by the member of Congress.

Typically, however, individual members of Congress are neither equipped nor disposed to conduct detailed investigations of alleged deficiencies or irregularities at the service academies that cadets, midshipmen, or their parents might bring to their attention. The groups upon which they rely for annual reviews of all facets of the academies, from physical plant to education and training, are the boards of visitors (one for each academy).

Boards of Visitors

The Board of Visitors to the Coast Guard Academy consists entirely of members of Congress. Its role has been supplemented since 1934 by that of a team of civilian educators who serve on an academy advisory committee. In contrast, the board of visitors to each DOD academy includes six persons designated by the President as well as four members from each branch of Congress. The former invariably include leading industrialists and college presidents or deans. (There are no exceptions to that generalization in the years since World War II.) Presidential appointees to the boards to West Point and Colorado Springs also always include prominent retired general officers (such as Brigadier General Charles A Lindbergh to the Air Force Academy in the late 1950s, General of the Army Omar Bradley to West Point in the early 1960s). Boards to the Naval Academy, in contrast, typically have not included flag-rank retired officers, although Annapolis graduates sometimes have been appointed in other roles. For example, billionaire H. Ross Perot (USNA 1953), president and chairman of the board of the Electronic Data Systems Corporation of Dallas, chaired the Board of Visitors to the Naval Academy in the early 1970s.

Representation to boards of visitors from the Senate tends to occur on a rotating basis among all members of the Senate, with the exception of the chairmen of the appropriations and armed services committees, who often have exercised their right to be represented rather than designate a proxy. In the House, in contrast, recurrent representation has tended to occur not only for the chairmen of the House counterparts to these two committees, but also for representatives whose districts include or adjoin a service academy. Thus, those who have had long periods of service on boards of visitors to the academies include not only the powerful chairman of the armed services and appropriations committees. They also include representatives with a constituency linked geographically, and sometimes economically and socially, to an academy. In addition, there have been some members of the House whose continuing concern with military affairs has led to relatively extended service on boards of visitors to the academies.

In the case of the Board of Visitors to the Coast Guard Academy, as well as of that to each of the DOD academies, a minimum of one visit per year is required by law. During each visit, board members receive formal briefings from the Superintendent and key members of his command. They visit sample classes, inspect facilities, and often solicit views from samples of cadets or midshipmen. Parades or athletic events often have been included on the itinerary.

Rarely, however, have the boards served as sources of important advice or as agents of significant reform. The *de facto,* as distinguished from *de jure,* function that the boards of visitors perform is roughly midway between that which Representatives Les Aspin of Wisconsin and F. Edward Hébert of Louisiana have attributed to the House Armed Services Committee. "The Armed Services Committee looks on its role as the protector of the uniformed military officer," Aspin has observed; "The committee thinks its role is to find out what the military wants and then try to get it for them." Hébert, as committee chairman, has described Aspin's view as "a typical example of irresponsibility in negative statements. . . . Yes, I'm a friend of the military," Hébert confessed, "but I'll take them to the woodshed and spank them any time."[26]

The boards of visitors have shown a paternalistic readiness to take academy officials "to the woodshed," but only when gross mismanagement or organizational crises have become public knowledge (for instance, in the aftermath of an honor scandal). Characteristically, a symbiotic relationship exists between board members and academy officials. After a visit to the academy, board members, armed with the latest data on academy affairs that is supplied in carefully prepared briefings on each visit, serve as emissaries of good will on behalf of the academies. Congressional members have been especially valuable friends of the academies when funds have been needed for new facilities or programs, and congressional board members have been able to proselytize among fellow members of Congress for needed appropriations. For example, in their 1957 report, which was entered in the *Congressional Record*, the Board of Visitors to the Coast Guard Academy was moved to "express on the record its alarm and amazement at the use of wartime wooden buildings for many of the barracks, classrooms, laboratories and so forth. . . ." The board noted that they and Superintendent Leamy were in agreement on the urgency of the matter. In the 1958 report of the meeting of the board, the Superintendent was reported as expressing "his gratitude to the Members of the Senate and House who by floor amendment added the specific replacement cost [for the wartime wooden buildings], as a special item in the 1959 appropriation bill."[27]

Congressional Investigation of the Academies

To the extent that Congress has made in-depth probes of problems at one or more of the academies, the inquiries have been special investigations rather than the work of boards of visitors. However, formal investigations have their distinct limitations also. One recurrent stimulus to congressional concern with the academies, for instance, has been allegations of hazing. Even in the few instances in which the serious injury or death of a cadet or midshipmen seemed to be linked to irregularities or abuses within an academy, it has been difficult for Congress to ascertain the facts of the case. The reluctance to "tattle" to "outsiders," which Cadet Douglas MacArthur displayed toward the congressional investigation that eventually resulted in the legislation proscribing hazing, continues to be part of the informal code of group solidarity among cadets and midshipmen.[28]

In response to a probe initiated in 1967 in the wake of cheating scandals at the Air Force Academy and the grade quota dispute at the Naval Academy, Representative Hébert received five letters from Air Force Academy cadets expressing resentment at congressional "interference." Hébert's reaction is typical of the indignation which members of Congress feel when in the performance of their investigative role they are perceived as "outsiders." "I thought Congress had established [the Air Force Academy]," Hébert observed.

I thought we provided the money. I thought it was a creature of the Congress. But, maybe I'm wrong. So I wrote the young men and told them I would be very happy to have them among the first witnesses here to explain this. Maybe I am wrong and they are right. If I am wrong, I want to get set right.[29]

That congressional investigation should be regarded as meddling does not necessarily imply that investigation is an ineffectual tool for gathering information and inducing change. However, promoting major change at the service academies rarely has been a goal to which a sizable portion of the membership of Congress has subscribed. Given a congressional bias on behalf of the maintenance of the status quo at the academies, investigation has tended to be a device to enhance organizational legitimacy rather than to threaten it. As Hébert explained to an Air Force cadet during the 1967-68 hearings:

The impression is we are coming in trying to change the [honor] code, maybe or change something here. This is the furthest from our intent. . . . But the criticism [of the academies] is coming so fast and so hard, from many places, that we want to acquaint ourselves and make ourselves knowledgeable with what the actual facts are.

We are not out to disrupt anything. We are out to strengthen everything we possibly can.

Mr. [Charles S.] Gubser (California]: Sometimes a congressional committee is a great buffer between an ill-informed public and those that are being hurt by bad public relations.

Mr. Hébert: It is, yes.[30]

The subcommittee's findings generally defended the academies against those who alleged that recent honor scandals were evidence of much more serious problems of structure and priorities at the academies, and those (such as Rickover) who argued that such practices as the grueling indoctrination of "plebes" and "doolies" were anachronisms that ought to be eliminated.[31]

Only in their final three findings did the subcommittee find reason to question existing practices at the academies. They were unhappy with the lack of uniformity among the three DOD academies regarding the term of office for an academic dean, and even in the terminology used to designate the position. Secondly, they suggested that the review of curriculum at the academies had been left too much in the hands of those who were concerned with academics, to the neglect of professional and military training. The service academies existed primarily to produce fighting men, not Rhodes Scholars, in the view of committee members. This bias of the subcommittee in favor of the Spartan component of the academies' mission had been evident from the first day of the investigations, when Hébert emphasized the point by observing, "When you hire Mickey Mantle you hire him to hit home runs. You don't care if he doesn't know who invented the atomic bomb."[32]

The final finding of the subcommittee reflected the stamp of the chairman's preconceived preferences to an even greater extent. Arguing that "intercollegiate athletics are an indispensable adjunct to service academy training and have a distinct morale impact on all of the men and women in our career military services," and noting that intercollegiate athletics are not only completely self-supporting but provide revenues that support intramural and other sports programs at the academies, the subcommittee insisted that

> when the opportunity presents itself, athletic teams at the Academies should be encouraged to participate in postseason intercollegiate athletic contests; and . . . the decision to participate in future recognized postseason intercollegiate athletic contests should, in accord with long-established service and Department of Defense policy, be made at the Academy concerned.[33]

Specifically, Hébert was a staunch advocate of participation by academy football teams in postseason bowl games. He was miffed that West Point, on orders from Secretary of the Army Stanley Resor, had declined an invitation to the Sugar Bowl in New Orleans on New Year's Day 1968. It was a decision in which Hébert, in whose district New Orleans is located, had taken an intense interest. Hébert acknowledged that some persons would find his own enthusiasm for service academy participation in bowl games difficult to understand when the American armed forces were fighting in Vietnam. But when it came right down to it, he did not see that patriotic concern for the war effort should lead a service academy to decline a bowl bid. To the contrary, in Hébert's view, support for football and patriotic spirit were virtually synonymous. "I am a very strong believer in intercollegiate sports, and of course particularly in football," he said during the session of the subcommittee at West Point, "because I believe in the American way of life. . . ."[34]

The General Accounting Office

The study of the service academies by the General Accounting Office (GAO), that was begun in 1973 and rendered to Congress in three reports in 1975 and 1976, provided a much more comprehensive review of academy programs and structures than was provided by the 1967-68 hearings of the House Special Investigating Subcommittee.[35] Budgeted at some $800,000, and doubtless costing the taxpayer at least twice that when the demands on personnel at the various academies are figured in, the study mobilized the services of twenty to twenty-five fulltime GAO agents as well as a team of high-powered consultants that included a number of former academy officials.

The study had been initiated in response to a request from Senators William Proxmire and Birch Bayh, eventually augmented by other members of Congress, that the GAO assess the cost of maintaining the academies in view of mounting cadet and midshipman attrition—which was of special concern in the light of incidents such as drug use, at least one suicide, and alleged irregularities in the conduct of some training (i.e., hazing). In addition to the four armed service academies, the Merchant Marine Academy was included in the GAO study at the direction of the Comptroller General, Elmer B. Staats, although the mission of the Merchant Marine Academy, most of the graduates of which go directly into private industry, is quite different from that of the academies of the four services.

The GAO decided to focus its inquiry on three issues: attrition, curriculum, and financing. However, these three topics provided a rationale for probing into virtually all facets of the "academy environment," through extensive library research, interviewing, and surveys of the opinions of current members of the student body of each academy as well as of former cadets and midshipmen.

Data gathering on the financial operations of the academies and on the curricula proceeded during the months that were spent in discussions of the survey instrument. Thus, reports on these two topics were the first completed by the GAO, released in February and in October 1975. The final report, on attrition, proved to be the most time-consuming in preparation because it involved the analysis of interrelationships among approximately 250 variables that had been used in the survey. However, a draft of the attrition study was completed by October 1975, circulated to the Departments of Defense, Transportation, and Commerce for comments, and released with these comments incorporated in March 1976.

Findings of the GAO Studies

GAO personnel were able to bring their most impressive experience and expertise to bear on the assessment of the cost of operating the academies. After assigning a dollar figure to even the most detailed facets of education and training, and to the support and maintenance that were required to keep the academies running, the GAO concluded that operational costs could be substantially reduced in two areas. First, more then five hundred support positions (clerks, band personnel, and the like) could be replaced by civilians, at a total savings to the government of $1.6 million per year. Second, potentially large savings (of an amount that could not yet be ascertained) could be attained if the services were to comply with a previously issued policy circular from the Office of Management and Budget requiring governmental organizations to rely on private enterprise in obtaining services such as food services and custodial services. (The Coast Guard was in compliance; the other services were not, according to the report.)

When the GAO moved from cost accounting to analyses that required the interpretation of historical experience, judgments about policy, and the weighing of alternative philosophies of education and training, they proved to be on much more slippery terrain. In preparing the report on academic and military programs, GAO personnel relied heavily upon previous reports by academic accrediting agencies and by boards of visitors to the academies, as well as upon judgments made by consultants to the GAO. The result was that most of the recommendations made by the GAO were ones that had been made previously by some other monitoring agency or body. Moreover, the analytical burden for appraising the three DOD academies and their military programs had been partly lifted from the GAO by virtue of the concurrent analysis by the DOD Committee on Excellence in Education.

The attrition problem, which the GAO study examined in its final report, was one which the academies themselves had been examining for many years (although the magnitude of the problem had been increasing at the DOD academies in recent years). Indeed, one of the most useful contributions which the GAO attrition study made was that of compiling and synthesizing the results of earlier internal studies, thirty-seven of which the GAO cited and described.[36] When they presented the data derived from their own analysis (mostly from the survey they had conducted), GAO personnel had little to say that academy officials did not already know. Most of the regression analyses, factor analyses, and other efforts to provide a definitive explanation of attrition proved inconclusive, leading the Comptroller General in his presentation of the report to Congress to caution that

> If there is one observation that emerged from our attrition study, it is that what motivates people to continue in or drop out of an organization is complex and understanding it requires long and methodical study. Changes to reduce attrition should not be hastily made without careful consideration of their effects on the quality of the graduates. While we believe that we have narrowed the possible causes of student attrition at the academies, more needs to be done, and we have encouraged the academies to use our findings and conclusions as a focal point for action and further examination.[37]

It is not surprising that the GAO displayed decreasing confidence the farther it ventured into fundamental issues of organizational structure and philosophy. The previous experience and training of the GAO personnel who conducted the inquiry and the time constraints under which they operated made them heavily dependent, on the one hand, upon interpretations that could be supported by statistical tests of significance (which were applicable to few important issues), and, on the other hand, upon their consultants (the views of whom reflected close association with the academies, in most cases, and therefore had to be regarded with caution by the GAO because of the possibility of bias).

Despite the numerous serious limitations of the GAO reports, the investigation had an impact on the process of organizational change at the academies in at least two important respects. First, the GAO's *act of doing* the study undoubtedly led academy officials to be more attentive to those facets of their organization and operations that were the object of special attention by the GAO than the officials otherwise would have been. Much of the time and energy that academy personnel had to expend in order to satisfy the incessant demands of GAO field representatives or the Washington office for information, documentation, interviews, and rationale for various programs was regarded by academy officials (with some justification) as an enormous waste—a distraction from more important organizational activities. Still, by having to explain to the GAO the academy mission, and the purported contribution which various endeavors made to the mission, academy officials were forced to scrutinize relative priorities among organizational goals and means more carefully themselves. Moreover, although the GAO was limited in its power to see that its various recommendations were carried out, academy officials were confronted with the obvious fact that Congress promoted a GAO investigation beginning in 1973, and it can do so again.

Secondly, the GAO reports have provided ammunition for those within the academies or in other parts of the task environment who are proponents of change. The reports have served this function not by the novelty of the ideas that are expounded in the GAO recommendations, but rather by augmenting and helping to legitimize ideas that many persons had been advocating for years, such as elimination of petty harassment in the plebe systems.

Similarly, the reports highlight widespread dissatisfaction among cadets and midshipmen with remaining rigidities in the academic programs. Totally prescribed curricula have been abandoned. However, in varying degrees, each academy continues to require a core of courses, some of which cadets and midshipmen find useless and unreasonable. GAO emphasis on this complaint will augment the voices of those who have been contending for many years that further reform in the academic realm is required.

Limits of the Responsiveness of the Academies to the Task Environment

We have described a variety of ways in which elements of the task environment have exerted important influence upon the service academies. The overall pattern in recent decades is one of reduced autonomy by the academies. We observed (and shall emphasize again in the concluding chapter) that reduced autonomy does not necessarily increase the likelihood of major change. However, reduced autonomy has meant that academy officials have been more responsive than in the past to external cues, whether the cues called for reform or for maintenance of the status quo. Sometimes the response has been direct and immediate. For example, the GAO advised the Secretary of Defense "to direct the Military Academy to reexamine the need for the current level of drill and ceremonies." Even before the GAO report was published, the Military Academy has reduced the number of drills and ceremonies by 35 percent for the fall semester of 1975, with further reductions anticipated in the following semester.[38] More often, stimuli from the task environment have generated a response that was indirect and gradual, for at least three reasons.

First, in contrast to signals such as legislation requiring the academies to admit women, often the cues are ambiguous, and in the aggregate provide mixed or conflicting guidance. For example, the major cheating scandal that came to light in 1951 made the Military Academy the focal point for widespread concern and criticism, much of it confusing and contradictory. Some members of Congress, some senior military officers, and some segments of the press charged publicly or privately that the Academy ought to de-emphasize football, thereby reintegrating football players into the corps of cadets and restoring the integrity of the honor system. However, the Army football team had been in the national limelight for several years, bringing prestige to the Academy, to the team, and to the nationally renowned coach, Red Blaik. When the cheating scandal broke, many sports columnists and a number of Academy graduates argued that football players who were being ousted for honor violations were being made the scapegoat for fundamental deficiencies in the honor system, or in top Academy leadership. Some influential members of Congress, as well as some senior military officers (including a former Academy Superintendent), also came to the defense of Blaik and the football team. The immediate changes at the Academy that resulted from the 1951 incident, which might have been major ones had uniform advice been received, were modest. The 1976 cheating incident at West Point resulted in a similar barrage of conflicting signals; but as preceding discussion has revealed, the Secretary of the Army, in response to congressional pressures, took policy matters largely out of the hands of the Academy Superintendent.

A second, closely related, point is that as clear and forceful as persons in the task environment might think they are being in expressing their views to academy officials, it is the perception of these views by academy officials that determines the action that will result. These perceptions, in turn, reflect the biases and predispositions that those who are in positions of authority at the academy have developed over a period of many years. For example, a succession of superintendents at the Naval Academy was subjected to criticisms by Vice Admiral Hyman Rickover, and to his sweeping proposals for reform of the service academies in general and of his alma mater, Annapolis, in particular. The early objects of Rickover's wrath tended to view him, at best, as a meddlesome nuisance. There was a tendency, therefore, to discount his advice, which they received either from Rickover directly or from published accounts of his testimony before congressional committees. However, Rear Admiral (later Vice Admiral) James Calvert had worked under Rickover in the nuclear submarine program. By no means did this experience convince Calvert that Rickover's views on education were infallible, nor was Calvert prepared as

Superintendent to continue to regard Rickover as "the boss." However, the fact that the two men had had a long and close relationship, and that Calvert respected Rickover even when he disagreed with him, meant that recommendations made by Rickover would be viewed more favorably than had been true previously.

Third, even in instances in which service academy officials have accurately perceived unambiguous signals from the task environment, sometimes they have felt that the cues must be disregarded or the message reinterpreted, even at some cost in terms of support from that element of the task environment. A classic example of such an instance was provided in 1963 when the Naval Academy received an order from the civilian Secretary of the Navy, Fred Korth, that all commissioned officers on the faculty (nearly all of whom lacked academic qualifications beyond the baccalaureate) were to be replaced by civilians possessing at least a master's degree. The Naval Academy Superintendent, assisted in Washington by the Chief of Naval Personnel, responded with a crash program of upgrading the academic qualifications of officers (although the ratio of civilians to officers was increased). A loophole in the wording of the Korth memorandum, which laid emphasis on the need to upgrade the qualifications of the faculty, made this strategy defensible, although it was clear that Korth expected his order to be carried out literally. In this instance, the lack of direct responsiveness to an order from superiors probably was explicable both in terms of the deep conviction which the Superintendent shared with many senior naval officers that an all-civilian faculty would be a disaster for the Academy, and in terms of resentment at an unprecedented "intrusion" by a civilian Secretary into policy matters heretofore left to the Academy and the Chief of Naval Personnel. It is a bit of historical irony that little more than a decade later, after complying with the Secretary at least to the extent of increasing the civilian-to-military ratio on the faculty to 50:50, the Naval Academy was ordered by the Pentagon to cut back on the number of civilians on the faculty.

Yet if the perceptions or disposition of academy officials sometimes leads them to disregard demands from the task environment, or to respond to such demands only indirectly, the task environment has some influence even over these perceptions and dispositions. That is, officials in the parent arm of the service, sometimes incorporating suggestions from other elements of the task environment, make the selection of top officials to the academies in the first place, and often the promotion and reassignment of such officials subsequently. (However, the superintendents may be consulted about the selection of a commandant of cadets or a dean.) These are powerful tools of influence, and, in the final analysis, they may be the most important sources of influence upon the direction of the academies that the task environment wields. Even those elements of the task environment such as the parent arms of the service that monitor the academies continuously, rather than spasmodically, cannot keep abreast of all activities and all decisions that are made at the academies. Thus, the choice of those who make the key decisions on a day-to-day basis becomes vital.

10

Growth and Bureaucratization

Expanding on findings from the case studies of Part 2, the analyses of chapters eight and nine have shown the mutual constraints which academy internal leaders and elements of the task environment impose on one another in the change process. The analyses have highlighted an exchange process between producers (the academies) and consumers (the task environment) that often amounts to bargaining, as elements of the task environment seek to impose priorities and preferences upon the academies, and academy officials in turn attempt to exert at least a measure of control over their environment.

The emphasis on bargaining is a useful corrective to images of organizational change which view such change as being simply a matter of reasoned analysis followed by decisive action. However, even this refined picture may lead one to exaggerate the element of rationality that is involved, if one assumed that both academy leaders and those in service headquarters, in Congress, and elsewhere in the task environment have acted fully by design, with a clear understanding of the probable consequences of their actions. The present chapter is designed to counteract any such tendencies toward viewing the change process as totally rational.

This chapter focuses on elements of change in the postwar decades over which even the most creative among academy leaders have had only the semblance of control: organizational growth and bureaucratization. The major task of Part 3, which is that of presenting the major analytical findings of the study, is brought to completion here. This chapter also brings the discussion full circle by returning to themes that were introduced at the outset of the first chapter and were developed in Part 1. That is, especially as one considers the consequences of growth and bureaucratization at the academies, one becomes sensitized to the implications of the transition that has been occurring as the academies have become "modernized," leaving behind selected elements of their seminary-academy past.

The continuing mission of the service academies has been not merely that of developing the skills and imparting the knowledge which cadets and midshipmen will find useful as career military officers. Central to the mission, now as in the past, is the task of professional socialization. The socialization process involves the transmission not only of skills and information, but also of organizational values and traditions. If socialization is successful, the academy graduate will have received not only the "tools of his trade" but also the motivation and feeling of identification that links him to earlier generations of military professionals and to the society which he serves.

There is no evidence that academy officials in the postwar decades have been less committed to professional socialization than were their prewar counterparts. Indeed, modern academy officials have tended to be far more explicit about this commitment than were prewar officials. However, growing enrollments in the postwar years have served as a stimulus to organizational complexity and to increased specialization of roles. Values that once could be transmitted from one generation to another informally have had to be codified. Rules of discipline which had been relatively few and maintained largely through verbal commands have proliferated in written form. In short, the context of socialization at the academies has been altered profoundly in the postwar years, making the task far more difficult.

The Martial Virtues

The Spartan ideals of the seminary-academy were those of duty, honor, and loyalty. Even in the pre–World War II years, the identification of cadets and midshipmen with these ideals was rarely as overtly enthusiastic or as unquestioning as academy officials might have liked. Large numbers of young men had been attracted to the service academies during the Depression by the opportunity to obtain a four-year tuition-free education. Neither patriotism nor romanticized images of the challenges of a military career had been relevant to their decision to attend the academies. Moreover, a grumbling acquiescence in routines and rituals characterized the attitude of nearly all cadets and midshipmen some of the time, and of some of them all of the time.

Still, there had been a keen and pervasive sense of community at each of the academies in the prewar years. It was the intimate sense of community, even of organic unity, that gave meaning to otherwise abstract ideals. In relatively abstract terms, for instance, loyalty meant being ready to fight for America and the American way of life. Formal attempts to instill a sense of loyalty in these terms rested on the assumption that twentieth-century young men would still be inspired by invocations such as that of Nathan Hale to "pray God to bless that flag. . . . Stand by Her, boy, as you would stand by your mother."[1] For the cadet or midshipman, however, it was loyalty to others with whom he was closely associated on a daily basis that made the concept meaningful. Being ready to "fight for America" meant being ready to share an obligation that his close companions bore as well, and one which his academy predecessors had borne before him. The routines of academy training also reinforced the ideal of loyalty in personal terms. One grew to believe in the importance of maintaining the watch faithfully at sea, for example, not so much because the country might be in peril if one were negligent, but because one's shipmates' lives might be imperiled. That is why loyalty was so closely tied to duty and honor.[2]

Put simply, duty especially meant accepting responsibility for your share of the load. This in turn meant sticking to it when you were assigned a task, down to the last detail, and not letting your buddies down. If nevertheless you should foul up a particular assignment, a sense of duty meant that your only reply to a superior who wanted to know why the task had not been accomplished to his satisfaction would be, "No excuse, sir." Such a reply could be painful to give sometimes; there might well be persuasive reasons why you had been unable to perform to expectations. However, although politicians might get away with making a thousand excuses for sloppy performance or for failing to fulfill campaign promises, for the military professional there should be no alibi—no "puny b-ache" [belly-ache], as cadets would say. Once one learned this, he could take real satisfaction in having become a person who could be counted on to do a job correctly, and to offer no excuses if this proved impossible.

Of course, one frequently would chafe at the ubiquity of regimentation and discipline. Dwight D. Eisenhower's description of his cadet days before World War I recalls a milieu that cadets and midshipmen for generations would find familiar.

> Offenses were possible everywhere. Dust on the window sills of the room. Improperly folded garments in the clothes locker. A few seconds late for formation. A badly prepared lesson. An unbuttoned jacket. An improper element of the uniform. Negligence of almost any kind. Each had its prescribed demerits and if in any month the total exceeded a certain level, the victim was required to walk the area—an expression used to describe punishment [marching back and forth within an area with a rifle on the shoulder] inflicted during free hours.[3]

Still, if life at the academy was severe, it was not impersonal; and in times of enlightened leadership, the rules were applied with reason and fairplay. As Eisenhower notes:

> The discipline was not so much harsh as inexorable. If one was guilty of an offense, report was automatic and the number of demerits to be received was exact. Justice was evenhanded, even though at times it seemed too swift.[4]

Fundamental to the acceptance of a Spartan life and its associated demands was the sense of mutual trust that was integral to the community. The concept of honor thereby retained a vitality in the lives of cadets and midshipmen, not because it was so prevalent in formal indoctrination lectures, but because its importance was comprehensible in terms of everyday experience. To be "honorable" meant that you must be deserving of the trust which others placed in you, just as you expected others to live up to your trust in them. At West Point, for example, one could walk his evening rounds as cadet-in-charge of quarters and rather than inspect each of two or three dozen rooms to make sure that all occupants were in bed and all was in order, get an "all right, sir," from the occupants, which served as their word that all were in bed and all was in order. As noted in earlier chapters, in the 1950s, the 1960s, and the 1970s, first the Military Academy and later the Air Force Academy (which had modelled its honor system after the one at West Point), and then the Military Academy again, would be wracked by honor scandals. Some came to question whether the use of the honor code to enforce regulations (as in the "all right" system of inspections) had contributed to the cynicism that some cadets had apparently developed regarding honor codes. But in the prewar academy, the "all right" was seen as simply one of the means of simplifying a required routine that men with common responsibilities for carrying out that routine had devised. Saying it or hearing it was to invoke the bonds of shared trust; and like the clear, sad notes from the bugle blowing taps each night, the "all right" was an end-of-the-day reaffirmation of one's membership in a very special sort of community.

Moreover, one recognized that the habits of duty and integrity that one acquired as a cadet or midshipman, even in fulfilling chores such as making a room check as charge-of-quarters, or in performing boring tasks such as standing the watch, would stand one in good stead in meeting the demands of military life that lay ahead. What former Secretary of War Newton Baker had said of the vital relevance of a sense of honor for West Pointers surely applied equally to graduates of the other service academies:

> Men may be inexact or even untruthful in ordinary matters and suffer as a consequence only the disesteem of their associates or the unconvenience of unfavorable litigation, but the inexact or untruthful soldier trifles with the lives of his fellow men and with the honor of his government, and it is therefore no matter of pride but rather a stern necessity that makes West Point require of her students a character for trustworthiness that knows no evasions.[5]

Cadets and midshipmen were told of the achievements and heroic exploits of academy graduates and of earlier military heroes from the day they entered the academy. Thereafter, there were ever-present symbols—songs, portraits, murals, statues, battle trophies—to serve as reminders to the cadet or midshipman that he was following in the footsteps of noble forebears. Cadets and midshipmen were enveloped in this romanticized past to a greater or lesser degree at each academy. Yet, because the myth as well as the reality had been nurtured over several generations, the legends had acquired a legitimacy which cadets and midshipmen in the prewar era were not disposed to challenge, any more than they challenged the legitimacy of actual past accomplishments. On the contrary, the aura of romance was part of what was special about the academies. The feeling of being part of institutions that were special was a source of pride.[6]

Quantitative and Qualitative Changes

On the face of it, the martial virtues which constituted the vital core of the seminary-academy tradition have been retained intact even into the 1970s, in spite of other changes which have occurred at the academies in the decades since World War II. Sensitive observers know better, however.

The commitment of academy officials to the cultivation of the traditional ideals of duty, honor, and loyalty remains undiminished. Discipline is still pervasive and severe. The milieu at each academy continues to evoke the memory of venerated heroes whose lives and deeds are linked to the traditional ethos, and to recall through ritual and routine the continuity of present with past. Formal indoctrination in the Spartan values has increased rather than decreased in intensity.

But the intensification is evidence of failure, not success. The redoubling of efforts to instill in modern-day cadets and midshipmen an identification with traditional ideals typically has come in the wake of an organizational crisis in which (as in instances of widespread cheating) core values seem to have been rejected or abused on a large scale. Idealistic appeals that had an authenticity in an earlier era have lost their resonance amidst organizational growth.

The seminary-academy had been sustained especially by intimacy. When each member of the community of an academy came into daily contact with every other member on a face-to-face basis, bonds of camaraderie were forged and tradition and custom developed naturally.

Even on the eve of World War II, the Coast Guard Academy remained tiny. West Point and Annapolis had expanded somewhat. As Figure 4 reveals, however, the period of significant growth at each academy (including the newly created Air Force Academy) has come since the onset of World War II. West Point and Annapolis each experienced rapid expansion during the war years, with enrollments falling off briefly at the end of the war. From the late 1940s to the early 1950s, the Naval Academy grew rapidly, experiencing gradual enrollment increases, with some fluctuations, thereafter to the 1970s. The size of the Military Academy remained relatively constant at about 2,400 from the late 1940s to the early 1960s, after which enrollments increased rapidly until they peaked in the early 1970s.[7] The Air Force Academy experienced rapid and unrelenting growth from the admission of the first class of 306 cadets in the summer of 1955 to a tapering off at approximately 4,000 in the 1970s.

By the 1970s, each of the three DOD academies was graduating classes of 800 to 900 students from a student body of about 4,000. If these figures still seem small relative to the student populations of most modern state universities, the growth of the academies nevertheless had introduced structural complexity, specialization, codification of custom, and the proliferation of formal rules to a degree that represented a qualitative change in the life of each institution. Only the Coast Guard Academy, which by the 1970s was graduating classes of some 200 annually from a student body of approximately 1,000, remained small enough for most members of the community to be acquainted personally with nearly all other members. Yet, to a lesser degree, even the Coast Guard Academy was beginning to experience the trends which the other academies were experiencing toward bureaucratization.

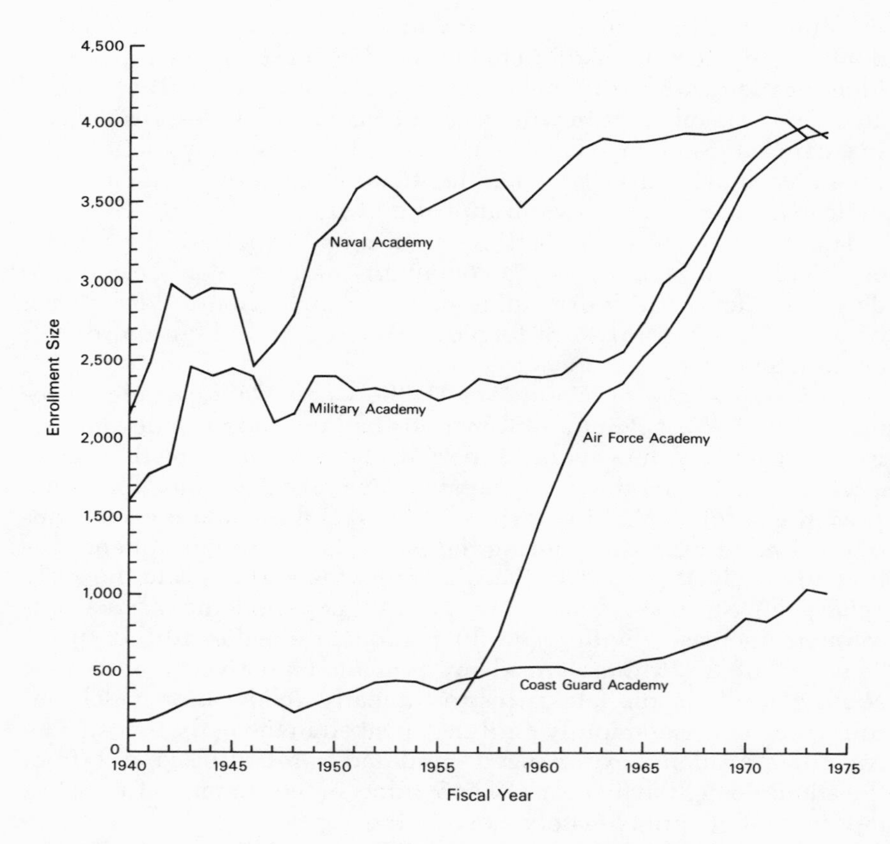

As at any institution of higher education, enrollment figures for any given year are necessarily an approximation, because of student withdrawals from school at various points throughout the year. Enrollment figures for the academies were computed by combining the figure for the size of the freshman class at entry with figures for the size of the sophomore, junior, and senior classes, respectively, at graduation. The method exaggerates the size of the freshman class, where a substantial amount of attrition will already have occurred even after two months, and underestimates the size of the sophomore and junior classes, which will be larger in the estimate-year than at graduation. However, a cross-check of data computed by this method with those reported on occasions when the academies have provided annual enrollment figures reveals the reliability of this method.

Figure 4. Service Academy Enrollments, FY1940–FY1974

Complexity and Specialization

In the nineteenth and early twentieth centuries, the reasonably well-defined organizational roles at each academy were those of Superintendent, Commandant of Cadets or Midshipmen, Protestant chaplain, faculty member (who frequently doubled as a disciplinarian and drill instructor), and cadet or midshipman. With growth, a division of labor among staff and faculty members was developed to the point that it is accurate to speak of the modern service academy as consisting of a set of interrelated, functionally specialized subsystems (Figure 5). Staff and faculty members throughout each academy still can preserve a sense of organizational identity through their dedication, at a lofty level of abstraction, to a common mission: the education and training of prospective career officers in the military service. At more mundane operational levels, however, choices must be made in the allocation of organizational time and resources to the various subsystems that perform discrete tasks which, in varying degrees, contribute to the overall mission. These choices have the effect of defining the priorities among subsidiary goals, such as recruitment, public relations, athletics, military training and discipline, and academics. It is hardly surprising that individuals who represent different subsystems often disagree with one another as to what the priorities should be.

The modern Superintendent is not just a father figure, he is also manager, who must coordinate the efforts of, and sometimes referee disputes among, a galaxy of subordinate units. Indeed, especially at the three Defense Department academies, each of which includes a sizable plot of land and an extensive housing development, the Superintendent is a community "mayor," as well as the man in charge of the operations that directly affect cadets or midshipmen. He is assisted by a chief of staff (at USMA or USAFA), an executive assistant (at USNA), or an assistant superintendent (at USCGA), as well as by one or more aides-de-camp. In addition, at the three larger academies, specialized staffs have proliferated, most of them headed by an officer of the rank of Army or Air Force colonel or Navy captain. In many instances (for example, admissions, public relations, personnel, communications) the staff handles responsibilities that prior to World War II were performed as additional duties by members of the academy faculty or staff. The office of admissions, especially, has become professionalized, with far greater organizational resources committed to the recruitment of cadets or midshipmen and to the processes of selection and record-keeping than was true before World War II when

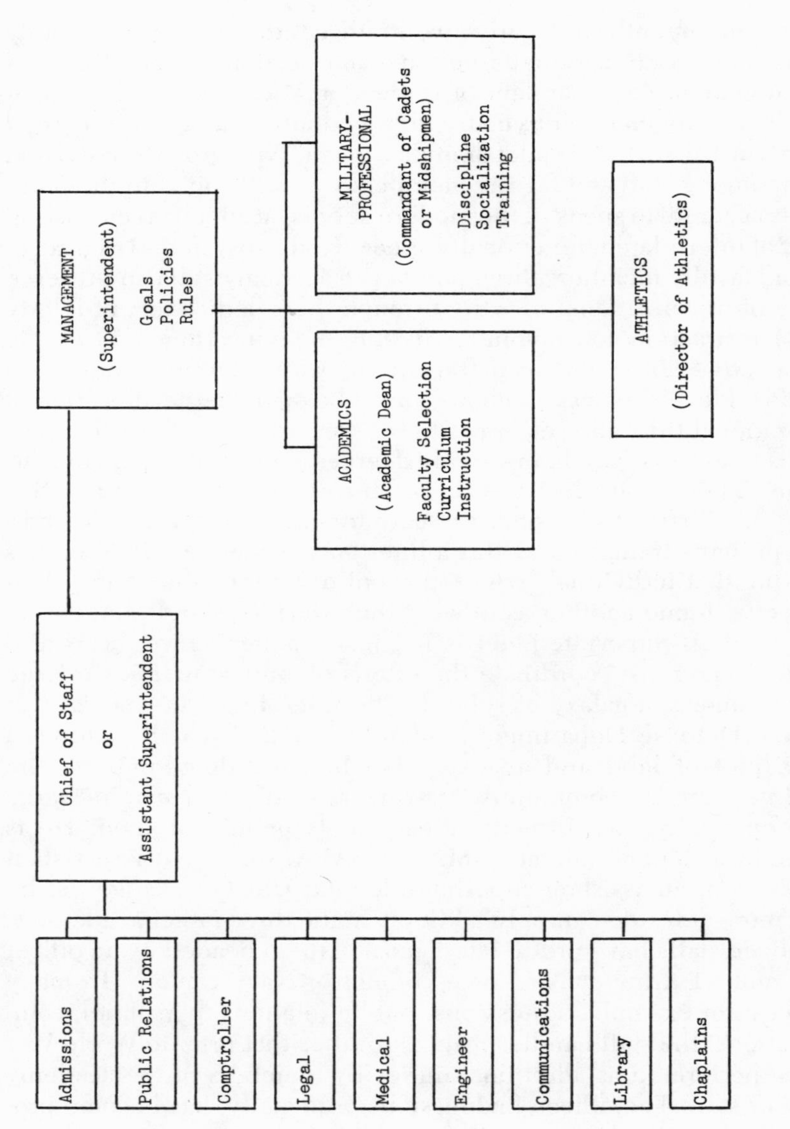

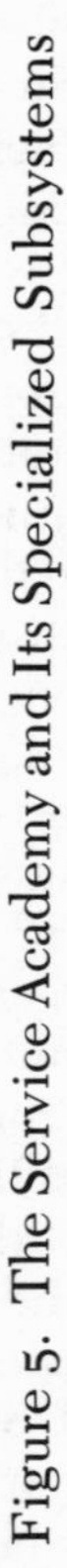
Figure 5. The Service Academy and Its Specialized Subsystems

senior faculty members typically assisted the Superintendent in making admissions decisions. Each admissions office now is headed by a dean or director of admissions, assisted at the three larger academies by a large staff within the academy and by field representatives throughout the country.

As the case studies revealed, the two key operational subsystems of the academies—academics and military-professional training—also have become more highly specialized in recent decades. The result has been a heightening of tensions between them.

From the nineteenth century until the eve of World War II, the trade school orientation of the academies, and the emphasis on "building character" and instilling discipline as the primary mission, insured that a commitment to Athens would remain comfortably subordinated to a commitment to Sparta. However, the combination of accreditation requirements generated by the authorization to award the baccalaureate degree, beginning in the 1930s, and the widespread recognition at the end of World War II of the complexity of professional demands in the postwar environment, led to an increase in emphasis upon the academic component of the academy mission. Thus, when Vice Admiral James Calvert spoke in 1970 about the need to walk "the fine line between Athens and Sparta," he identified the tightrope act that had become required of all service academy superintendents in recent decades.[8] They have managed the act with varying degrees of dexterity, amidst increasing specialization within and emphasis upon the academic subsystem, which have aggravated Sparta-Athens tensions.

Specialization and an institutionalization of roles within the academic subsystem have taken several forms. We noted earlier that a formal position of Academic Dean has been created since World War II at each academy. Related postwar phenomena have been an expansion of the number of faculty positions that are assigned on a permanent, tenured basis, as contrasted to those assigned to officers on short tours of duty, and an upgrading of the academic qualifications of faculty members.

The concept of the professional responsibilities of the faculty member who has acquired a relatively advanced degree of expertise in an academic discipline is more complicated than it was in the days when most faculty members had only a general working familiarity with the subject matter they were assigned to teach. As late as the mid-1950s, the majority of faculty members at the academies had had little career experience outside of military assignments. Therefore, they were likely to define their professional responsibilities almost exclusively in terms of prevailing norms of conduct within the military establishment. The faculty member expected to be judged on those terms. ("Major Smith is prompt, courteous, well-groomed, and carries out instructions efficiently and with dispatch.") However, the increasing numbers of military officers and civilians who have joined service academy faculties in the 1960s and 1970s after having spent two or more years in civilian graduate schools pursuing advanced degrees often have perceived their responsibilities differently from the traditional formulation. Being able to claim advanced expertise, the faculty member has expected to be judged accordingly. ("Major Smith is a first-rate officer and a first-rate chemist; he has been imaginative and successful in the development of laboratory exercises that have advanced the research skills of his students.")

It is not always the case, however, that the expectations of the faculty members are shared by those with whom or for whom they work. Questions that might seem niggling loom large when viewed through the lens of the academies' efforts to reconcile the commitment to Athens with the commitment to Sparta. Should not the faculty member of a service academy, however impressive his academic credentials, display a dedication to Spartan concerns that is at least equal to his dedication to Athenian goals? Should he not, for instance, make certain that his students always say "Sir" when addressing him, that they sit up straight in their chairs, and that they always come to class with shoes that are shined to a high gloss?

The conviction that the appropriate answer to such questions is an emphatic yes has tended to prevail especially among commandants of cadets or midshipmen and their officer cadres. Frictions are generated by incidents, real or imagined, which lead such officers to suspect that standards of military discipline which they demand of cadets or midshipmen get lowered in the academic classroom. Civilian faculty members, who make up approximately half of the Naval Academy faculty and roughly one-third of the Coast Guard Academy faculty, have been particularly suspect. (Until very recently, both the Military and Air Force Academies maintained an all-military faculty. In response to a directive from the Deputy Secretary of Defense in April 1975, the two academies have begun to add a small number of civilians.) Conversely, civilian and officer members who have been the object of such suspicions have tended to resent them and to view with a quiet contempt those who would derogate the development of the cadet's or midshipman's intellect.

A trend toward increasing specialization within the domain of the Commandant of Cadets or Midshipmen has been evident also, most notably in the realms of leadership training and of counselling. In recent years, each of the academies has sent some of its prospective leadership instructors to graduate school for exposure to sophisticated and up-to-date training in subjects such as social psychology and management. A related trend is to make sure that at least some of those officers who have the most direct contact with cadets or midshipmen in disciplinary or counselling roles (as company officers, squadron officers, or tactical officers) have had similar training.

Officers who lack such training, however, continue to be dubious of its value. The anomalous position in which the specialists in the field often find themselves is illustrated by the discomfort long experienced by members of the Office of Military Psychology and Leadership at West Point. Although an integral part of the Commandant's domain since the office was created in 1946, its members were considered by many tactical officers to be "too bookish" for their role in leadership training to be fully acceptable. In addition, until 1977, when they were incorporated into the new Department of Behavioral Science and Leadership in response to a recommendation of the Borman Commission, they were too closely identified with the Commandant to be accepted fully as faculty members.

Those individuals at the academies whose special training is in counselling, as distinct from leadership, occupy an even more precarious existence. There long has been a deep-seated feeling among many military officers at each of the service academies that no cadet or midshipman with the requisite amount of "guts" to become a reliable officer in combat is going to need professional counselling services during his days at the academy.[9] After all, it is often said, the academy chaplains and the academy social hostesses are there to let cadets and midshipmen pour their hearts out, and thus to "punch their tickets" of grievances. Moreover, the company officer lets cadets and midshipmen know that "his door is always open."

If on occasion (such as the joint cadet-officer sensitivity sessions which were instituted for a brief period at the Coast Guard Academy) the trend toward increasing complexity and specialization has brought cadets or midshipmen into closer contacts with the academy "power structure," the more typical consequence has been to broaden the gap between students and academy authorities. Layers of organizational authority gradually have been introduced between those who make the rules that govern academy life and the cadets and midshipmen who typically are most severely affected by the rules. One should not exaggerate the remoteness of academy superintendents, commandants, or deans from "the masses," however. Probably more than most administrators in comparably sized civilian institutions, academy officials typically make a special effort to become faces and not just names to cadets and midshipmen. Still, the difficulty of succeeding in that effort is far greater today than it was when the entire student population was no more than a couple of hundred young men. Moreover, as noted in chapter nine, it has become much more common than it once was for key policy decisions regarding the academies to be made in Washington rather than by academy officials on site.

The growing complexity means also that the sense of group cohesion among cadets or midshipmen is a much rarer commodity than it once was. The number of primary units (companies or squadrons) has proliferated as components of a battalion or regiment expanded to a corps, brigade, or wing. The peers whom a cadet or midshipman gets to know well enough to foster a sense of camaraderie no longer constitute the entire student body, but rather are a fragmentary component of it—the company or squadron; a team or club to which one belongs; and to some extent one's graduating class, with the members of which one shares important trials and triumphs.

Paradoxically, academy officials tend both to cultivate and to fear the loyalties that develop naturally out of subgroup associations. On the one hand, the development of a fierce competitive spirit is seen as essential to the preparation of officers for military careers that may involve combat. Thus, team spirit is encouraged and intercompany or intersquadron competition in athletics and in drill is fostered. On the other hand, academy officials have come painfully to recognize that subgroup loyalty may compete with and even supplant the loyalty that academy officials try to nurture to the academy itself and to the values it represents.

Athletics

Organized athletics have been an integral component of the socialization process at the academies only since the turn of the century, but the rationale for the heavy emphasis that athletics have received in recent decades links athletic training to the martial virtues. It is argued that the skills and attitudes developed on the athletic field—especially in contact sports—will prove beneficial to the cadet or midshipman in the more critical later trials of combat. He will learn to give and take; he will learn team spirit, and the need to "go all out" for the team. Physical training and athletics help to cultivate a "can-do" spirit that is said to be vital to a military leader. The good military professional is action oriented rather than idea oriented, academy officials remind cadets and midshipmen—adding, however, that this is not to say that the military professional has no ideas. (As academy officials in the postwar period have put it, the point is that the military man is a Dwight Eisenhower rather than an Adlai Stevenson, a Curt LeMay rather than a Eugene McCarthy. Unlike the latter individual in each comparison, it is said, the military professional is not paralyzed by contemplation. Like the former, he prefers practical problem solving to abstract theorizing, and when confronted with a task, he wants to "get on with it" and get the job done.) Athletics are an important part of the preparation of cadets and midshipmen because on the athletic field, as on the battleground, it is action—results—not theorizing, that counts. Contact sports receive special emphasis at the academies for their perceived "playing fields of Eton" payoffs. And among contact sports, football has been the undisputed king. However, the cultivation of "big-time" athletics has had its price at the academies, as it has at many civilian colleges and universities. It is of interest that before 1890, West Point cadets were not permitted to engage in athletic competition with teams from other schools.[10] In that year, challenged by the Naval Academy to a game of football, West Point intercollegiate athletics began—badly, with a 24-0 defeat. The result, as recalled by an officer who had been a cadet at the time, was that

> Even the most hidebound conservatives conceded that the Army must settle its score with the Navy. Contributions came from every regiment in the Army. Dr. H. H. Williams, a former Yale player and future creator of great Minnesota teams, was employed as coach and we had our first series of games with outside teams.[11]

The following year, Army defeated Navy. Dennis Michie, cadet captain of the victorious team (and son of Colonel Peter Michie, West Point professor), is perhaps remembered more for his contribution to Army football than for giving his life in battle on San Juan Hill seven years later. (The present football stadium at West Point is named after young Michie.) The ferocity of emotions engendered by Army-Navy games led President Grover Cleveland to place a ban on them; but under President McKinley in 1899 they were revived.[12]

By the 1920s, intercollegiate athletic competition had become an integral part of the subculture both at West Point and at Annapolis. The Coast Guard Academy, because it was much smaller than the other academies, still remained aloof from such competition. With the move to the present site in the 1930s, however, the new academy at New London also began to schedule athletic competition with other schools. It became a point of pride that most of these schools were ones with enrollments several times that of the Coast Guard Academy.

The recruitment of athletes had become commonplace at colleges and universities by the eve of World war II. At the service academies, such recruitment was justified in terms of the alleged superiority of varsity athletes as combat leaders. A victorious athletic coach would receive plaudits at the Academy not only because a winning record enhanced institutional prestige and morale, but also because it was believed that athletics were an integral part of the process of instilling a commitment to team spirit and a will to win. As Lieutenant General Robert Eichelberger wrote to Army football coach Red Blaik shortly after Pearl Harbor (by which time Eichelberger had left the West Point superintendency for a combat-training assignment):

> Keep driving ahead on your job, for it's a top-flight project in our war effort, and don't let anyone tell you anything else. And you can tell 'em I said so—and I'm seeing every day fifteen thousand damn good fighting men who'd be a damn sight better if they'd spent their last year in school under you—and never entered a class room, if that had been necessary.[13]

Blaik was spectacularly successful in producing nationally acclaimed football teams at West Point. It took a major organizational crisis to alert Army and academy officials to the possibility that team spirit and a will to win football games were not necessarily synonymous with ideals of loyalty to the corps as a whole and to its honor code.

A cheating ring was uncovered just before graduation at West Point in 1951. The Class of 1951, not implicated in the initial reports available to Academy officials, was permitted to graduate while the investigation continued. However, ninety cadets from the two succeeding classes ultimately were charged with violations of the cadet honor code, and in the glare of nationwide press coverage were discharged or resigned under the taint of allegations of complicity in the affair. Some of the cadets had used the officially sanctioned system of tutoring (whereby cadets who were doing well in a subejct would tutor others who were not doing well—or would tutor members of an athletic team who were hard pressed for time) to pry out answers in advance on a quiz they were going to take. Others were implicated because they had provided such answers (for example to a quiz on

Tuesday identical to the one the tutors had taken on Monday). Still others were violators of the "no toleration" commitment. Even if the cadet had not cheated himself but had learned that another cadet had cheated, to fail to report it (thereby tolerating it) was to violate the honor code himself. Nearly the entire football team—thirty-seven varsity players—had been implicated in the scandal.[14]

General Maxwell Taylor later would recall that even by the time of his superintendency (1945-1948), "Big-time college football as it developed after World War II did not fit readily into the West Point pattern of education. . . ."

> The players were heavily burdened by the load of the academic schedule to which was added the need to prepare themselves to face the best teams in the country each Saturday. To utilize every minute of a working day, they followed a schedule quite different from that of other cadets. They practiced as a group, ate as a group, and often studied in groups under cadet tutors who volunteered to help them. This distorted pattern of life isolated the football squad from the rest of the Corps and eventually led to a spiritual alienation which, I am sure, contributed to the cheating scandal of a few years later, resulting in the dismissal of virtually the entire football squad. It has always been a cause of deep regret that as Superintendent I did not perceive clearly or soon enough the potentially baneful influence of big-time football on West Point.[15]

Taylor's regret perhaps is compounded by the knowledge that it was he who contributed to the expansion of Coach Blaik's fiefdom by naming him athletic director (in addition to his post as football coach) in 1946, and then appointing him to chair the athletic board late in 1948.

Coach Blaik, understandably, had an interpretation of the honor scandal that was vastly different from that of General Taylor (who was Deputy Chief of Staff for Operations of the Army at the time) and the authorities at West Point. Blaik was outraged at the way in which the incident was handled. The lives of fine young men were being ruined, he contended, when in most cases the fault lay in the system rather than in individual misbehavior. Academy officials had badly overreacted, Blaik charged, and had allowed themselves to act rashly and hastily through pressures, if not blackmail, from immature tactical officers and a small group of cadets threatening to resign if those implicated in the investigation were not dismissed. Ninety cadets were being made the scapegoats for the unwillingness of Academy officials to face up to their own failures and to the need to overhaul the system.[16]

There was some truth in Blaik's allegations. Army and Academy officials were willing to acknowledge only (1) that ninety cadets had failed to meet the requisite standards of honorable conduct, and (2) that divisive subgroup loyalties had developed in the football team. The emergence of other subgroup loyalties was scarcely acknowledged; but with growth, the corps was losing the sense of cohesion that had sustained informal norms of conduct on a corps-wide basis. Even in the minds of many of the vast majority of cadets who continued to observe its precepts, the formal honor system had an artificial quality in important respects, especially as it became entwined with the web of regulations that had come to intrude into minute details of the cadet's life.

In the aftermath of the 1951 scandal, the West Point honor code was preserved intact, with renewed emphasis given to the importance of the code in the indoctrination of incoming cadets. The practice of giving identical quizzes on alternate days was curtailed, but otherwise little change was made in the Thayer system practice of giving frequent quizzes; nor was the heavy reliance on "objective" rather than essay-type examinations altered. The system of discipline was maintained with little alteration, except for periodic changes of phraseology which cadets were required to enter into their copy of the Blue Book of regulations.[17]

As noted in chapter four, the West Point graduates who founded the Air Force Academy made sure that the first contingent of cadets who entered the new academy in 1955 would adopt an honor code virtually identical to that at West Point. Clearly no lessons from the 1951 incident at their alma mater had been absorbed beyond those of the official diagnosis of the problem. The irony was compounded by a major honor scandal at the Air Force Academy in 1965, also heavily implicating the football team. As at West Point, Air Force Academy officials proved unable to cope with the underlying problems, as the outbreak of another cheating scandal two years later revealed.

Codification of Once Informal Codes of Honor

In the pre–World War II academies, the hallmark of the success of the process of socialization by which young men had been assimilated into the organization had been the high degree of congruence that existed between values that were officially espoused and those that were internalized by cadets or midshipmen. It is true that from the days when cadets Jefferson Davis and Edgar Allan Poe went "over the wall" from West Point to seek diversion in Buttermilk Falls, defiance of official regulations has been accorded prestige in the cadet or midshipman subculture. However, there always has been a tacit but emphatic distinction made between merely "outfoxing" the authorities and acting in such a manner as to impugn one's character. The fact was that academy officials themselves often tacitly expressed their admiration for cadets or midshipmen who were clever enough to sneak out undetected after taps, or otherwise violate regulations and avoid demerits. To transgress one's honor was quite another thing, however. To cheat, for example, was to take advantage of classmates who were attempting to satisfy academic standards honestly. To lie was to call into question one's courage in giving truthful answers even when such answers adversely affected one's position.

Parameters of "permissible deviance" still are maintained informally among cadets and midshipmen. However, the likelihood that these parameters will be uniform throughout an academy has declined with the increase in enrollments. Thus, incidents of actions by some cadets or midshipmen that are at great variance with the officially endorsed norms of acceptable conduct but which are sanctioned by norms that prevail in a segment of the student body have become more common. In the investigation of the 1967 honor scandal at the Air Force Academy, for example, it was discovered that one or more of the squadrons in which most of the academic cheating had taken place had been organized into a "cool group" that took pride in its deviance from officially prescribed norms of conduct.[18] Similarly, in the investigation of the massive honor scandal at West Point in 1976, it was disclosed that a "cool-on-honor" counterculture had emerged among many cadets.[19]

The operational norms that in fact govern behavior are those that have been internalized. These are transmitted most effectively informally. Such was the mode of transmission of standards of honorable conduct at the academies historically, into the twentieth century. Beginning at West Point during the MacArthur superintendency, however, the academies have codified the honor system. Codification has the apparent benefit—from the perspective of authorities—of assuring uniformity of indoctrination, to the point that no offender can plead that formally sanctioned interpretations differ from his understanding of the code. However, codification also has increased the likelihood that the code will be viewed as merely an additional dimension of the externally imposed system of organizational control.

An officer-cadet special study group appointed by the Superintendent of the Military Academy to study the honor system noted perceptively in its report in 1975 that, over the years, Academy officials had found it necessary "to specify in greater detail those acts that constitute honorable or dishonorable behavior. . . ."

> As cadet life has become vastly more complicated, the number of situations demanding interpretation and legalistic decisions have so multiplied that there are relatively few solid points of consensus, and there are numerous areas left open to self-serving rationalizations. The inevitable drift is toward an increasing listing of specifics. This trend tends to obscure the spirit of the Code and exacerbate the conflict that cadets conjure up between honor and regulations.[20]

Less than a year after the study group had rendered its report, a major honor scandal erupted at West Point. In the months that followed, both internally and externally sponsored investigations of the widespread violations of the honor code suggested that the integrity of the institution itself had become questioned. As one of the officers at West Point who had been assigned to a board studying the case observed, he had come to recognize that the revelation of extensive cheating on a home-study problem in an electrical engineering class was merely "the tip of a much larger, more complex iceberg."

> The diffuse, unconnected, nonconspiratorial character of the cheating incidents indicates to me [that] we happen to have lighted on one particular skeleton in our academic closet. Statistically, it is unreasonable to assume [that] the Class of 1977 is anomalous, an unhappy convergence of reprobates and bounders. That simply does not make sense given our admissions procedures. Moreover, I find it difficult to believe that Fortune guided us to 21 percent of a class the first and only time it ever cheated so that we could purge the miscreants and maintain unsullied the purity of the institution.[21]

Still another officer observed that "Cheating, to certain degrees has become a way of life and cadets aren't sure what is cheating and what is not." Another affirmed that "The Honor System is not alive and well at West Point. In truth it is very sick."[22]

However complex and diverse the individual motives of cadets who were implicated in the 1976 incident, it is clear that the honor system itself had fallen victim to the problems which the institution as a whole was experiencing in adapting to growth and bureaucratization. As noted by the Borman Commission,

Far from being a statement of immutable principles, the Honor Code as defined has become a compendium of changing rules. The body which has been entrusted with the primary responsibility for interpreting and applying the Code—the Honor Committee—annually changes its leadership, thereby precluding development of a stabilizing institutional memory.

Equally troublesome is the fact that the Honor Code has been exploited as a means of enforcing regulations—a view shared by 76 percent of the Cadet Corps in 1974. Cadets and officers have taken the shortcut of placing a cadet on his honor rather than themselves assuming necessary responsibility for the enforcement of regulations. Consequently, the Honor Code, by merging with the extensive Academy regulations, has lost much of its unique meaning. It has become part of the "system to be beaten."[23]

Formalization of the Rite of Initiation

Although the recent major honor scandal was peculiar to West Point, the other academies have not escaped the adverse consequences of the codification of norms and standards of behavior which once were transmitted informally. Problems which all of the academies have experienced in effecting reform of the most intensive phase of the socialization process, which occurs during plebe year, are illustrative. (For simplicity, we use only the West Point–Annapolis terminology in subsequent discussion, recognizing that first-year students are termed "swabs" at New London and "doolies" at Colorado Springs.)

Over a period of decades, academy officials have perceived the need for providing increasingly explicit guidance regarding the conduct of the plebe system. Not only growth but also the susceptibility of the system to abuse made it imperative to codify the operative rules. However, few officials anticipated the costs of codification in terms of the loss of authenticity of the system to its participants.

Like tribal initiation rites, plebe systems at the academies developed not only as means of indoctrinating new members but also as tests of their worthiness and desire to join an elite community. Severe demands on plebes by upperclassmen were thereby legitimized, making the systems vulnerable to chronic abuses. Physical hazing was a recurring problem, instances of which were reported throughout the nineteenth century. However, when two young men died from causes that appeared to have been linked to hazing which they experienced as plebes at West Point, Congress (after an investigation) passed legislation in 1901 proscribing physical hazing at the service academies. Additional limits were imposed at West Point when MacArthur became Superintendent after World War I. He insisted that seniors draw up a formal statement of what was expected of plebes. Designed to curtail bullying and verbal abuse as well as to carry out the congressional ban on hazing, the statement was distributed to all cadets in pamphlet form, as "Fourth Class Customs."[24] In time, the other academies followed the example.

However, the line between legitimized actions which serve to "put a plebe through his paces" and unauthorized hazing has remained difficult to draw. The frustrations which officials at the various academies have experienced in subsequent reform efforts has been noted in the case studies. As enrollments have increased, officials have sought to maintain standardization of practices and centralized control over upperclass behavior which no longer could be monitored simply by the Commandant and an assistant or two. Thus, there has been less reliance on the discretionary judgment of upperclassmen in administering the system, and progressively more on detailed regulations and on supervision by a bevy of commissioned officers at the company and battalion levels working at the direction of the Commandant.

The logical extension of this trend had been envisaged many decades earlier by MacArthur when he was superintendent at West Point. He had entrusted the first several weeks of intense indoctrination of plebes (a phase known as "Beast Barracks") to commissioned officers rather than to upperclassmen. The rationale was that the former would have the maturity which the latter lacked, and would be more likely to set an example in the use of leadership techniques which had applicability to officer/enlisted relations subsequent to the cadet's graduation. However, MacArthur's successor as Superintendent[25] returned the conduct of "Beast Barracks" to upperclassmen, where it remained thereafter—although with increasing officer supervision, as noted. As with the honor systems, academy officials have sought to perpetuate the myth that the plebe systems "belong" exclusively to the students; but officials have been unwilling to accept the risks associated with relinquishing control.

To the extent that upperclassmen retain the responsibility for the day-to-day administering of the plebe systems, despite the intrusion of centralized control and directives, plebe indoctrination remains not only an initiation rite but a game. Demands are imposed which the upperclassman knows the plebe cannot meet fully. He requires the plebe to commit to memory enormous amounts of detailed information from the "plebe bible" (a small handbook for plebes), and to recite even whimsical passages with the precision and solemnity of bringing the message to Garcia. Faltering or failure brings a merciless reprimand from the upperclassman.

"Aren't you required to have that information committed to memory, Mister?"

"Yessir."

"Then why can't you recite it verbatim?"

"Sir, I . . ."

"What's that?!"

"No excuse, sir."

"That's right; there is no excuse! Now, pop up that puny chest. Suck up that monstrous gut. Roll out that horrible sway in your back. Keep those glittering globules straight to the front. And now let's hear you recite the information again; and get it right this time!"

In part, the function served by this unreasonable-demand/harsh-reprimand process is that of mortification. As Erving Goffman notes, the process is characteristic of the initiation into institutions which assume almost total control over the lives of their initiates. The process assures "a deep initial break with past roles and an appreciation of role dispossession," and thus a readiness to assimilate new roles.[26] But this initiation is mock-serious drama at the academies, in which verbal abuse is part of an art form that is cultivated. The norms of the art require that the upperclassmen reproach the plebe so as to sting but not bludgeon; that he be indelicate but not profane. No self-respecting upperclassman would call a plebe a "stupid idiot," for instance—and certainly never a "bastard" or a "son-of-a-bitch" (terms which the upperclassman assumes the plebe is silently according to him). Through a process of trial and error, the academies have learned to reduce demeaning aspects of the plebe system, giving greater emphasis to those designed to instill pride. Moreover, the admission of women has imposed even greater restraints upon upperclassmen in their lexicon of confrontation with plebes. Insults and rebuke remain essential components of the art, however, provided they are introduced with *style*. Thus, alliteration in the command to "Pop up that puny chest" hyperbole in the reference to a "monstrous gut" and the inventive use of the term "glittering globules" rather than "eyes" are evidence of an upperclassman playing his role with flair.

Conversely, if a plebe is to avoid becoming an emotional casualty of the process, he must learn to pick up the cues, knowing when to be reticent and submissive and when to be brash and play for the galleries. A sense of humor and even of the absurd are as essential to his role as is being in physical trim.[27]

Yet in the process of formalizing what had once been an informal rite of passage, much of the spontaneity of the game of plebes matching wits with upperclassmen has been sacrificed. Moreover, requirements for the memorization of vast amounts of information, much of it nonsensical trivia, which in an earlier time were tolerable as part of the game, have become increasingly intolerable as cadets have become subjected to more severe demands in the academic realm. The GAO study of student attrition at the academies, released in 1976, found that "the outline of knowledge requirements at one academy was 30 pages long and referred to four other official documents for specific details on the information to be memorized."[28] Even third-classmen, who bore the major responsibility for drilling plebes on their knowledge, complained about the massive amounts of trivia that plebes were required to commit to memory. At all the academies, the plebe systems often had an effect opposite the one intended, the GAO found. They exerted a demotivating rather than a motivating influence.

The Proliferation of Written Rules

The trend toward formalizing in detail the plebe systems of the academies is closely related to the proliferation that has occurred of formal rules and regulations generally. Whereas a relatively small set of written regulations once sufficed as the formal guidepost of permissible behavior, to be supplemented by verbal injunctions at the discretion of the officers who were in positions of authority, the trend has been one of expanding ad infinitum the list of formal regulations.[29] Over the decades, less and less has been left to chance or to the discretion of academy personnel, as new articles, paragraphs, and subparagraphs are written to anticipate each contingency and to reduce the list of excuses that inventive cadets and midshipmen might mobilize on behalf of behavior that is contrary to official expectations.

In recent years, officials at each academy have begun to recognize that the endless proliferation of formal rules is self-defeating. In action that received front-page attention in many newspapers and coverage on national TV in 1973, West Point officials (most notably the Commandant of Cadets, Brigadier General Philip Feir) issued a revised "Blue Book" of regulations that was actually shortened in length. However, the Superintendent, Lieutenant General William Knowlton, was quick to point out that "The new regulations were *not* designed to liberalize cadet life. They are *more,* rather than *less* demanding."[30]

Griping about rules is no recent development. Cadets and midshipmen have chafed at the regimentation of their lives since the earliest years of the academies. The proliferation of formal rules has provided new sources of aggravation, however. In the first place, academy students correctly perceive that regulation has been extended into minute details of their lives. Secondly, despite the laudable intent in the formalization of many of the rules of reducing opportunities for the arbitrary exercise of authority, the growing length and complexity of formal rule-books has had the ironic effect of reducing the probability that rules will be applied fairly and uniformly. One of every three Coast Guard Academy third-classmen who were surveyed by the General Accounting Office reported that in their view, rules at the Academy were not being applied uniformly *any* of the time. Only 10 percent of those surveyed thought that rules were applied uniformly most of the time. Although the GAO did not give precise data for the other academies, the report cited as an attitude typical of students at each of the academies "that those who survive . . . learn to cut corners without getting caught; those who try to live totally within the regulations rarely make it through an academy."[31]

The large numbers of those who resign rather than endure the regulations, and the cynicism among those who stay and "survive," are equally sources of concern about the health of the academies. Student attrition has risen to above 40 percent at the three DOD academies in the 1970s, and had been above that figure for many years at the Coast Guard Academy. Among those who have remained to graduate from the academies there have been disturbing signs of malaise. For example, a survey of the graduating class of the Air Force Academy in 1970 showed that more than half would not attend the Academy if they had the choice to make again.[32] Trend data were available for West Point, which had been asking a similar question for years to those about to graduate. The date (Table 8) revealed that whereas approximately 90 percent of those graduating from the Military Academy in the late 1950s and early 1960s had indicated that they would make the same choice again if they could reconsider the decision to come to West Point as cadets, by 1963 the affirmative responses had declined to slightly over 60 percent. In 1970 and 1971, fewer than half of those in the graduating class gave responses that indicated that they had made the correct decision in coming to West Point. More recently, there has been a slight upturn in favorable responses; but only 54.7 percent of the Class of 1976 gave responses of "definitely yes" or "probably yes" to the question.

Sources of malaise were varied. It seems probable that the high negative responses from the Classes of 1970 and 1971 in part reflect ambivalence if not regret at being associated with a high-visibility military institution at a period when societal disenchantment with American military involvement in Southeast Asia had reached major proportions. However, it also seems evident that feelings of negativism toward the Academy in the 1970s are rooted in part in widespread cynicism regarding the maze of rules that govern the lives of cadets. For example, only half of the first-classmen of the West Point Class of 1975 were convinced that the Spartan life of the Academy "promotes" or "strongly promotes the development of self-discipline." One-fourth were convinced, on the contrary, that the Spartan life "hinders" or "strongly hinders the development of self-discipline." Perhaps the most telling indicator of the pervasiveness of discontent, however, was the fact that 68 percent of cadets were somewhat dissatisfied or very dissatisfied with official explanations that were being provided of existing procedures and practices.

Table 8 Feelings of Satisfaction or Dissatisfaction with the Decision to Attend the USMA, as expressed by First-Classmen of the Classes 1957–1976.[a]

If you could reconsider your decision, would you now come to the U.S. Military Academy?[b]	CLASS									
	1957	1958	1959	1960	1961	1962	1963	1964	1965	1966
Yes	88.2	89.7	90.0	81.6	88.3	73.7	63.4			
Undecided	—	—	—	—	—	11.6	16.3	c	c	c
No	11.8	10.3	10.0	18.4	11.7	14.0	20.3			
same question	CLASS									
	1967	1968	1969	1970	1971	1972	1973	1974	1975	1976
Yes				47.5	47.3	54.9	49.3	63.0	66.3	54.7
Undecided	c	c	c	18.4	11.6	14.8	17.8	12.1	13.1	13.4
No				32.4	41.1	29.6	32.9	24.1	20.5	31.7

[a]Data are from those reported in USMA, Office of the Director of Institutional Research, *The First Class Questionnaire, Class of 1975* (West Point, N.Y.: USMA, June 1975), item 7; and *The First Class Questionnaire, Class of 1976* (West Point, N.Y.: USMA, Aug. 1976), item 20.

[b]Response categories were as indicated here for the Classes of 1957, 1958, 1959, 1960, 1961, 1962, 1963, 1970, and 1973. Response categories have been collapsed here from those used for the Classes of 1971, 1972, 1974, 1975, and 1976; respondents could answer "Definitely yes," "Probably yes," "Undecided," "Probably no," or "Definitely no."

[c]The question was not posed to these classes.

Conclusions

The simplicity of pre–World War II academy life, when the academies remained small and relatively intimate, must seem to many modern-day academy officials to represent an idyllic era. However, it must be emphasized that such dismantling as has occurred of past practices and approaches (as with the termination of marching to classes, and the abandonment of the lockstep curriculum) typically has been amply justified. Moreover, growth has had its benefits. Only with larger enrollments, for instance, could increases in staff and faculty be justified. Because a qualitative upgrading accompanied the increase, a richer diversity of advanced course offerings became possible. Similarly, larger enrollments necessitated some expansion of physical plant and facilities, but also justified expenditures for sophisticated laboratory, computer, and other facilities to a greater extent than would have been true with fewer persons using the facilities.

The problems which have been described in this chapter have arisen not from growth *per se*. Even bureaucratization *per se* is an insufficient explanation, in the sense that it is possible to imagine the academies having acquired many of the classical bureaucratic characteristics (complexity, specialization, formalization) without having developed problems of the magnitude of the ones that have been experienced in recent years. These problems are more accurately attributable to failures to anticipate and compensate for the probable consequences of the bureaucratization that proved necessary. Instead, efforts were made to perpetuate traditional values (such as discipline and honor) and routines (academic and military) with insufficient recognition (1) of the altered response to indoctrination and socialization that was bound to occur in a transformed organizational context, and (2) of the diversity of perspectives and organizational interests that would develop even among staff and faculty members with growing complexity.

The unforeseen adverse consequences of bureaucratization have not been totally uniform among the four academies. Generally speaking, the Military Academy has experienced the most severe problems, the Coast Guard Academy the least severe ones. Furthermore, there have been notable fluctuations within each academy in the skill and creativity which its top leaders and their subordinates have demonstrated in adapting to organizational growth and other changing requirements.

For the foreseeable future, it appears that academy leaders will be spared the necessity of coping with additional expansion of enrollments or facilities. However, unresolved problems such as continuing high student attrition, and new requirements such as the admission of women make the present era one of intense challenge.

PART IV

Conclusion

11

The Future of the Service Academies

In recent years, the American service academies have been the object of scathing criticism, such as that levelled at West Point by some members of Congress in the wake of the 1976 honor scandal, and of detailed and disturbing reviews of performance, such as that of all the academies by the GAO. They have been subject to far-reaching demands, sometimes shaking the foundations of the organizational subculture, such as the requirement that women be admitted to these previously all-male sanctuaries. Moreover, the academies have faced even more sweeping proposals (several of which will be discussed in this chapter) for alteration of their structures, practices, and goals.

In short, the academies have entered a new era of adaptive challenge. The challenge, it should be noted, is posed not only to those officials and their faculties and staffs who bear the day-to-day operational responsibility for the academies. The challenge also faces those persons in the task environment who are charged with representing the interests of the nation, which the academies were created to serve.

However, the challenges of adaptation are subtle and complex. Organizational adaptation implies responsiveness to demands from the task environment—but not to *all* demands. Organizational leaders must learn to respond to some demands while insulating the organization from others, lest the organization remain in such a state of flux and turmoil that morale and performance are jeopardized. Similarly, from the perspective of those in the task environment who monitor organizational performance, the appropriate question is not: has the organization responded to demands for change? Rather, the pertinent questions are: what kinds of changes have been made, and which practices and structures remain unchanged? There is no inexorable law of adaptation that insures that the structures and practices which an organization discards in the process of change invariably are obsolete ones and that new practices which are adopted necessarily are improvements. Major periods of change in the life of an organization therefore typically also are periods of conflict and anguish both within the organization and within the task environment; proponents and opponents of particular changes are able to argue with equal conviction that the proposals under consideration (1) are essential to the attainment or maintenance of organizational legitimacy and excellence of performance, or (2) if adopted, would undermine organizational vitality and effectiveness.

Periods of important ferment and change at the American service academies have been characterized by a clash of views of precisely this sort, as illustrated by the case studies. Consensus internally and externally prevailed as to the need to meet the requirements of the post–World War II environment. Nearly everyone in positions of authority at the academies and in their task environment came to recognize the need for important departures from the traditional seminary-academy model. But there was seldom a broad consensus that particular changes would constitute progress, any more than there was an unfailing consensus that the maintenance of particular traditions constituted wisdom. Moreover, by no means is there full confidence at the academies or in their task environment currently that the blend of change and tradition that has evolved in the optimal one.

The key dilemma remains that of reconciling the Athenian goals of a four-year undergraduate educational program with the Spartan goals of military training and professional socialization. Faith remains strong among academy leaders that the synthesis must be achieved, whatever the deficiencies of the present system. As Vice Admiral James Calvert has observed,

> . . . if [the service academy formula] has any genuine relevance [it] has to depend upon that element of discipline; and if you take that element of discipline out, then that synergism between discipline and academic endeavor which has been the essence of the West Point and Annapolis idea for 100 years is gone. So . . . this is what I mean by the balance between Athens and Sparta. Without the academic effort there, it's just Parris Island, you know, dressed up. . . . Without the discipline, it's just Harvard with blue suits.[1]

The troubles which the academies have experienced in recent years, however, raise a nagging question. In their quest to be both Sparta and Athens, both Parris Island and Harvard, have the academies become an imperfect mixture that lacks the essential elements of either? If the answer to the question is yes, then mere tinkering with curricula and reshuffling of organizational routines will not be sufficient for the preparation of professional military officers for the complex demands of the late twentieth century. More fundamental changes in academy structures, programs, and perhaps goals will be needed.

What is desirable is not necessarily what is probable, however. One cannot be certain what the pace, scope, magnitude, and direction of change at the academies will be—even within the next five to ten years. There are too many imponderables, both in terms of unforeseen contingencies that affect national security demands, and in terms of personalities who may assume positions of authority with an impact on the academies that is not fully predictable. However, the analysis in the preceding chapters of patterns of continuity and change at the academies, supplemented by relevant findings from studies of change in other organizations, provides a basis for assessing the probability or improbability of each of a number of alternative scenarios of policies that might be adopted by or imposed upon the academies in the coming decade.

In the remainder of this chapter, a wide range of alternative scenarios is examined. Some options are too implausible, however, to merit detailed consideration. Turning the clock back to the seminary-academy model is simply not feasible, for instance. Nor is it desirable, nostalgic mythology to the contrary notwithstanding. The seminary-academy had the virtues of intimacy, cohesion, and camaraderie. But the lockstep curriculum, the juvenile pranks and more serious abuses of the plebe system that were chronic, and the effective isolation of academy staff and faculty members from civilian higher education were elements that make the seminary-academy an inappropriate model, given the complex needs of the late twentieth century.

At the other extreme, the abolition of the academies is not a realistic policy alternative. It is true that high rates of student attrition, and performance failures such as the 1976 honor scandal at West Point, have been of sufficient severity to call into question the raison d'etre of the academies. It is also true, and important, that the academies are but one source of armed forces officers. Precommissioning education and training are provided through OCS programs and ROTC programs at much less cost than the per-student cost of academy education. Conceivably, the physical plants and facilities of the academies could be sold or leased, although portions of the grounds and buildings might be maintained as historical sites or museums. Alternately, one or more of the advanced service schools, such as a command and staff college or a war college, might be moved to the academy site and take over the facilities.

Proposals such as these have cropped up with predictable regularity in periods of intense, popular antimilitary sentiment throughout the life-span of the academies. But proposals to abolish the academies completely have less plausibility now than they did, say, in the nineteenth century or even as recently as the end of World War II. Within the past ten to twenty years, the Congress (and thus the American taxpayer) has invested millions of dollars in new buildings and modern facilities at each of the four academies. Despite the per-student cost that the maintenance and operation of such facilities entails, the sunk cost of the initial investment represents a persuasive argument on behalf of "seeing the investment through."

Moreover, aside from the issue of the economic investment that has been made, an important question which would surface if a proposal for abolishing the academies were to be considered seriously would be: what would the effects of such action be on the morale of the armed forces and on public confidence in the military establishment? One of the most visible aftereffects of the prolonged American involvement in Southeast Asia, followed by the revelations of high-level duplicity that were associated with the Watergate scandal, was a profound loss of popular confidence in governmental institutions. A modicum of popular skepticism regarding government, and especially its military component, may be essential to the maintenance of democratic practices. But deep and prolonged mistrust is likely to have cumulative effects that not only weaken the nation's defenses but also undermine democratic institutions. A military profession which, over an extended period, is the object of widespread popular hostility not only will have difficulty recruiting able and dedicated new members; it also will lose through resignation many of its talented and committed current members. Those who remain are likely to be estranged from the society which they are expected to serve.

By the late 1970s, public confidence in its governmental institutions had risen slightly beyond the low ebb of the second Nixon administration. However, the outright abolition of the service academies surely would be interpreted among the public and among members of the military profession as a judgment by the nation's legislature that a key element of the military profession no longer warranted trust. Quite apart from the function which the service academies perform as tourist attractions and historical shrines (which could be maintained to a degree even if the operational activities of the academies were abolished), the academies are a symbol of the American military profession. In making such an argument, it is not necessary to subscribe to the view which zealous proponents of the academies sometimes advance that academy graduates diffuse the key values that have been instilled in them to their nonacademy colleagues throughout the service. Rather, one need only observe, minimally, the linkage that is maintained in popular thinking between the ideals which the academies represent and the values of the military profession as a whole. To abolish the academies would be perceived, therefore, not merely as a signal from Congress that the academies themselves no longer were worthy of support; it also would seriously call into question the continuing legitimacy of the military profession as a whole.

Although the abolition of the service academies is extremely implausible, it is worth noting that the total maintenance of the status quo at the academies is an equally untenable policy option in the coming decade. The seeds of at least minimal further change have already been planted. As the ratio of women to men at the academies increases, for instance, further modifications of programs are highly probable. Similarly, continuing changes in technology and in demands upon the military profession inescapably will lead to requirements for further modifications of academic coursework and military training.

Change is inescapable. But plausible alternative scenarios of the future of the academies range from ones in which their structure and mission are altered greatly to ones in which present programs and practices remain largely intact with change proceeding incrementally. Although a number of scenarios might be postulated, each having some plausibility, it is useful to consider the relative likelihood that each of the plausible options might become a reality. In doing so, it is unnecessary to create alternative futures out of whole cloth. A number of specific proposals for change have been made in recent years by participants in the policy process or by concerned observers of the academy. The discussion that follows expands upon or highlights particular features of the various proposals in order to explain the reaction which such proposals have elicited, or are likely to elicit, from academy officials and from persons in authority in the task environment. Even those whose probable reactions are described below may object that the policy consequences of such reactions would not be fully consonant with the vision which they have of the contribution of the service academies to American society. What follows is an assessment of the relative probabilities. It is intended neither as a declaration of the author's preferences, nor as a deterministic argument of inevitability. Hope lies in the belief that clarification of the probable consequences of alternative courses of action will serve to enhance the likelihood that choices that are made among alternatives will be ones that are responsive to the needs of the nation.

Scenario One: A Combined Services Academy

The revolutionary changes in technology which occurred during World War II, the unprecedented involvement of the United States in world affairs in the early postwar era, and discussions of the unification of the armed services combined to generate a number of far-reaching proposals for change at the service academies. The Service Academy Board put most of such proposals to rest. However, it is testimony to the severity of the problems which the academies have experienced in recent years that a number of change proposals similar in form and magnitude to those advanced after World War II have been advanced.

In a book on the American military establishment published in 1971, Edward Bernard Glick proposed that prospective career officers in the armed services spend the first three years of their undergraduate education and training in a combined services academy, with two additional years spent in one of the currently existing academies associated with a particular arm of service. Glick repeated the proposal in another source in 1977, suggesting that the sequence might require five or even six years.[2]

Proposals such as Glick's were relatively abundant during the period of planning for the postwar era. For example, in 1944, Harold Smith, the U.S. Director of the Budget, called for the creation of a combined services academy which would replace those at West Point and Annapolis, and would obviate the necessity for creation of a new academy for the soon-to-be independent Air Force.[3] However, even the more modest proposal by General Dwight Eisenhower that West Point cadets spend a year of study at Annapolis, in exchange for a comparable number of midshipmen who would be spending a year at West Point, was rejected by the Service Academy Board in favor of its endorsement of the existing system, to be supplemented by creation of a separate Air Force Academy. As noted in chapter three, West Point and Annapolis did endeavor through curricular emphasis on interservice cooperation, through student exchanges for periods of several days, and through combined training exercises during the summer following the sophomore year, to reduce service parochialism. It may be argued that such efforts were too limited to counteract the effects of living for four years within a milieu where service pride is cultivated and rivalry with the other academies, especially in athletics, often is intense. Still, as the analysis in this book has shown, the service academies today are experiencing a number of problems far more serious than that of service parochialism. Merger proposals, or variations on the theme such as the Glick plan, are largely irrelevant to such pressing problems as student alienation and cynicism. Indeed, if the cadet or midshipman role were to be extended into a fifth or sixth year, such problems probably would become even more acute.

However, Glick's proposal reflects not what he feels is optimally desirable but rather what he feels might be a politically tenable means of remedying continuing tendencies at the academies toward a narrow, technical approach to education, which large numbers of cadets and midshipmen reject. Thus, Glick suggests that the first few years of academy education might be spent in a National Service Academy, which would include as students not only prospective military officers but also young men and women training for jobs in the State Department and the civil service.

Laurence Radway has identified a similar option.[4] Generally supporting a "civilian tilt," he notes that one possibility would be to transform the academies into "universal service institutions charged with the function of preparing young people for a range of public careers—in diplomacy, the police, the civil service, and international organization, as well as the armed forces." Radway is keenly aware of the plausibility of alternative change scenarios but, like Glick, favors the "civilian tilt" approach as being preferable to the present system and probably feasible politically. Radway's views carry special weight because of his pioneering studies in military education and his long and continuing contacts with the service academies as a consultant.[5]

Scenario Two: Mixed Civilian-Military Collegiate Experience

On balance, however, the probability that a universal service or combined services academy will be created in the coming decade appears low. Radway himself notes that the major objection to the "civilian tilt" is that the academies might thereby

> lose the advantage of product differentiation. The more they resemble civilian universities, the more likely that critics will ultimately ask whether it is necessary to have them at all, especially since they are so expensive. A graver risk, should the threat of war increase, is that they may appear inadequate in preparing men for what must still be regarded as an essential social function.[6]

A compromise solution to that problem is to include civilian education in the precommissioning format, while retaining a role for the academies. Such a compromise has been suggested by Richard F. Rosser, president of DePauw University and a retired Air Force colonel. Rosser, who formerly headed the political science department at the Air Force Academy, would have prospective career officers in the armed forces spend three years in a civilian institution of higher education, followed by two years in a service academy. The rationale would be that students would gain a better understanding of civilian outlook through the three years spent on a civilian college campus, and as solid an educational experience, if not more so, than they would have gained during their first three years at the academy. Not only could the process of professional socialization be accomplished within the two years that cadets and midshipmen spent within the academy milieu, but probably it would be more effective because proportionately more of those present would have made an enduring career choice by virtue of having until completion of their junior year to make the commitment. Moreover, motivation would be sustained more effectively with the formal cadet or midshipman role lasting only two years rather than four. Furthermore, the cost to the taxpayer of funding academy education and training would be greatly reduced (although not halved, since Rosser anticipates doubling the number admitted to each class from present policies). Anticipating the admission of women into such a program, Rosser's 1973 proposal provided that

> At the end of the fifth year the individual would receive a bachelor's degree from his or her parent institution and a professional degree from the academy. This practice long has been common with the so-called 3-2 programs some of the best liberal acts colleges have had with colleges of engineering.[7]

Like the Glick suggestion for including experience in a combined services setting, Rosser's proposal is one that has precedents in the postwar planning of the 1940s. For example, in 1948 a group of Air Force officers who had been assigned the task of proposing the best format for academy education for Air Force officers recommended a five-year program consisting of two years of study at civilian institutions followed by three at a newly created Air Force Academy. The Air Force Chief of Staff, General Hoyt Vandenberg (USMA 1923), rejected the recommendation and directed instead that planning for the

Air Force Academy be based on the assumption of a four-year course of undergraduate instruction.[8]

Rising costs of service academy education, among other concerns, perhaps give such proposals for a system of education that would combine civilian college and academy experience more chance of serious and sustained consideration at high policy levels than they had in the early postwar years. However, if the 3-2 plan such as the one Rosser proposes would require the designation of selected cooperating civilian colleges and universities, it would run the risk of relying on an unrepresentative institutional base, a problem that already plagues ROTC programs to a degree. In any event, the trend elsewhere in professional education (for instance, in law, medicine, engineering) seems to be away from 3-2 plans, in favor of requiring students to complete their undergraduate education before entering the professional school. If the armed services were to follow this trend, the mission of the service academies would be changed to that of providing postgraduate education and training.

Scenario Three: The Academies as Postgraduate Institutions

One of the early advocates of such a change of mission for the academies was Representative Melvin J. Maas, a member of the Board of Visitors to the Naval Academy on the eve of World War II. In a minority report that was included with the Board of Visitors report in 1940, Maas argued that it was impossible to reconcile satisfactorily the task of providing a solid undergraduate education with the task of providing the specialized training and discipline which future naval officers require. He suggested that the Naval Academy accept only applications from young men who had completed their undergraduate education, and that Annapolis provide a two-year professional program devoted primarily to technical studies.[9]

The notion that the service academies should provide postgraduate rather than undergraduate education was rejected, not only by the majority of the Board of Visitors on which Maas participated, but also in emphatic terms (along with other alternatives to the existing system) by the Service Academy Board in a preliminary report in 1949 and again in its final report in 1950. The twofold argument which the Service Academy Board made was (1) by recruiting young men no more than a year or two out of high school, the academies got students at an age when they "are most receptive to training, to motivation and to learning [the essential] qualities of leadership and service . . ."; and (2) that the academies should concentrate on providing a general educational background, which would enable graduates to pursue more specialized postgraduate studies in other institutions in any of a number of fields.[10]

Although the report of the Service Academy Board sufficed for many years to head off renewed advocacy of the transformation of the academies into postgraduate institutions, in the 1970s a flurry of such advocacy has reappeared. The quantity as well as the quality of recent proposals suggest that a postgraduate mission for the service academies may be an idea whose time has come.

Each of the various arguments that have been made in recent years for making the academies postgraduate institutions has its own distinctive elements. However, an illustrative change scenario incorporating some features of each of several presentations, would look like this:

—Provisions would be made to enable students to get a taste of military training, and an orientation to the demands and opportunities associated with a military career, through ROTC programs, some of which might serve as regional centers for a number of colleges and universities rather than being affiliated with just one.

—Those students who displayed aptitude for military service through ROTC would be encouraged to consider seeking a regular service commission through the postgraduate education and training at the academy. However, provisions would be made also for lateral entry into the pre-academy program, even after graduation from college (with remedial programs available to compensate for requisite training that was missed).

—One or more eight-week periods of summer training at the academy would be provided to prospective applicants, during or immediately after their undergraduate experience.

—Postgraduate education and training of twelve to eighteen months' duration would be provided at the academy. The petty harassment that currently is associated with the distinction between upperclassmen and fourth-classmen, and with the concern of the academies as undergraduate institutions with *in loco parentis*, would be eliminated. However, the program of professional academic and military training would be rigorous, with high standards maintained. Those who completed the program successfully would be commissioned and granted a professional degree.

There are at least three reasons why it seems probable that proposals for such a change in the mission of the academies will receive more serious attention in the near future than they have received in the past: (1) Increased reliance on objective assessment criteria, rather than on subjective judgment and on the mere invocation of historical accomplishments, makes it more difficult for military officials to dismiss without consideration arguments for a change to a postgraduate mission. (2) The most salient of recent proposals for having the academies transformed to postgraduate institutions have been advanced by persons who have had close association with one or more of the academies over a period of years; some are well known and respected within the military profession. Thus, the familiar rationale that often is mobilized in the dismissal of radical-change proposals by "outsiders"—"they don't really understand the problem"—is inapplicable. (3) The authors of recent change proposals have identified a disturbing number of serious problems at the academies which might be remedied or at least greatly alleviated by a change of mission. Members of Congress and military officials might shy away from the radical step proposed to remedy the problems; but precisely because the problems are serious, it seems likely that the proposals will receive high-level attention. Each of these three factors requires brief further elaboration.

Performance-Assessment Criteria

Throughout most of the post–World War II era, military officials were able to ward off any critical queries about the performance of the service academies by letting the historical record "speak for itself": Pershing, MacArthur, Eisenhower, Bradley, Patton, Arnold, Dewey, King, Nimitz, Halsey, Leahy, Burke, Waesche. . . . Merely to invoke the names, and thus the memory of celebrated wartime performance, often was sufficient for military officials in the 1940s and 1950s to convince themselves that no further critical assessment of the academies was needed, and to squelch those who thought otherwise.

However, in the 1960s and 1970s (roughly the era since McNamara), more often than not military officials have been required by civilian superiors in the Executive branch and by Congress to justify their programs and institutions in terms that permit concrete comparisons with policy alternatives. Many military officials have complained, with some justification, that to yield to such demands completely is to provide but the illusion of precision. With regard to the academies, for example, one might accurately assert that some of the most important effects of academy training and of the process of socialization that occurs over the four years are not reliably reducible to mere statistics. Efforts to identify the beneficial results of the academy experience, weighing these benefits against the cost of academy education, and comparing the cost-benefit ratio to that of ROTC or OCS programs must confront the intangibility of key performance goals, such as instilling in would-be professional officers a sense of duty, personal integrity, and loyalty.

A related point is that whereas Annapolis and New London, especially, once were concerned primarily with preparing midshipmen and cadets to assume positions as junior officers, in recent years all four academies have defined their mission in terms of "the long haul." Thus, their effectiveness or ineffectiveness is ascertainable only through the measurement of performance throughout a career; and even if one had indicators for measuring such performance in which one had confidence, how could one be sure that results obtained were applicable to comparison of the effects of the academy experience with that of ROTC and OCS?

There is no need to belabor the point. The measurement of effectiveness of a process which includes intellectual development, character building, and military training is elusive. Although frequently frustrated by the apparent desire of the General Accounting Office and other monitoring agencies to reduce organizational performance measurement to an exact science, the truth is that over the past two decades, academy officials themselves often have succumbed to the manipulation of sometimes meaningless data in the quest for a demonstration of "progress."

James D. Thompson's discussion of the foibles of organizational assessment is relevant. As he notes, the problem of assessment for an organization is particularly acute when ambiguity exists regarding the end product most highly desired by salient elements of the task environment (for example, military officers instilled with a keen sense of discipline, or with a high degree of intellectual creativity?) and when knowledge of causality is incomplete (what *are* the long-range effects of a "tough" plebe year?). In such situations, organizational leaders tend to rely on reference group comparisons ("We are competitive with the other academies," or "GRE scores of our graduates were above the mean for graduates of a large national sample of civilian colleges and universities"). Similarly, they try to demonstrate that the organization "scores well" according to criteria that are highly visible to important elements of the task environment. Thompson notes, for example, that

> Public schools are sensitive to measures of pupil hours in various activities, to student-teacher ratios, to drop-out rates, or to dollars spent per pupil per year. None of these is necessarily a reflection of intrinsic achievement or preparation for future intrinsic value, but they are visible criteria from which improvement can be demonstrated.[11]

Similarly, extrinsic standards have been the guide to which service academy officials have tended to turn in recent decades. The percentage of an entering class who had been high school class presidents or valedictorians, those who had been Eagle Scouts or varsity athletic letter winners, and performance on College Boards (SATs) demonstrate the degree of success from one year to the next in recruitment. The excellence of the academic program is demonstrated by measures such as the performance of members of the graduating class on the nationally standardized Graduate Record Examination, by the number in a given class who were awarded Rhodes Scholarships, and by the number of publications by faculty members during the preceding academic year. Student participation in an activity such as airborne training on a voluntary basis is cited as evidence of professional motivation on the part of cadets and midshipmen. The particular measures which academy officials mobilize vary somewhat according to the audience (increases in the percentage of Ph.D.'s on the faculty for boards of visitors, for instance; the won-lost record in varsity sports for alumni gatherings). Of course, the data which academy officials provide may be determined by the dictates of a review group rather than by academy initiative. Thus, academy officials have had to provide data such as those on student attrition to the GAO and other investigating bodies despite the adverse implications of such data.

Ironically, now that academy officials themselves have become so reliant on extrinsic measures of the "worth" of the academies, they are more susceptible than in the past to criticisms which rest heavily on quantifiable indicators of performance, despite the limitations of such measures. Especially salient are indicators that are highly visible to the task environment, such as student attrition, postgraduation attrition of academy graduates (especially as compared to attrition among officers of other commissioning sources), and per-student costs of educating academy graduates (again, as compared to the comparable costs of ROTC and OCS programs). Thus, the persuasiveness of the recent proposals for transforming the academies into postgraduate institutions derives in part from the use of these and other indicators to demonstrate the magnitude of the problems which the academies face in their current format.

Critic Credibility

But the probability that the transformation of the service academies into postgraduate institutions will be considered seriously within the military establishment itself, rather than dismissed out of hand, derives also from the credentials of those who have expressed variants of the idea. Although the idea is one which has been broached in some popular writings and in informal discussions among military men and students of military education, recent proposals which merit special attention are ones which have appeared in military or scholarly journals by authors who have had close ties to the academies and/or to military education.

Of particular note have been proposals in 1970 by Paul R. Schratz; in 1972 in separate publications by Frederick C. Thayer and by Fred Harburg; in 1975 by Paul M. Regan; and in 1977 in a co-authored piece by Roger A. Beaumont and William P. Snyder.[12] Most of these authors are academy graduates. Schratz (USNA 1939) retired from the service as a Navy captain, Thayer (USMA 1945) as an Air Force colonel, and Snyder (USMA 1952) as an Army colonel, each leaving a military career after extensive service to pursue an academic career. Each had gained wide respect among military colleagues through highly rated performance in military assignments. Each brought to his analysis of military education the added credentials of a Ph.D. (from Ohio State, Denver, and Princeton, for Schratz, Thayer, and Snyder, respectively) and experience in the civilian academic world. Beaumont, Snyder's co-author, a graduate of the University of Wisconsin with a Ph.D. from Kansas State, is a historian who has written a number of articles and books on contemporary and historical military experience. Although Harburg, who was writing as a cadet in his fourth year at the Air Force Academy, lacked the status of Schratz, Thayer, Snyder, or Beaumont, his analysis had the authenticity in discussion of cadet motivation that comes from immediate first-hand experience. Similarly, Regan could support his argument with recent and direct observation from his three years of service as a member of the officer-crew of the cadet training ship *Eagle*, and as a junior faculty member at the Coast Guard Academy.

Each of the analyses gained some further credibility among military readers because of the outlet that was utilized for publication. None of the authors published his recommendations in a popular magazine or newspaper. Obviously, none was headline-hunting. Instead, each was concerned primarily with laying his argument before an audience the vast majority of which would be sympathetic to and extremely knowledgeable about the academies. The articles by Schratz and by Regan appeared in academy alumni publications. That by Harburg appeared in the cadet magazine of the Air Force Academy. The Thayer piece first appeared in *Air Force Magazine*, and was reprinted in the Air Force Academy alumni magazine. The Beaumont and Snyder article appeared in the scholarly *Journal of Political and Military Sociology*.

Disturbing Diagnosis

The authors have in common affection for the academies, but also a conviction that the present undergraduate format is not working well. The latter is a conviction that each communicates forcefully. Although the emphasis of argument and the analysis on which it rests differ somewhat from one author or set of authors to another, a synthesis of the reasons advanced for effecting a transformation of the academies from undergraduate institutions to postgraduate ones is as follows:

—Professional training in the modern era requires a degree of specialization that is appropriate to the postgraduate more than to the undergraduate model. (Schratz; Harburg)

—The education that the academies do provide currently is, as Regan puts it, "thoroughly average—satisfactory but not excellent."[13]

—Relative to alternatives to the academy model for providing undergraduate education to prospective career officers (such as college-based ROTC programs), the cost of academy education is excessive. (Harburg; Regan; Beaumont and Snyder)

—Although in an earlier era it could be argued that the undergraduate preparation provided by the academies was so distinctive that it could not be replicated at a civilian institution, that argument has been greatly undermined by the "civilianization" of academy curricula. (Schratz; Beaumont and Snyder)

—Contrary to the argument advanced by the Service Academy Board and perpetuated by service academy officials that it is essential to "get 'em young" in order to tap youthful energies and enthusiasm and instill career motivation, recruiting young people just out of high school to the academies insures that the academies will suffer relatively high student attrition. Young people of college age typically are still developing career ideas and frequently change their minds. (Thayer; Beaumont and Snyder)

—By maintaining an undergraduate program, with the necessary regimentation associated with military discipline and the felt need to maintain all vestiges of *in loco parentis*, the academies are forced to cope with problems that are less attributable to faulty leadership than to youthful rebellion and the search for individual identity. Such problems would less frequently reach headache proportions if the academies were dealing with young people at a more advanced stage in the maturation process. (Regan)

—Moreover, the obsessive measurement by academy officials of virtually all facets of the behavior of cadets and midshipmen to determine "aptitude for the service" is bound to be largely an exercise in futility when those whose behavior is assessed are so young and inexperienced. (Thayer)

—A related point is that the commissioning system requires an arbitrary determination that *all* of those who graduate from a service academy, however marginal their performance may have been, should be given regular (rather than reserve) commissions in the armed services, whereas graduates from ROTC or OCS programs who by every other measure may be superior in qualifications to some academy graduates may be denied a regular commission because insufficient slots are available. (Thayer; Beaumont and Snyder)

—The effort to combine a four-year program of undergraduate education with a process of military training and socialization leads inexorably to conflicts (Athens-Sparta tensions) which reduce the effectiveness of both processes. Moreover, habits of thought and behavior are engendered which are antithetical to the professed goals of the academy. To wit, many cadets and midshipmen learn over time to find ways to "get by" in meeting the various requirements, rather than striving for excellence; many cultivate a highly legalistic approach not only to regulations but also to honor codes, rather than internalizing the spirit of each; many take a perverse delight at finding ways to "beat the system," complying when they must rather than developing a deep sense of self-discipline and identification with the academy's purposes. (Harburg; Regan)

—Finally, one might add the argument that Maas made in 1940, that the undergraduate academy format, by taking students away from the civilian sector at age seventeen or eighteen, makes it less likely that prospective military officers will acquire a sensitivity to civilian perspectives than would be true if such candidates had completed college before beginning their academy training.

Scenario Four: Continued Incrementalism

For the reasons enumerated, proposals for changing the mission of the service academies from an undergraduate to a postgraduate one of education and training will receive high-level attention in the Executive branch, in Congress, and perhaps in the mass media in the years immediately ahead. It is conceivable that the Coast Guard, being relatively free of the quest for standardization that makes autonomous major change in the academy of one of the other services unlikely, would actually institute such a change. The responsibilities of the

Coast Guard as a whole are expanding, including such challenging assignments as enforcement of the 200-mile fisheries limit, and of regulations for the control of pollution from ships and tankers. A sense of urgency could lead the Coast Guard to redefine the mission for its academy; moreover, this could be accomplished with somewhat less agony and turbulence than at West Point, Annapolis, or Colorado Springs because of the relatively small number of personnel and facilities to be redistributed at New London.

However, the change scenario that is most probable over the next five to ten years at all four academies is one of incrementalism, cautious or bold, rather than the more radical scenario of change of mission. Both our analysis of the pattern of change at the academies in the past and findings from studies of the change process in other complex organizations support such a conclusion.

As noted above, the benefits to be derived from altering the military educational system so that prospective career officers receive undergraduate education at civilian colleges and universities (possibly with summer military training at the academies), followed by postgraduate education and training at the service academies, may be many. But in the first place there are many important questions which even the more detailed of the change proposals (Regan; Beaumont and Snyder) do not answer. For example, what specifically would be the content of the curriculum of a postgraduate academy? Which among the current personnel billets would be needed in a postgraduate academy? In particular, how many current staff and faculty members would have to be released or reassigned with the change of mission? How many new faculty members would be needed, and what would be the cost of hiring them? Which current academic facilities would be needed? What would be done with facilities that were not needed? Similarly, what would be done with expensive athletic facilities, such as football stadiums, which presumably would not be utilized by the postgraduate academy?

Staff studies might be done which could provide reasonable estimates in response to the preceding questions; but there are other important questions that are even more elusive. What would be the effects of such a change on the recruitment of prospective officers into the armed forces? Would the opportunity for all those who would receive regular service commissions to become academy graduates increase the numbers who would seek professional careers in military service? Or might the number of those seeking military commissions decline precipitously? How effective would the combination of military training during the summers for prospective officers who were undergraduates followed by a postgraduate academy experience be in providing socialization into the military profession? Would the sense of duty and integrity among academy graduates be less than among academy graduates under the current system? Would the commitment of academy graduates to a military career be lower under the post-graduate format than under the present one?

In the relevant policy arenas (the headquarters of the parent arms of service; the Departments of Defense and Transportation; committees of Congress), among those who do not ignore these questions altogether, one can expect three general types of response. First will be those vehemently opposed to any change from the current system, for whom the above questions will serve to trigger a parade-of-horribles: "the services would not attract qualified recruits if the change of mission of the academies were made"; "a postgraduate academy would fail to instill a sense of duty, honor, and loyalty"; "the vital link to the past that has come from generations of academy graduates would be severed"; and so forth. At the opposite end of the spectrum, those who have been convinced by other arguments that the change is needed will be inclined to speculate optimistically about the numbers of young men and women who will find attractive the new format of precommissioning training, linked to postgraduate academy experience, and about the quality of those sent through such training and experience. In between will be those (the "swing vote," if you will) who simply admit to uncertainty regarding the answers to these questions. But incremental change, by definition, leaves intact most of the structures and procedures with which one is familiar, whereas a change of mission opens a pandora's box of uncertain consequences. Therefore, doubt is likely to be resolved in favor of the lower-risk incremental strategy.

With organizations steeped in tradition and possessing a concept of mission deeply rooted in the quasi-religious terrain of the seminary-academy, the burden of proof on those who would propose a departure from the current organizational mission is especially great. Academy officials and others in positions of authority who have had long association with the academies tend to regard flaws that are detectable in the existing system in the manner that Edmund Burke viewed defects in the institutions of government, like "the wounds of a father," to be approached only "with pious awe and trembling solicitude."[14]

Even among policy makers whose view of the academies is basically pragmatic and governed by secular considerations, the tendency to resolve any doubt regarding the probable consequences of change away from comprehensive proposals in favor of incremental ones will be compelling. For the policy maker, hope lies in the prospect that if the first increment of change that is instituted proves insufficient for coping with existing organizational difficulties, the second increment, or the third, will do the trick. In the meantime, most of the existing academy structure and most procedures will be retained intact, rather than incurring the risk (associated with comprehensive change) that the virtues of the present system, whatever its defects, would be scrapped in favor of an untried system with unknown results.

Moreover, proponents of incrementalism are bolstered by the inertia created by the considerable investment that Congress has made in the recent past in improving the academies as undergraduate institutions. Inertia derives not simply from the sunk dollar costs. Rather, each major investment also represents a set of policy expectations regarding the conditions under which the investment will be realized. The return that society and the military profession already have realized from the recent investment helps to sustain the argument for "seeing the investment through" by adhering to a strategy of incremental rather than comprehensive change. The return on the investment is evident not only in some splendid buildings and some superb educational facilities that were developed at Colorado Springs for the new Air Force Academy and at the other three academies with the expansion and modernization programs of the past ten to fifteen years, although these are one important component of the investment. Computer centers, closed-circuit television capabilities, well-equipped laboratories for the basic and applied sciences, expanded library resources, and similar developments provide the wherewithal for a first-rate educational experience. Similarly, the latest in weaponry, electronic equipment, military vehicles and vessels, and training aids are available for military training. But the investment also had been made in improving the quality of human resources, most notably by upgrading the permanent cadre of faculty members at each academy through the funding of postgraduate education for them, in a high percentage of cases to the Ph.D.

The transformation of the academies into postgraduate institutions might still permit utilization of most of the physical and human resources that have been developed. However, proponents of such transformation will face the argument that the dividends are just beginning to be realized on the investment in expanded undergraduate educational opportunities. Consequently, it will be argued, this is a most inopportune time for the abandonment of the undergraduate format. If important deficiencies are detectable in the educational experience which cadets and midshipmen derive, despite these expanded opportunities, it would be far better (it will be said) to make incremental changes in order to better utilize existing resources within the undergraduate format. For example, the talents of the faculty might be utilized more effectively if faculty members at all ranks were given a greater voice in academic policy making.

Of course, proponents of the change of mission to postgraduate education for the academies (at least those proponents cited above) dispute neither the excellence of the educational and training resources and facilities of the academies, nor the quality and dedication of the staff and faculty. Several do note the lack of teaching experience of a relatively high percentage of faculty who are assigned on a tour-of-duty basis. Furthermore, proponents of a change of mission do argue that, considering such phenomena as the relatively high pregraduation and postgraduation attrition rates, the performance of the academies is inadequate to justify maintaining them as four-year undergraduate institutions, given the cost of this system relative to alternatives. Moreover, some argue that the effort to reconcile the demands of four years of undergraduate education with those of military discipline, training, and professional socialization ensures that neither of the dual missions will be performed optimally.

But the probability that a "can-do" view of the Athens-Sparta dilemma will prevail is high. The can-do component of the professional ethos of the American military is manifested in a tendency to define organizational problems as ones that are reducible to leadership failure. Conversely, it is assumed that existing problems can be remedied by the application of the appropriate leadership technique—more aggressiveness, more openness, more clarity of communications, more decisiveness, or what have you. To the extent that the military has become more sensitive in recent years to the human relations movement in management theory, such views are augmented still further.[15] The can-do view tends to minimize or neglect altogether the possibility that the "givens" of organizational mission and structure (such as a four-year time frame for carrying out the professional socialization process) may impose inherent limitations on the effectiveness of leadership, regardless of ability and style.

Of course, one might argue that despite the proclivity of military leaders for incrementalist, can-do approaches to coping with problems such as those which the service academies currently are experiencing, Congress could adopt a more sweeping approach to reform. After all, the authority to change the mission of the service academies rests with Congress, not with the military. Congress used such authority when it mandated the admission of women to the academies, despite considerable opposition on the part of academy alumni and military officials to such a policy change.[16]

However, if on the one hand the legislative act granting admission to the academies to women demonstrates the capability of Congress for relatively radical reform, on the other hand the act is further evidence of the expectation on the part of Congress that for the foreseeable future the undergraduate format of the academies will be maintained. The arguments that were made above regarding the inertia that has been created on behalf of the undergraduate system at the academies by recent investments in it apply with special force to Congress, where appropriations are the key to the oversight function.

Moreover, in their approach to reform, members of Congress, like military officials, tend to prefer disconnected incremental remedial measures to overall designs for radical surgery. This preference stems in part from the human predilection in problem solving to opt for simplicity over complexity. Appraisal of the probable consequences of an incremental solution involves the assessment only of a limited number of variables and of uncomplicated causal relationships, as opposed to the more complicated appraisal of the probable consequences of more comprehensive reform measures.[17] But the preference for incremental solutions stems also from political experience and know-how. Workable compromises in support of particular incremental reform measures often can be put together even among groups and factions with widely disparate perceptions of the nature and magnitude of the problem at issue, and with sharply contrasting value priorities. Comprehensive measures, in contrast, are ones to which a wide variety of groups may object for differing reasons. As the risks from change becomes broadened and magnified, the feasibility of achieving a consensus becomes reduced.

Bold Incrementalism

Nevertheless, to say that changes at the service academies are likely to be incremental over the next five to ten years is not to say that they will be trivial, nor is it to assume that they will prove totally ineffectual. If a sense of urgency is felt at the academies or in Washington within the next several years in response to new organizational crises such as cheating scandals or to alarming trends such as a sharp increase in student attrition, bold action might be taken and might achieve demonstrable results.

Cautious incrementalism, the typical strategy for change when a sense of urgency is lacking, is applied only to matters of structure or procedure about which the probability of eliciting organizational consensus is high. Bold incrementalism, in contrast, is applied to facets of organizational endeavor where the need for reform appears to be most acute, despite evidence that the reform efforts will antagonize sizable numbers of organizational personnel. The following are examples of bold incremental changes that are plausible at one or more of the academies over the next five to ten years, if the sense of urgency is felt.

—The Academic Board at West Point would be reduced in size. Rather than the current system in which each department head has a permanent seat on the board, the change would provide for representation from one of several departments in a common academic division (such as the basic sciences; the social sciences and humanities; the engineering sciences) on a rotating basis among chairmen. A change of this sort surely would be opposed at least by the more senior members of the board, since it would reduce their power. Precisely for that reason, a similar proposal (for creating a smaller policy board as the operational arm of the Academic Board) was advanced by the Army's West Point Study Group in mid-1977 as being a major structural reform essential to assuring adaptability to needed change. More powerful over the years than comparable top-level academic policy-making bodies at the other academies, the Academic Board at West Point also has proven to be more resistant to innovation. The report of the Borman Commission, upon which the Army's West Point Study Group drew in its follow-up analysis, highlighted the failure of Military Academy top officials and senior faculty members to remain abreast of changing values and attitudes not only among cadets but also among junior faculty members. A major restructuring of the Academic Board would give somewhat greater freedom of action to reform-minded superintendents. But at the same time, the change would be accompanied by measures such as creation of a faculty council (also proposed in 1977 by the West Point Study Group) designed to insure that the voices of junior faculty members as well as those of senior ones are heard in discussions of matters of academic policy and governance.

—A civilian academic dean would be appointed at the Coast Guard Academy, the Air Force Academy, or the Military Academy. There is no evidence at this point in time of support internally or externally for such a change. However, it is conceivable that circumstances such as those that led the Secretary of the Navy to mandate the appointment of a civilian dean at Annapolis, despite the resistance of many of his uniformed subordinates, could be replicated in one or more of the other services. Especially as the number of civilians who join the faculties of the Military Academy and the Air Force Academy increase, and as civilians at the Coast Guard Academy rise to positions of greater influence, and as arguments are made (as they will be) for more clearly distinguishing the nine-month academic program from the demands of military training and discipline, the idea of appointing a civilian educator rather than a military professional to the dean's post may gain greater attention.

—The members of the cadet or midshipman first class (seniors) would be housed separately from the rest of the corps, brigade, or wing, and would be treated essentially as junior officers. They would be exempt from most military drill and ceremonies, and would be free from most of the rules and regulations that apply to cadets and midshipmen of the other classes, except for ones that also would be applicable to officers. They would dress in officer uniforms, but with distinctive insignia (as cadet officers in ROTC wear). As with commissioned officers assigned to the academies, first-class cadets or midshipmen would be free to leave the military reservation whenever they were not in the academic classroom or involved in other duties; they would be permitted to wear civilian clothing during off-duty leisure time.

Such changes would be opposed vehemently by some of the permanent cadre at each academy, as well as by many concerned alumni. The argument would be made that the various forms of separating the first class from the rest of the student body would be injurious to organizational cohesion, and thus to *esprit de corps*. Expanded privileges such as letting first-classmen wear civilian clothes when they were off duty, many would argue, would represent a lowering of standards that would imperil the maintenance of discipline in other respects.

Contrary to these arguments, those who favored these bold but incremental changes would expect benefits to accrue in the form of higher morale not only among first-classmen but also among the freshmen, sophomores, and juniors. The latter three classes would be given responsibilities of leadership in the corps, brigade, or wing at an earlier stage than currently is the practice, with attendant privileges which presently are accorded only to first-classmen. The four-year "survival test" (as many cadets and midshipmen perceive it) would be reduced, in effect, to three.

—Provisions would be introduced for any cadet or midshipman to take a year's leave of absence at any time after completion of his or her first year. Morris Janowitz advanced a proposal of this sort in a paper published in London in 1973; he repeated the proposal the following year in a banquet address at West Point.[18] As Janowitz, among others, has noted, the facts of contemporary life are that fewer college-age Americans than in the past are willing to pursue a four-year uninterrupted program of studies at a single institution of higher education. Many begin at one institution, then transfer to another. Many others drop out for a year or more of work experience or travel. The academies would acknowledge this trend by establishing procedures that legitimize such behavior.

Many academy officials and members of the permanent cadre are sure to regard leave-of-absence proposals as ominous. Deep-down, the suspicion would be that the vast majority of cadets and midshipmen who took advantage of such a plan, once having tasted the wild fruit in the orchards of civilian campuses, would never return to the regimented life of the academies. However, overtly the issue would be posed in terms of logistics, and in terms of the threat to class cohesion that would be posed as sophomore and junior classes had their ranks depleted by students taking advantage of the plan.

Although no groundswell of support for a leave-of-absence plan is evident currently, it is of some interest that the cadets who were expelled from West Point in 1976 during the investigations of widespread cheating were permitted to resume their cadet status at the Academy in 1977. Willy-nilly, the Army had granted them a year's leave of absence. Moreover, contrary to the dire predictions of some that few would return, approximately two-thirds of the 151 cadets who were evicted in 1976 returned in June 1977.

Cautious Incrementalism

Changes such as the proposed leave-of-absence plan will be adopted, however, only if at the academies or—more crucially—within the Congress or the service headquarters a sense of urgency is felt regarding problems which the academies are experiencing. In the meantime, cautious incrementalism will be the characteristic strategy. If a scenario of cautious incrementalism prevails—and this seems the most likely—one can expect changes of the following sort to occur over the next five to ten years.

—Tinkering with the curriculum will occupy much of the time of academy officials and faculty members. Efforts will be made to provide larger numbers of students than in the past with opportunities to design their own programs of study, with emphasis on research during the senior year. The trend toward including in the curriculum coursework that focuses on policy issues of current prominence, such as global energy problems, environmental protection, and the law of the sea, will continue.

—After years of debating the issue, West Point at last will move to a program of academic majors for all students, in lieu of the current program of less specialized "concentration" areas for each cadet.

—The Military Academy and the Air Force Academy will add a number of civilians to their faculty, mostly on a visiting or short-term basis. An effort will be made each year to attract nationally known scholars as visiting professors, who will be accorded great deference during their stay. The other civilian appointees also will be treated courteously, but will have little impact on academy policy making.

—Female cadets and female midshipmen will be utilized in prominent roles on student advisory boards and in cadet/midshipman leadership positions. Partly as a result of the recommendations of female cadets and midshipmen, and partly as a response by academy officials to problems associated with their initial efforts to utilize methods of discipline and training with women identical to those used with men, the academies will alter their training and disciplinary procedures increasingly to accommodate the distinctive needs of women.

—As programs of women's athletics flourish and receive favorable publicity, they will receive added internal support. The academies will become more aggressive in recruiting female athletes.

—Each of the four academies will make further efforts to reform the indoctrination practices associated with plebe/swab/doolie year. The emphasis will be on utilizing positive leadership techniques, similar to ones that would be effective in regular units of the armed forces in training enlisted personnel. The severity of the distinction that has existed during the regular academic year between fourth-classmen and upperclassmen will be reduced.

—Programs of summer training which provide cadets and midshipmen with opportunities for leadership experience as "third lieutenants" or "apprentice ensigns" with regular units of the armed forces will be continued. Efforts will be made to provide such experience earlier in the cadet's or midshipman's four years than in the past, with special orientation programs devised to broaden his or her understanding of the life of enlisted personnel.

—Academy officials will continue to agonize over incidents of cheating, theft, falsehood, quibbling, and other manifestations of dishonorable conduct. Programs designed to cope with the problems will be spawned, modified, abandoned, and new programs instituted at a relatively rapid rate. Academic courses focusing on ethics will be introduced. In military leadership training, more attention than in the past will be paid to role-playing exercises which confront the student with ethically complex choices. Prominent guest speakers will be brought in to lecture on ethics.

—Especially through the expansion of privileges granted to cadets and midshipmen during off-duty time and on weekends, academy officials will attempt to keep attrition below levels that would trigger renewed investigation of the academies by Congress or the GAO. Periodic downward fluctuations in attrition rates will encourage academy officials to believe that reforms have been successful. However, intermittent upturns in attrition will renew the search for remedies. The expansion of liberties will prove insufficient in overcoming the sense of deprivation which most academy students feel relative to civilian peers. Their nagging complaint that the four-year system of regimentation inherently demands "capitulation" (as a dropout from Annapolis during the period of the Calvert reforms put it) will continue. The opposite attitude ("discipline has become too lax") among some staff members, alumni, and even some cadets and midshipmen will exert pressures for periodic tightening of regulations. Incidents in which cadets or midshipmen apparently have abused privileges (for example, through sloppy appearance while on liberty, excessive drinking, recklessness in driving automobiles) will provoke new restrictions or more severe punishments.

In short, many of the themes of the past will be themes of the coming decades, although novelty will be apparent in procedural and programmatic detail, and in the growing prominence of women in the activities of the academies. Some key academy officials will be imaginative leaders with visions of the future; others will be prisoners of rule books and their own past, struggling to maintain control. There will be "young turk" elements within the staff and faculty, working for reform; there will be "old guard" elements fearful of change, and working to thwart it. There will be times when the academy community is close-knit, proud, and confident, other times when it is divided, insecure, and confused. There will be periods of revelry and exaltation, periods of gloom and despair; periods of hard-earned organizational triumph, periods of bitter failure. The quest to be both Athens and Sparta will continue—but so also will the attendant frictions and frustrations. The traditional ideals of duty, honor, and loyalty to country will continue to be a wellspring of inspiration for nearly all members of the academy community, even on occasions when the ideals have not been fulfilled. Continuing external criticisms will evoke renewed defense of the service academy in terms of its ideals, in terms of its past responsiveness to changing societal and professional needs, and in terms of extrinsic measures of comparison with the performance of other institutions. But only the foolish among members of the academy community will never doubt that the changes made are adequate to the challenging demands of the future.

APPENDIX 1.

Interviewees

NOTE: All interviews were conducted on the date indicated in person by the author, unless otherwise indicated. Rank and position of interviewee are given as of the date of the interview (or of the latest interview, if more than one was conducted).

Albright, William E., Jr. COL USAF. Former faculty member, Dept. of Political Science, USAFA. 20 Aug. 1974 and 12 May 1977.

Bender, Chester R. ADM USCG. Commandant of the Coast Guard. 7 Feb. 1974.

Birrer, Ivan. Education Advisor, U.S. Army Command and General Staff College. 8 Feb. 1978.

Briggs, James E. LTG USA. Former Supt., USAFA. 24 Feb. 1973 by USAF Oral History Committee.

Buckley, Harry. COL USA. Director, Office of Military Psychology and Leadership, USMA. July 1971.

Calvert, James F. VADM USN (ret.). Former Supt., USNA. 5 June 1975 by phone.

Chancellor, George. LTC USA. Professor of Chemistry, USMA. 20 June 1973.

Clark, Albert P. LTG USAF. Supt., USAFA. 4 Oct. 1972.

Coltrin, William H. MSGT USAF. Noncommissioned Officer of the Wing, USAFA. 15 April 1963.

Conord, Albert E. Asso. Director for Educational Development, Academic Computing Center, USNA. 30 May 1972.

Cousland, Walter C. COL USA. Regimental Commander USCC, USMA. 21 June 1973.

Crowley, John D. Professor and Head, Dept. of Physical and Ocean Sciences, USCGA. 23 July 1973.

Crump, James B. LTC USAF. AOC 2d Cadet Group, USAFA. 18 April 1963.

Cutler, Elliott C. COL USA. Professor and Head, Dept. of Electrical Engineering, USMA. 20 June 1973.

Davidson, Bruce M. Academic Dean, USNA. 22 May 1972.

Davidson, John F. RADM USN (ret.). Former Supt., USNA. 25 May 1972.

Davis, Ray E. CAPT USN. Director, Division of U.S. and International Studies, USNA. 10 Aug. 1971.

DeMichiell, Robert. CDR USCG. Director, Computing Center, USCGA. 24 July 1973.

Erdle, Philip J. COL USAF. Vice Dean of the Faculty, USAFA. 1 April 1976.
Ferrari, Victor J. COL USAF. Wing Air Officer, Commanding, USAFA. 17 April 1963.
Foye, Paul F. CAPT USCG. Academic Dean, USCGA. 1 June 1972 and 24 July 1973.
Garigan, Thomas P. LTC USA. Asst. Information Officer, USMA. 19 June 1973.
Gruenther, Richard. COL USA. Director, Office of Military Instruction, USMA. 21 June 1973.
Harris, Boyd. MAJ USA. Company Tactical Officer, USMA. 23 April 1977 by phone.
Hurley, Alfred F. COL USAF. Head, Dept. of History, USAFA. 31 March 1976.
Huston, John W. Professor and Acting Chairman, Dept. of History, USNA. 30 May 1972.
Inman, E. E. CDR USN. Professor of Psychology, USNA. 17 Aug. 1971.
Jannarone, John R. BG USA. Dean of the Academic Board, USMA. 10 May 1972.
Janse, Donald A. Asst. Professor of Musical Activities, USCGA. 23 July 1973.
Jarrell, William R., Jr. COL USAF. Director of Admissions, USAFA. 5 Oct. 1972.
Johnson, Bruce. Associate Professor of Systems Engineering, USNA. 15 June 1972.
Johnson, Paul. Librarian and Archivist, USCGA. 24 July 1973.
Jordan, Amos A. COL USA. Head, Dept. of Social Sciences, USMA. July 1971.
Kirk, Neville. Professor of History, USNA. 16 June 1972.
Knowlton, William A. LTG USA. Supt., USMA. 12 March 1973.
Korb, Lawrence J. Asso. Professor of Political Science, USCGA. 14 July 1973.
McCormick, Michael E. Professor and Chairman, Dept. of Naval Systems Engineering, USNA. 5 June 1972.
McCormick, Robert F. Rector, Catholic Chapel, USMA. 19 June 1973.
McDermott, Robert F. BG USAF (ret.). President, United Services Automobile Assn.; former Dean of the Faculty, USAFA. 18 April 1963; 8 and 9 July 1974.
McDonald, William. LTC USAF. Acting Chairman, Dept. of Political Science, USAFA. 16 April 1963.
Martin, Ben. Head Football Coach, USAFA. 1 April 1976.
Megargee, Richard. Asso. Professor of History, USNA. 2 May 1972.
Melson, Charles. VADM USN (ret.). Former Supt., USNA. 25 May 1972.
Minter, Charles S., Jr. VADM USN. Deputy CNO for Logistics; former Supt., USNA; former Commandant of Midshipmen, USNA. 14 June 1972.
Moody, Peter R. Provost and Vice President for Academic Affairs, Eastern Illinois University; former Head, Dept. of English, USAFA. 24 Nov. 1975.
Morris, Max. RADM USN. Commandant of Midshipmen, USNA. 12 June 1972.

Nehrt, Lee. Professor, School of Business, Indiana University, and graduate, 1949, USCGA. 12 March 1974.

Nye, Roger H. COL USA. Professor of History, USMA. 19 June 1973.

Olvey, Lee Donne. COL USA. Professor and Head, Dept. of Social Sciences, USMA. 20 June 1973.

Peters, Wayne. ENS USN. Staff member, USNA. 26 Oct. 1971.

Pitt, A. Stuart. Professor and Chairman, Dept. of English, USNA. 26 May 1972.

Probert, John R. Professor and Chairman, Dept. of Political Science, USNA. 10 Aug. 1971.

Rosser, Richard F. COL USAF (ret.). Dean, Albion College; former Head, Dept. of Political Science, USAFA. 5 Oct. 1972 and 2 Sept. 1974.

Ruhe, Richard. LT USCG. Company Officer, USCGA. 5 April 1974.

Shinn, Allen Mayhew. VADM USN (ret.). Former Commandant of Midshipmen, USNA. 15 June 1972.

Smith, Stanley. CAPT USCG (ret.). Provost, S.E. Branch of the University of Connecticut at Avery Point; former Academic Dean, USCGA. 24 July 1973.

Smith, Willard. ADM USCG (ret.). Former Commandant of the Coast Guard; former Supt., USCGA; former Commandant of Cadets, USCGA. 14 and 15 July 1976.

Starling, Ira Caryl. Protestant Chaplain, USNA. 15 June 1972.

Tate, J. Philip. MAJ USAF. Asst. Prof. of History, USAFA. 15 Oct. 1973.

Taylor, Maxwell D. GEN USA (ret.). Former Army Chief of Staff; former Supt., USMA. 8 Dec. 1972 by Col. Richard A. Manion for the Army Military History Research Collection.

Taylor, William J., Jr. LTC USA. Asso. Prof. of Social Sciences, USMA. 22 June 1973.

Thompson, John F. RADM USCG. Supt., USCGA. 1 June 1972.

Tuneski, Robert S. CDR USCG. Asst. Commandant of Cadets, USCGA. 23 July 1973.

Viglienzone, Walter S. LT USCG. Instructor in Humanities, USCGA. 4 April 1974.

Wakin, Malham M. COL USAF. Professor and Head, Dept. of Political Science and Philosophy, USAFA. 31 March 1976.

Wells, Ronald A. CDR USCG. Professor and Head, Dept. of Social Sciences and Humanities, USCGA. 2 June 1972.

Williams, Chester Y. LTC USAF. Former Asst. Prof. of Political Science, USAFA. 15 Oct. 1973.

Williams, Malcolm J. CAPT USPHS. Director of Admissions, USCGA. 5 April 1974.

Woods, Jimmie D. CDR USCG. Professor and Head, Dept. of Mathematics, USCGA. 24 July 1973.

Woodyard, William T. BG USAF. Dean of the Faculty, USAFA. 31 March 1976.

Zierdt, William. LTC USA. Asst. Professor, Office of Military Psychology and Leadership, USMA. 12 Oct. 1973.

APPENDIX 2

Military Academy Superintendents, Commandants of Cadets, and Deans of the Academic Board, 1945-1978

(The date of assumption of office is indicated in parentheses.)

Superintendent	Commandant of Cadets	Dean of the Academic Board
	George Honnen (November 1943)	
Maxwell D. Taylor (September 1945)		Roger G. Alexander (Acting 1945; permanent 1946)
	Gerald J. Higgins (January 1946)	
	Paul D. Harkins (June 1947)	Harris Jones (June 1947)
Bryant E. Moore (January 1949)		
Frederick A. Irving (January 1951)	John K. Waters (June 1951)	
	John H. Michaelis (June 1952)	
	Edwin J. Messinger (September 1953)	
Blackshear M. Bryan (September 1954)		
	John L. Throckmorton (April 1956)	
Garrison H. Davidson (July 1956)		Thomas D. Stamps (July 1956)
		Gerald A. Counts (July 1957)

Superintendent	Commandant of Cadets	Dean of the Academic Board
	Charles W. G. Rich (September 1959)	William W. Bessell, Jr. (July 1959)
William C. Westmoreland (July 1960)		
	Richard G. Stilwell (July 1961)	
	Michael S. Davison (March 1963)	
James B. Lampert (June 1963)		
	Richard P. Scott (April 1965)	John R. Jannarone (July 1965)
Donald V. Bennett (January 1966)		
	Bernard W. Rogers (September 1967)	
Samuel W. Koster (June 1968)		
	Sam S. Walker (October 1969)	
William A. Knowlton (March 1970)		
	Philip R. Feir (September 1972)	
		John S. B. Dick (Acting 1973)
Sidney B. Berry (July 1974)		Frederick A. Smith, Jr. (July 1974)
	Walter F. Ulmer, Jr. (April 1975)	
	John C. Bard (January 1977)	
Andrew J. Goodpaster (June 1977)		

APPENDIX 3

Naval Academy Superintendents, Commandants of Midshipmen, and Academic Deans, 1945-1978

(The date of assumption of office is indicated in parentheses.)

Superintendent	Commandant of Midshipmen	Academic Dean
	Stuart H. Ingersoll (March 1945)	
Aubrey W. Fitch (August 1945)		
James L. Holloway, Jr. (January 1947)		
	Frank T. Ward (July 1948)	
	Robert B. Pirie (July 1949)	
Harry W. Hill (April 1950)		
	Charles A. Buchanan (January 1952)	
C. Turner Joy (August 1952)		
Walter F. Boone (August 1954)	Robert T. S. Keith (October 1954)	
William R. Smedberg III (March 1956)		
	Allen M. Shinn (August 1956)	
Charles L. Melson (June 1958)	William F. Bringle (August 1958)	

Superintendent	Commandant of Midshipmen	Academic Dean
John F. Davidson (June 1960)	James H. Mini (July 1960)	
	Charles S. Minter, Jr. (June 1961)	
Charles C. Kirkpatrick (August 1962)		
		Arthur B. Drought (Pro tem 1963; permanent 1964)
Charles S. Minter, Jr. (January 1964)	Sheldon H. Kinney (January 1964)	
Draper L. Kauffman (June 1965)		
	Lawrence Heyworth, Jr. (August 1967)	
James F. Calvert (July 1968)		
	Robert P. Coogan (September 1969)	
		Edward J. Cook (Interim 1970)
	Max K. Morris (July 1971)	Bruce M. Davidson (July 1971)
William P. Mack (June 1972)		
	Donald Forbes (August 1973)	
Kinnaird R. McKee (August 1975)		
	James A. Winnefeld (June 1976)	
	Jack N. Darby (January 1978)	
William P. Lawrence (August 1978)		

APPENDIX 4

Coast Guard Academy Superintendents, Commandants of Cadets, and Deans of Academics, 1945-1978

(The date of assumption of office is indicated in parentheses.)

Superintendent	Commandant of Cadets	Dean of Academics
James Pine (July 1940)	Henry S. C. Sharp (August 1944)	
	Carl B. Olsen (July 1946)	
Wilfred N. Derby (August 1947)		
Arthur G. Hall (September 1950)	Edwin J. Roland (August 1950)	
Raymond J. Mauerman (September 1954)	Leon H. Morine (August 1954)	
Frank A. Leamy (July 1957)	Willard J. Smith (June 1957)	Albert Lawrence (September 1957)
S. Hadley Evans (February 1960)	William B. Ellis (July 1960)	
		Stanley L. Smith (July 1961)
Willard J. Smith (July 1962)	Chester I. Steele (July 1962)	
	Austin C. Wagner (July 1964)	Paul F. Foye (July 1964)

Superintendent	Commandant of Cadets	Dean of Academics
Chester R. Bender (July 1965)		
Arthur B. Engel (June 1967)	Curtis J. Kelly (July 1967)	
John F. Thompson, Jr. (July 1970)		
	John B. Hayes (June 1971)	
Joseph J. McClelland (July 1973)	William S. Schwob (June 1973)	
William A. Jenkins (June 1974)		Roderick M. White (July 1974)
	Sidney B. Vaughn (June 1975)	
Malcolm E. Clark (June 1977)		

APPENDIX 5

Air Force Academy Superintendents, Commandants of Cadets, and Deans of the Faculty, 1954-1978

(The date of assumption of office is indicated in parentheses.)

Superintendent	Commandant of Cadets	Dean of the Faculty
Hubert R. Harmon (August 1954)	Robert M. Stillman (September 1954)	Don Z. Zimmerman (September 1954)
James E. Briggs (August 1956)		Robert F. McDermott (Acting 1956; permanent 1959)
	Henry R. Sullivan, Jr. (August 1958)	
William S. Stone (August 1959)		
	William T. Seawell (June 1961)	
Robert H. Warren (July 1962)		
	Robert W. Strong, Jr. (March 1963)	
Thomas S. Moorman, Jr. (July 1965)	Louis T. Seith (June 1965)	
	Robin Olds (December 1967)	
		William T. Woodyard (July 1968)
Albert P. Clark, Jr. (July 1970)		

Superintendent	Commandant of Cadets	Dean of the Faculty
	Walter T. Galligan (February 1971)	
	Hoyt S. Vandenberg, Jr. (February 1973)	
James R. Allen (August 1974)		
	Stanley C. Beck (July 1975)	
Kenneth L. Tallman (June 1977)		
	Thomas C. Richards (March 1978)	William A. Orth (July 1978)

NOTES

Chapter One

1. The Military Academy was founded in 1802, the Naval Academy in 1845, the Coast Guard Academy in 1876, the Air Force Academy in 1954 (with the first class entering the following year). These four academies constitute the focus of this book. A fifth federal academy, the Merchant Marine Academy, differs from the other four in its organizational mission and is not included in this study. The Merchant Marine Academy, founded in 1938 with instruction initiated in 1942, graduates young men and women into a variety of maritime-related careers, typically with the maritime industry. In contrast, unless medically disqualified, graduates of the other academies are commissioned into the armed forces.

2. I was in attendance that evening. The local paper devoted three front-page stories to the event the following day. Annapolis *Evening Capital*, 5 May 1972.

3. I had been invited to the Military Academy to participate in a workshop on contemporary problems in junior officer leadership, and attended the Dellums lecture, 12 May 1972. Representative Dellums directed his remarks primarily to black cadets, urging them to work hard and rise to top military leadership positions so the voice of blacks could be heard more clearly.

4. See J. William Fulbright, *The Pentagon Propaganda Machine* (New York: Vintage Books edition, 1971). CBS's "The Selling of the Pentagon" was critically reviewed by Claude Witze in *Air Force Magazine* (April 1971), pp. 10-12. A reply from the president of CBS, with other commentary from readers, appears in the May 1971 issue, pp. 12-15.

5. 1966 data cited by Robert F. McDermott, Brig. Gen. USAF (Ret.), "The USAF Academy Academic Program," *Air University Review* 20 (Nov.-Dec. 1968): 12.

6. USNA, *Catalog 1972–73*, p. 65.

7. Sidney B. Berry, LTG USA, Superintendent USMA, "The Superintendent's Letter," *Assembly* 36 (June 1977): inside front cover.

8. The concept of "task environment" was developed by William R. Dill, "Environment as an Influence on Managerial Autonomy," *Administrative Science Quarterly* 2 (March 1958): 409-443. It has been further refined by James D. Thompson, *Organizations in Action* (New York: McGraw-Hill, 1967).

9. Alone among the four academies, the Coast Guard Academy was not a

party to the *Anderson v. Laird* case that resulted in the 1972 finding that a chapel attendance requirement was unconstitutional. However, it was obvious to authorities at New London that the reasoning applied with equal force to their regulations; thus, they also eliminated the requirement. Likewise, the Coast Guard Academy was not included in the appropriations bill rider requiring the DOD academies to admit women. However, a similar bill applying to the Coast Guard Academy had been introduced in the House of Representatives, although no legislative barrier to the admission of women existed. They had been denied admission by provision of service regulations. Thus, the Commandant of the Coast Guard found it the better part of wisdom to preempt congressional mandate by decreeing in August 1975 that women would be admitted to the Coast Guard Academy. See "Women to Join Cadet Corps," USCGA *Alumni Bulletin* 37 (Sept.-Oct. 1975): 8-11.

Chapter Two

1. H. I. Marrou, *A History of Education in Antiquity*, trans. George Lamb (New York: Mentor Books, 1964), pp. xi-xii.
2. Cf. George P. Schmidt, *The Old Time College President* (New York, 1930).
3. *Board of Visitors to the USMA*, 1820.
4. See R. Ernest Dupuy, *Sylvanus Thayer: Father of Technology in the United States* (West Point, N.Y.: Assn. of Graduates, 1958), 23 pp. Sidney Forman, *West Point: A History of the United States Military Academy* (New York: Columbia University Press, 1950), pp. 36-60. Stephen E. Ambrose, *Duty, Honor, Country: A History of West Point* (Baltimore, Md.: Johns Hopkins Press, 1966), chaps. 4 and 5. Thomas J. Fleming, *West Point: The Men and Times of the United States Military Academy* (New York: William Morrow, 1969), pp. 3-87.
5. USMA, *Records, March 1814 to February 1838* (West Point: USMA, n.d.), pp. 314-315.
6. USMA, *Records, March 1814 to February 1838*, p. 110.
7. *Board of Visitors to the USMA*, 1820.
8. Swift, *Memoirs*, p. 123, quoted in Edgard Denton III, "The Formative Years of the United States Military Academy, 1775–1833" (Ph.D. diss., Syracuse University, 1964), p. 221.
9. The Military Philosophical Society is discussed in Forman, *West Point*, pp. 20-35.
10. Brevet Lt. Col. Robert H. Hall, ed, *Laws of Congress Relative to West Point and the United States Military Academy from 1786 to 1877* (West Point: USMA Press, 1888).
11. USMA, *Regulations U.S. Military Academy* (West Point, 1802–1816).
12. Daniel Hovey Calhoun, *The American Civil Engineer: Origins and Conflict* (Cambridge, Mass.: MIT, The Technology Press, distributed by Harvard University Press, 1960), p. 44.

13. Frederick Rudolph, *The American College and University: A History* (New York: Vintage Books, 1962), pp. 37-38.

14. W. H. Cowley, "European Influences upon American Higher Education," *The Educational Record*, 20 (April 1939): 165-190. Also, Samuel Eliot Morison, *Three Centuries of Harvard* (Cambridge, Mass.: The Belknap Press of Harvard University Press, 1964), pp. 222-245.

15. Morison, *Three Centuries*, p. 230; Forman, *West Point*, p. 139.

16. Ambrose, *Duty, Honor, Country*, p. 196.

17. Letter Jared Mansfield to John Michael O'Connor, 12 Feb. 1819, ms. USMA, quoted in Denton, "The Formative Years," p. 202. On French influence at West Point in the pre-Thayer period, see Denton, "The Formative Years," pp. 139, 151; Forman, *West Point*, pp. 79-80; Thomas Everett Griess, "Dennis Hart Mahan: West Point Professor and Advocate of Military Professionalism, 1830–1871" (Ph.D. diss., Duke University, 1969), p. 69. An excellent discussion of education at the École Polytechnique in this period, and its reputation, is provided in Frederick B. Artz, *The Development of Technical Education in France, 1500–1850* (Cambridge, Mass.: The Society for the History of Technology, and MIT Press, 1966), esp. pp. 100-101, 155, 243n.

18. Dupuy, *Sylvanus Thayer*, pp. 1-7.

19. Griess, "Dennis Hart Mahan," pp. 235-237.

20. Burton R. Clark, *The Distinctive College: Antioch, Reed & Swarthmore* (Chicago: Aldine, 1970), p. 259.

21. Ambrose, *Duty, Honor, Country*, chap. VII.

22. Rudolph, *The American College*, p. 37.

23. See Roger Hurless Nye, "The United States Military Academy in an Era of Educational Reform, 1900–1925," (Ph.D. diss., Columbia University, 1968).

24. Rudolph, *The American College*, pp. 157-164.

25. Samuel P. Huntington, *The Soldier and the State* (Cambridge, Mass.: The Belknap Press of Harvard University Press, 1959), p. 229.

26. Huntington, *The Soldier*, p. 228.

27. John Hope Franklin, *The Militant South, 1800–1861* (Cambridge, Mass.: The Belknap Press of Harvard University Press, 1956), p. 167.

28. Huntington, *The Soldier*, p. 206.

29. John A. Logan, *The Volunteer Soldier of America* (Chicago: R.S. Peal and Co., 1887). See also, Russell F. Weigley, *History of the United States Army* (New York: Macmillan, 1967), pp. 270-271.

30. Leonard D. White, *The Jeffersonians: A Study in Administrative History, 1801–1829* (New York: Macmillan, 1951), chap. 20.

31. Harold and Margaret Sprout, *The Rise of American Naval Power, 1776–1918* (Princeton, N.J.: Princeton University Press, 1939, 1967), chap. 8.

32. See Vice Admiral James Calvert, "The Fine Line at the Naval Academy," U.S. Naval Institute *Proceedings* 96 (Oct. 1970): 63-68. Captain Frank V. Rigler (USN), "The First Quarter Century," *Shipmate* 33 (Sept.-Oct. 1970): 14-17.

33. Florian Cajori, quoted by John A. Tierney, "William Chauvenet—Father of the Naval Academy," *Shipmate* 32 (Sept.-Oct. 1969), 6-11.

34. At that time, after a midshipman had passed the requisite examination that entitled him to a commission, he had to wait until a vacancy for lieutenant occurred. During the interim, he was designated a "passed midshipman."

35. Marcy, as quoted by John D. Hayes (USNA '24), "Influence of West Point on the Founding of the Naval Academy," *Assembly* 19 (Winter 1961): 8-11.

36. *A Guide to the United States Naval Academy*, compiled by the Writers Program of the Works Projects Administration (WPA) in the State of Maryland, sponsored by the U.S. Naval Academy (New York: Devin-Adair, 1941), p. 54.

37. Hayes, *Assembly* 19 (Winter 1961): 8-11.

38. Tierney, *Shipmate* 32 (Sept.-Oct. 1969): 7, 10.

39. Hayes, *Assembly* 19 (Winter 1961): 9n. Also, USMA, *Register of Graduates* (1970).

40. Biographical data about Commander Sydney Smith Lee were obtained from a file on Naval Academy Commandants maintained in the Office of the Commandant, U.S. Naval Academy.

41. Rigler, *Shipmate* 33 (Sept.-Oct. 1970): 17.

42. *A Guide to the United States Naval Academy*, p. 54.

43. Biographical sketch of Eleventh Commandant of Midshipmen, on file in Office of the Commandant, U.S. Naval Academy.

44. Actually, Rodgers served as Superintendent from Sept. 1874 to June 1878, and again for a five-month period in 1881. The second tour, although abbreviated, was a tribute to the success of the first.

45. The quote and data regarding Rodgers's career are from a biographical sketch on the Sixth Commandant of Midshipmen, on file in the Office of the Commandant, U.S. Naval Academy.

46. William E. Simons, *Liberal Education at the Service Academies* (New York: Teachers College, Columbia University, for the Institute of Higher Education, 1965), pp. 50-52.

47. Rear Admiral C. C. Kirkpatrick, "Rhodes Scholars? Yes! But Naval Officers First, A Century of Academic Achievements," *Shipmate* 26 (Sept.-Oct. 1963): 4-10, 18.

48. W. D. Puleston, *Mahan* (New Haven: Yale University Press, 1939), pp. 5-17, 56-61. For other insights into the early development of Mahan's thought see Rosa P. Chiles, ed., "Letters of Alfred Thayer Mahan to Samuel A'Court Ashe (1858-59)," *Duke University Library Bulletin* (July 1931), pp. vii-xvii, 1-121.

49. Captain Frank A. Rigler, "Superintendents of the Naval Academy," *Shipmate* 35 (April 1972): 19-22. Also "Long Blue Chain," *Shipmate* 33 (Sept.-Oct. 1970): 26-31.

50. Rigler, *Shipmate* 35 (April 1972), 19. Also, see Sprout and Sprout, *The Rise of American Naval Power, 1776–1918*, chap 12. During Sampson's superintendency, prominent scientists from other universities were invited to give lectures.

51. Not only was there extensive participation of the Naval Academy Alumni Association in the development of the Navy League, but also Horace Porter, a West Point graduate of the Class of 1860, assumed the presidency of the

League in 1905. See Armin Rapoport, *The Navy League of the United States* (Detroit, Mich.: Wayne State University Press, 1962).

52. *Reef Points 1970–1971* (Annapolis, Md.: USNA, 1971), pp. 58-60. Also, *The Chapel of the United States Naval Academy* (Annapolis, Md.: USNA, 1969), 32 pp. The last-cited pamphlet was written by a Naval Academy chaplain under the direction of his superior officer, also a chaplain. Because the pamphlet describes the admiral who laid the cornerstone of the chapel as "John Dewey," one must surmise that chaplains are somewhat less steeped in naval history and lore than are their line-officer counterparts.

53. Peter Karsten, *The Naval Aristocracy: The Golden Age of Annapolis and the Emergence of Modern American Navalism* (New York: Free Press, 1972), pp. 280, 285. Puleston notes that "During the naval stagnation between 1868 and 1898, Academy graduates were encouraged to resign." Puleston, *Mahan*, pp. 229-230.

54. Chandler and Clark are quoted in U.S., Congress, House, Committee on Economy and Efficiency, *House Doc. 670*, 62d Cong, 2d sess., 4 April 1912, pp. 304-305, 325-336; discussed in Leonard D. White, *The Republican Era: A Study in Administrative History* (New York: Free Press, 1958), pp. 116-117.

55. Stephen H. Evans, Captain (USCG), *The United States Coast Guard 1790–1915* (Annapolis, Md.: U.S. Naval Institute, 1949), pp. 86-94.

56. Quoted in Riley Hughes, *Our Coast Guard Academy, A History and Guide* (New York: Devin-Adair, for the Coast Guard Academy Alumni Assn., 1944), pp. 59-60.

57. The distinctive character is described in detail in a special 104-page centennial issue of *The Bulletin* (1976) of the USCGA Alumni Association. The issue was prepared by Paul Johnson, Academy archivist and librarian, and William Earle, editor of *The Bulletin*.

58. U.S., Congress, House, Committee on Appropriations, Subcommittee on Treasury, *Hearings, Treasury Department Appropriation Bill for 1941*, 76th Cong., 3d sess., 1940, pp. 552-554.

59. MacArthur went into the heat of battle without a helmet or gas mask and carrying a riding crop rather than a weapon. He describes his own World War I exploits in some detail in Douglas MacArthur, *Reminiscences* (New York, Toronto, London: McGraw-Hill, 1964), pp. 51-73.

60. U.S., Treasury Department, *Report of the Cruise of the U.S. Revenue-Cutter* Bear *and the Overland Expedition for the Relief of Whalers in the Arctic Ocean*, Treasury Document no. 2101, Division of Revenue-Cutter Service (Washington, D.C.: GPO, 1899).

Chapter Three

1. Roosevelt's affinity for the Navy was evident from his childhood. By the time he graduated from Groton in 1900, in the heyday of naval expansionism (with cousin Teddy preaching the virtues of Mahanism), Franklin had decided that he wanted to attend the Naval Academy at Annapolis. However, his

father persuaded him to go to Harvard instead. Karl Schriftgiesser, *The Amazing Roosevelt Family, 1613–1942* (New York: Wilfred Funk, 1942), chap. 21.

2. Coast Guard activities in the late 1930s and throughout the war are described in some detail in Malcolm F. Willoughby, *The U.S. Coast Guard in World War II* (Annapolis, Md.: U.S. Naval Institute, 1957), pp. 6-24.

3. U.S., Bureau of the Budget, *The Budget of the United States for the Fiscal Year Ending June 30, 1940;* and *The Budget of the United States for the Fiscal Year Ending June 30, 1941.* With the formal entry of the United States into the war, following the attack at Pearl Harbor, Coast Guard appropriations increased even more sharply, to $64.3 million for fiscal year 1942.

4. Riley Hughes, *Our Coast Guard Academy: A History and Guide* (New York: Devin-Adair, for the USCGA Alumni Assn., 1944), pp. 73, 86, 172.

5. Williams, widely regarded as the dean of revisionist historians, graduated from Annapolis in 1944 (with the Class of 1945). He recalls that he "learned much more history, literature, etc. than my civilian colleagues imagine." Williams to author, 27 Dec. 1971.

6. "War Hits the School," *Shipmate* (July 1942): 4-5, 35; and "The Academy at War," *Shipmate* (Oct. 1943): 8-9, 71-73.

7. *Annual Report of the Superintendent USMA*, 30 June 1942.

8. F. B. Wilby, Maj. Gen., Superintendent USMA, "West Point at War," *Assembly* 2 (July 1943); 4-5.

9. Ronan C. Grady, *The Collected Works of Ducrot Pepys* (Newburgh, N.Y.: Moore, 1943), p. 56. *The Collected Works* purports to be a diary maintained by Cadet Ducrot Pepys. The name of this fictitious character combines "Ducrot"—or "Do Crow," one of several epithets used by upperclassmen in addressing plebes—with the surname from *The Diary of Samuel Pepys,* with which cadets in the 1940s had become familiar in English classes.

10. In cadet usage, the term *soirée* evolved from the original French meaning to refer to almost any activity, social or otherwise, that cadets were required to attend, thus representing another unwelcome demand on their time.

11. Roger Hilsman, foreward to Grady, *Ducrot Pepys.*

12. Kendall Banning, *Annapolis Today*, 6th ed. rev. by A. Stuart Pitt (Annapolis, Md.: U.S. Naval Institute, 1963), p. 300.

13. Anonymous critics, cited by John W. Masland and Laurence I. Radway, *Soldiers and Scholars: Military Education and National Policy* (Princeton, N.J.: Princeton University Press, 1957), p. 106.

14. Vincent Davis, *Postwar Defense Policy and the U.S. Navy, 1943–1946* (Chapel Hill: University of North Carolina Press, 1962), p. 135.

15. "Holloway, James L(emuel), Jr.," *Current Biography* (1947), pp. 313-314. Also, Robert Greenhalgh Albion and Robert Howe Connery, *Forrestal and the Navy* (New York: Columbia University Press, 1962), pp. 245-246.

16. Maxwell D. Taylor, *Swords and Plowshares* (New York: Norton, 1972), p. 114.

17. U.S. Coast Guard, Special Board for Conducting a Study of Procurement and Education of Coast Guard Officers, *A Plan for the Procurement and Education of Coast Guard Officers,* submitted 2 January 1946.

18. U.S., Secretary of Defense, *A Report and Recommendation to the Secretary of Defense by the Service Academy Board*, January 1950, App. A., pp. 17-18. (Cited hereafter as *Report of the Service Academy Board*.)

19. The board made forty specific recommendations, which appear in the *Report of the Service Academy Board*, pp. 9-40.

20. U.S., War Dept. General Staff, Director of Organization and Training, *Survey of the Current Situation at the United States Military Academy*, report prepared by USMA Headquarters, 16 July 1946, p. 6. Also, *Annual Report of the Superintendent USMA*, 30 June 1946, p. 22. My knowledge of these events has been enhanced by discussions with Professor Neville Kirk, Department of History, U.S. Naval Academy, who served as one of the exchange faculty to West Point in the early postwar period.

21. Nehrt interview. Nehrt recalled the incident with Professor Hoag firsthand, as a 1949 graduate of the Coast Guard Academy.

22. "Electronics Laboratory," *Assembly* 6 (April 1947): 4-5.

23. "Fitch, Aubrey (Wray)," *Current Biography* (1945), pp. 190-192. Also Davis, *Postwar Defense Policy*, pp. 128-129.

24. For the Taylor-Stilwell relationship, see Barbara W. Tuchman, *Stilwell and the American Experience in China, 1911-45*, paperback edition (New York: Bantam Books, 1972), pp. 215-218, 317-318, 390.

25. Anthony L. P. Wermuth to author, 17 Oct. 1973. Wermuth graduated from West Point in 1940, and was serving as a faculty member in the Department of English at the Military Academy during the period under discussion.

26. S. E. Gee, Lt. Col. (USA), "The Department of Military Psychology and Leadership, U.S.M.A.," *Assembly* 8 (Jan. 1950): 10-11.

27. Wermuth to author, 1973, notes the frictions that developed between the Office of Military Psychology and Leadership and the Department of Military Art and Engineering.

28. H. B. Seim, Lieutenant Commander (USN), "USNA—1948," *Shipmate* (Nov. 1948): 7-9, 18.

29. The psychologist who was hired was Malcolm Williams, who became a captain with the U.S. Public Health Service while serving as Director of Admissions at the Coast Guard Academy. Williams came to the Coast Guard Academy from wartime service as a psychologist with the Army Air Corps.

30. Colonel John D. F. Phillips, "The Course in Social Sciences at the U.S. Military Academy," *Assembly* 9 (Oct. 1950): 3.

31. Colonel G. A. Lincoln and Major S. H. Hays, "Program of Studies, U.S.M.A.," *Assembly* 9 (April 1950): 6-8.

32. Lt. Col. John S. Harnett, "SCUSA—II," *Assembly* 9 (Jan. 1951): 10-11.

33. Captain Harold S. Walker, Jr., "The Department of English, U.S.M.A., 1802–1950," *Assembly* 9 (Jan. 1951): 6-9.

34. Figures are based on data reported in USMA and USNA, "Curricular Comparison," a two-page typewritten report signed by Maj. Gen. Bryant E. Moore, Supt. USMA, and by Rear Adm. J. L. Holloway, Jr., Supt. USNA, and authenticated 8 Sept. 1949 by William S. Shields, educational advisor and Asst. Secy. Academic Board USNA. Some academic subjects have been re-

classified here among the three major categories (social-humanistic; scientific-engineering; military) in the interest of consistency. For example, all history courses are classified as "social-humanistic," including naval history, which is so-classified in the original report, and military history, which is treated as a "military" subject in the original report.

35. The Coast Guard Academy Superintendent told cadets that "if we consider mathematics, science, and engineering as one subject then we will find that this combination takes about 60% of your time as compared to 25% for professional subjects other than engineering, and 15% for general studies such as English and history." Rear Admiral Wilfrid N. Derby, Superintendent USCGA, "Education and Training of Cadets," *Alumni Association Bulletin* 9 (Dec. 1947): 458-469.

36. USCGA, Superintendent, *Report to the Board of Visitors* (1948).

37. Seim, *Shipmate* (Nov. 1948): pp. 9, 18.

38. *Report of the Service Academy Board,* p. 48.

39. *Report of the Service Academy Board,* App. D, p. 48.

40. U.S., *Report of the Board of Visitors to the United States Military Academy,* 1950.

41. *Report of the Service Academy Board,* p. 49.

42. Stanley Smith interview.

43. Sanford M. Dornbusch, "The Military Academy as an Assimilating Institution," *Social Forces* 33 (1955): 316-321.

44. Kirk interview.

45. U.S., *Report of the Board of Visitors to the U.S. Naval Academy,* 19 April 1949, statement of the Superintendent USNA, pp. 20-21.

46. Kirk interview.

47. William E. Simons, *Liberal Education at the Service Academies* (New York: Teachers College, Columbia University, for the Institute of Higher Education, 1965), p. 117.

Chapter Four

1. See Edward A. Miller, Jr., "The Struggle for an Air Force Academy," *Military Affairs* 27 (Winter 1963–64): 163-173.

2. Preliminary Report of 4 April 1949 quoted in U.S., Secretary of Defense, *A Report to the Secretary of Defense by the Service Academy Board,* Washington, D.C., Jan. 1950.

3. See C. W. Borklund, *Men of the Pentagon: From Forrestal to McNamara* (New York: Praeger, 1966), chap. 5.

4. Memorandum from Johnson to the Secretaries of the Army, Navy and Air Force, and the JCS, 13 May 1949, cited in Miller, *Military Affairs* 27 (Winter 1963–64): 167-168.

5. The board also rejected various other alternative proposals. *Report, Service Academy Board,* pp. 4-6.

6. John W. Masland and Laurence I. Radway, *Soldiers and Scholars: Mili-*

tary Education and National Policy (Princeton, N.J.: Princeton University Press, 1957), p. 120.

7. Letter to Commanding General, Air University, from General Hoyt S. Vandenberg, Chief of Staff, USAF, 1 Sept. 1948, cited in Miller, *Military Affairs* 27 (Winter 1963–64): 167.

8. U.S., Congress, House of Representatives, Committee on Armed Services, *Hearings on H. R. 5337 to Provide for the Establishment of a United States Air Force Academy,* 82nd Cong., 2d sess., 1954, p. 3015.

9. USAFA, Office of Information Services, Historical Div., *History of the United States Air Force Academy 27 July 1954–12 June 1956,* prepared by LTC. Edgar A. Holt, Dr. M. Hamlin Cannon, and Dr. Carlos R. Allen, Jr. (USAFA, 1 Aug. 1957), II: 890-891.

10. Harmon to Col. W. C. Fowler, 9 Dec. 1953, quoted in *History of the USAFA 27 July 1954–12 June 1956,* II: 908.

11. USAFA, "United States Air Force Academy Honor Code, 1955–1956," mimeographed (USAFA, Colo., n.d.), pp. 2-3. See also Pete Todd and Lou Tidwell, "Year in Retrospect," *Talon* 1 (June 1956): 8.

12. Harmon to Fowler, quoted in *History of the USAFA 27 July 1954–12 June 1956,* p. 909.

13. William Truman Woodyard, "A Historical Study of the Development of the Academic Curriculum of the United States Air Force Academy" (Ph.D. diss., Univ. of Denver, 1965), p. 102. John L. Frisbee, Col. USAF, "Educational Program of the Air Force Academy," *Higher Education* 13 (Dec. 1956): 64-65.

14. Biographical data on key members of the initial staff and faculty are compiled from the following sources: (1) biographical sketches on file at the USAFA Archives (these sketches are the primary source of data on academy officials described later in the chapter also); (2) Woodyard, "A Historical Study," pp. 299-305; (3) *History of the USAFA 27 July 1954–12 June 1956,* II: Annex A. For West Point graduates on the Air Force Academy staff and faculty, supplemental data were obtained from the *Register of Graduates USMA* (West Point, N.Y.: Assn. of Graduates USMA, updated annually).

15. As of June 1948, academy graduates constituted 30.2 percent of regular Army officers, 38.0 percent of regular Navy officers, and 10.0 percent of regular Air Force officers. However, the pool of college graduates with commissions from sources other than West Point was considerably less than 90 percent of all Air Force officers, since only 41 percent of all Air Force officers, including academy graduates, had earned the baccalaureate degree by mid-1948. In the Army and Navy, approximately three-fourths of all officers had the baccalaureate. *History of the USAFA 27 July 1954–12 June 1956,* I: 369.

16. McDermott interview. Others with whom I talked who knew Harmon contend that McDermott is exaggerating here; but none denies that Harmon held West Point in very high esteem.

17. Harmon to Col. G. A. Lincoln, 2 March 1953, and Zimmerman to Harmon, 3 May 1954, cited in *History of the USAFA 27 July 1954–12 June 1956,* I: 390-394.

18. *History of the USAFA 27 July 1954–12 June 1956,* I: 408-410.

19. Woodyard, "A Historical Study," p. 60.
20. Talbott had told the director of athletics, "You find the kids, I'll appoint them." Briggs interview. Briggs was USAF Deputy Chief of Staff for Operations under Talbott. Details of the Academy Board meeting have been provided by McDermott, who represented the Dean at the meeting; interview. Plans for congressional appointments to the Academy were discussed in congressional *Hearings on H. R. 5337*, pp. 3018-3027, 3035-3036, 3054.
21. Pattillo to author, 17 Sept. 1977. Also, McDermott and Woodyard interviews. Both had talked with Pattillo about the accreditation problem.
22. McDermott interview.
23. Pattillo to author, 17 Sept. 1977. Pattillo learned, to his considerable irritation, that his letter to Harmon, which had included criticisms of Zimmerman's performance as Dean, had been referred to Zimmerman for reply.
24. Briggs interview.
25. McDermott interview.
26. Briggs interview.
27. Briggs interview.
28. Briggs interview.
29. *New York Times,* 6 March 1962, p. 14. Stillman at that time was commander of the 313th Air Division. His remarks were made to a gathering of the Society of American Engineers.
30. USAFA, Office of Information Services, Historical Div., *History of the United States Air Force Academy 13 June 1956–9 June 1957,* prepared by LTC. Edgar A. Holt, Dr. M. Hamlin Cannon, Dr. Victor H. Cohen, and Mr. Emory H. Dixon (USAFA, 1 Nov. 1958), I: 173.
31. Briggs interview.
32. Briggs interview.
33. Martin Mayer, "The Air Force's Egghead Factory," *Saturday Evening Post* 236 (23 Nov. 1963): 92-94.
34. USAFA, Office of Information, "Biography of Brig. Gen. Robert F. McDermott, Dean of the Faculty," n.d.
35. McDermott interview.
36. Woodyard, "A Historical Study," pp. 139-140.
37. Col. Robert F. McDermott, Dean of Faculty, Memorandum, subject: "Enrichment of the Curriculum," typewritten (USAFA, 5 Dec. 1956).
38. McDermott, Memorandum, 5 Dec. 1956, p. 2.
39. McDermott and Briggs interviews.
40. Erdle interview. The importance of the roles played by Higdon and Moody, in the early years of the Academy has been stressed to the author by many other interviewees also, including McDermott.
41. Moody interview.
42. Moody to author, 10 Nov. 1975.
43. McDermott interview.
44. Dean of Faculty, Memorandum, subject: "Absenteeism and Interference with Cadet Study Time and Free Time," 15 Oct. 1956, quoted in *History of the USAFA 13 June 1956–9 June 1957,* I: 173.
45. Data on the USAFA curriculum for various years are from the following

sources. Woodyard, "A Historical Study," p. 134. "Astronautics Course Set at Academy," *Army, Navy, Air Force Journal* 95 (15 Mar. 1958): 820, 844. Robert F. McDermott, Brig. Gen. USAF, "Educating Cadets for the Aerospace Age," *Air University Quarterly* 13 (Summer 1961): 3-17. USAFA, Academic Advisory Committee, "Report" (USAFA, 24 Feb. 1964). The Academy equated a "semester-hour" with "150 minutes of cadet effort per week, including both classroom work and outside preparation, throughout the semester of 17 or 18 weeks."

46. McDermott, in transcript of U.S. Academies of the Armed Forces, Conference of Superintendents, *Record of Proceedings* (Annapolis, Md.: USNA, March 1960), pp. 38-39.

47. Martin interview.

48. The other academies also decided that "overload" had unfortunate connotations and should be replaced by a euphemism. U.S. Academies of the Armed Forces, Conference of Superintendents, *Record of Proceedings,* 1961, p. 39.

49. See William E. Simmons, *Liberal Education in the Service Academies* (New York: Teachers College, Columbia University, for the Institute of Higher Education, 1965), pp. 134-135. Also *Air Force Times,* special issue devoted to the Academy, 3 April 1963, p. 7.

50. McDermott interview.

51. Woodyard interview. Also Archie Higdon, Brig. Gen. USAF, to author, 10 May 1976.

52. Peter R. Moody to author, 5 Jan. 1976.

53. Briggs interview.

54. Robert F. McDermott, Brig. Gen. USAF, Dean of the Faculty USAFA, "Some Answers to Our Critics," *The Talon* (June 1960), reprinted as a pamphlet for general distribution.

55. "Accreditation of Engineering Science Scores as Another 'First' at Academy," *Air Force Times,* 3 April 1963, p. 4. Presumably relying on information provided by the Air Force Academy, the *Times* erroneously claimed that the USAFA was the first service academy ever to receive ECPD accreditation. In fact, the Coast Guard Academy received ECPD accreditation as early as 1939, but had lost it in 1957 (see chapter 6).

56. McDermott, "Some Answers to Our Critics," Also Conference of Superintendents, *Record of Proceedings* (March 1960), p. 39.

57. Conference of Superintendents, *Record of Proceedings* (March 1960), p. 41.

58. President Dwight D. Eisenhower, quoted by John Eisenhower, his Assistant Staff Secretary, in a memorandum to the Secretary of Defense, excerpted by Robert F. McDermott, Brig. Gen. USAF, Dean of the Faculty USAFA, memorandum for record, 11 Aug. 1960.

59. Robert F. McDermott, Brig. Gen. USAF, Dean of the Faculty, USAFA, memorandum to Col. J. S. Hudson, USMC, Office of the Asst. Secretary of Defense (Manpower, Personnel and Reserves), subject: "Further Information on the Air Force Academy's Proposed Master's Program" [27 Dec. 1960].

60. American Council on Education, Committee on Relationships of Higher Education to the Federal Government, "Minutes of the Meeting," typewritten (Washington, D.C.: 13 April 1961), appendix: "The Air Force Academy's Proposed Master's Degree Program."

61. McDermott interview.

62. LeMay, quoted by McDermott, interview.

63. Complaints of this sort were communicated to me during a research visit to the Academy in April 1963.

64. The area that separates the academic classrooms (the domain of the Dean) from the cadet barracks (the domain of the Commandant) is of terrazzo surface. Thus, at the Air Force Academy the Athens-Sparta rift is known as the "terrazzo gap."

65. "Experience, Motivation, Character, Leadership Stressed," special issue of the *Air Force Times* devoted to the Air Force Academy, 3 April 1963, p. 9. Mayer, *Saturday Evening Post* 236 (23 Nov. 1963): 92-94.

66. Memorandum, "House Committee," 21 Feb. 1962, Tab B. p. 14.

67. Coltrin, Crump, and Ferrari interviews.

68. Thomas Thompson, "Ex-Pilot Guides 3-Phase Program," *Amarillo* (Texas) *Globe-Times*, 11 Sept. 1963, pp. 40-41. Thompson, editor of the *Globe-Times*, was a guest of Warren's for the visit, along with Texas oil men and insurance executives.

69. Rosen '67, "The Commandant of Cadets," *The Talon* 10 (June 1965): 11.

70. [Cadet] Rosen, *The Talon* 10 (June 1965): 11.

71. U.S., Dept. of the Air Force, "Report of the Special Advisory Committee on the United States Air Force Academy," mimeographed (Washington, D.C.: Dept. of the Air Force, 5 May 1965). Cited hereafter as "White Committee Report."

72. USAFA, *Annual Report of the Superintendent, FY 1965*, excerpted at length in J. Arthur Heise, *The Brass Factories* (Washington, D.C. Public Affairs Press, 1969), chap. 3.

73. Erdle and McDermott interviews.

74. "White Committee Report," p. 54.

75. It was revelations of new abuses of athletic recruiting, McDermott is convinced, that triggered the replacement of the Superintendent and the Commandant. Interview.

76. "White Committee Report," pp. 85-86.

77. "White Committee Report," pp. 28-29, 85-88.

78. "White Committee Report," pp. 85, 40-43.

79. Morgenstern, quoted in Heise, *The Brass Factories*, pp. 53-54.

80. "White Committee Report," pp. 90-91.

81. Charles Konigsberg, Lt. Col. USAF, to Rep. F. Edward Hébert, 4 Sept. 1967, reprinted in U.S., Congress, House, Committee on Armed Services, Special Subcommittee on Service Academies, *Report and Hearings*, 90th Cong., 1st and 2d sess., 1967–1968, p. 10899. Konigsberg's own testimony as a faculty member before the White Committee had been followed by an order to have him "thrown off the base within 24 hours." The intervention of sup-

portive colleagues led to the rescinding of the order. Konigsberg to author, 25 Aug. 1976.

82. *Report and Hearings,* 1967–1968, p. 10900.

83. Martin interview.

84. Brown to Moorman, quoted in USAFA, Directorate of Historical Studies, *History of the United States Air Force Academy 1 July 1965–31 July 1970,* prepared by Dr. M. Hamlin Cannon and Henry S. Fellerman (USAFA, July 1970), p. 83.

85. See William Snead, Jr., with Jack Shepherd, "Air Academy's Cheating Scandal," *Look* 31 (24 Jan. 1967): 23-25.

86. U.S., Congress, House, Committee on Armed Services, *Report and Hearings of the Special Subcommittee on Service Academies,* 90th Cong., 1st and 2d sess., 1967–1968, p. 10909.

87. *Report and Hearings,* p. 10912.

88. U.S., President, *Report of the Board of Visitors to the USAFA 1967,* USAFA, 1967, p. 16.

89. *Report and Hearings,* pp. 10224b-10224e.

90. McDermott became president of the United Services Automobile Association.

Chapter Five

1. USMA, Committee on Curriculum Survey (Green Board), *Report and Recommendations to the Dean of the Academic Board* (West Point, N.Y.: USMA, 1 April 1954), p. 6. Cited hereafter as *Green Board Report.*

2. *Green Board Report,* App. A, p. A-27.

3. Irving Katenbrink, "The Results of a Recent Survey," *The Pointer* 31 (5 March 1954): 4-7.

4. USMA, Superintendent, *Annual Report* (1954), p. 4. USMA, Superintendent, *Report of the Working Committee on the Historical Aspects of the Curriculum for the Period 1802–1945,* prepared by the Superintendent's Curriculum Committee chaired by LTC. C. E. Covell (West Point, 31 July 1958), Appendices 22–23.

5. Katenbrink, *The Pointer* 31 (5 Mar. 1954): 7. Superintendent, *Annual Report* (1954), pp. 3-4.

6. Davidson to author, n.d. [July 1974].

7. Davidson to author, 6 March 1974.

8. Davidson to author, n.d. [July 1974]. See also "Academy's New Chief Played Soldier as a Boy in the Bronx," *New York Times,* 18 March 1956, p. 7.

9. The account of these events is based upon descriptions, consistent with one another in basic detail, provided in correspondence with Davidson and in Earl "Red" Blaik, *The Red Blaik Story* (New Rochelle, N.Y.: Arlington House, 1974), chaps. 6-8. The latter incorporates and supplements *You Have to Pay the Price,* which Blaik wrote with Tim Cohane, foreword by Douglas MacArthur (New York: Holt, Rinehart, and Winston, 1960).

10. "Next Superintendent, USMA," *Assembly* 15 (April 1956): 1.

11. Davidson to author, 25 Feb. 1976.
12. Birrer interview.
13. Bryan to Davidson, 19 March 1956.
14. USMA, Superintendent, *Annual Report*, 1957, pp. 3-5; 1958, pp. 2-4; 1959, pp. 4-5. Garrison H. Davidson, "Plain Talk," *Assembly* 19 (Summer 1960): 1. "In Retrospect, General Davidson's Four Years as Superintendent, 1956-1960," *Assembly* 19 (Summer 1960): 22-23.
15. Davidson to author, n.d. [July 1974].
16. USMA, Superintendent, *Annual Report*, 1957, p. 12.
17. Davidson to author, n.d. [July 1974].
18. Davidson, "Plain Talk," *Assembly* 16 (April 1957): 1, and *Assembly* 17 (Winter 1959): 1.
19. "Intercollegiate Athletic Policy," *Assembly* 16 (Fall 1957): 16; emphasis in the original text. The policy statement was issued on 26 August 1957.
20. I am convinced by Davidson's letter to me of 14 July 1974 and by Blaik's comments in *The Red Blaik Story* that each of the two men, recognizing the reservoir of ill will that had built up from their association in the 1930s, was studiously correct in his dealings with the other after Davidson became Superintendent, however determined each might be not to yield to the other's views on matters of athletic policy.
21. Blaik, *The Red Blaik Story*, pp. 394-397. Substantiation of Blaik's claim that the remainder of the athletic program was underwritten by income derived from Army football games is provided by the "Army Athletic Association Financial Statement," *Assembly* 17 (Winter 1959): 7.
22. Blaik, *The Red Blaik Story*, p. 394.
23. Davidson, "Plain Talk," *Assembly* 17 (Winter 1959): 1.
24. Blaik, *The Red Blaik Story*, p. 391.
25. Davidson, "Plain Talk," *Assembly* 18 (Spring 1959): 1.
26. Davidson to author, 25 Feb. 1976. Also, Bryan to Davidson, 19 Mar. 1956.
27. Details in this and the succeeding two paragraphs on Throckmorton's career are from "New Commandant of Cadets," *Assembly* 15 (Oct. 1956): 14.
28. USMA, Superintendent, *Annual Report* 1958, p. 6. Also, *Assembly* 16 (Summer 1957): 1.
29. Throckmorton to author, 11 April 1976.
30. Throckmorton to author, 11 April 1976.
31. The report formed the basis for a two-part article on "The 4th Class System," *Assembly* 21 (Winter 1963): 6-9; *Assembly* 22 (Spring 1963): 10-13.
32. Bean to author, 18 Feb. 1976.
33. George M. Bean, *Be Not Conformed to this World: Sermons Preached to the Class of 1955 by their Chaplain* (New York: Morehouse-Gorham, 1955), pp. 11-20.
34. Katenbrink, *The Pointer* 31 (5 March 1954): 4-7.
35. USMA, Office of the Registrar, Research Division, "Over-all Tabulation, First Class Questionnaire, Class of 1960," Project No. 6005-01-2, mimeographed (West Point, N.Y.: USMA, 20 Oct. 1960), p. 37.
36. Bean to author, 18 Feb. 1976.

37. Bean to author, 18 Feb. 1976.
38. Throckmorton to author, 11 April 1976. Emphasis in the original text.
39. Davidson to author, 25 Feb. 1976. Bean and his family frequently were invited to the Bryan home, and the Bryans in turn came often to the Bean's to "relax and let down their hair." This pattern was not maintained under Davidson. Bean to author, 18 Feb. 1976.
40. "Clerics Charge West Point Bias," *New York Times*, 25 April 1958, p. 15.
41. Bean to author, 18 Feb. 1976.
42. Davidson to author, 25 Feb. 1976.
43. Throckmorton to author, 11 April 1976.
44. Davidson to author, 25 Feb. 1976.
45. USMA, Superintendent, *Annual Report to the Army Chief of Staff*, 1957, p. 7.
46. Cutler and Knowlton interviews. Both were members of the Ewell Board.
47. USMA, *Superintendent's Curriculum Study* (West Point, N.Y.: USMA, 1958), App. A: "The Army Officer of the Future."
48. USMA, Superintendent, *Annual Report 1957*, p. 7.
49. USMA, *Superintendent's Curriculum Study*, "Report of the Evaluation Committee."
50. USMA, *Superintendent's Curriculum Study*, "Report of the Evaluation Committee."
51. USMA, Superintendent, *Annual Report 1959*, p. 11.
52. Davidson to author, n.d. [July 1974].
53. See Col. Charles P. Nicholas, Professor of Mathematics USMA, "Preparing the Weapon of Decision," *Assembly* 17 (Winter 1959): 11-13.
54. Jordan interview.
55. The quote is from an unpublished, typewritten autobiographical sketch that General Davidson is preparing for his grandchildren, and made available to me with a letter 23 Jan. 1976.
56. Cutler interview.
57. Davidson autobiographical sketch.
58. USMA, Superintendent, Memorandum to Army Chief of Staff, "Modification of the Academic Curriculum, USMA," 29 Feb. 1960, Incl. 1.
59. "Modification of the Academic Curriculum, USMA," Incl. 1.
60. When typed by his secretary in final form, the memorandum to Lemnitzer came to thirty-two pages. Davidson to author, 14 July 1974. Lemnitzer to Davidson, 16 June 1960.
61. Davidson to Lemnitzer, 12 Nov. 1959.
62. Davidson to Gruenther, 4 April 1960.
63. Lemnitzer to Davidson, 2 May 1960. Emphasis in the original text.
64. "June Week 1960," *Assembly* 19 (Summer 1960): 4-11.
65. David Halberstam, *The Best and the Brightest*, paperback ed. (Greenwich, Conn.: Fawcett, 1972), p. 666. For his own views on the war, see General William C. Westmoreland, *A Soldier Reports* (Garden City, N.Y.: Doubleday, 1976).

66. For the use of such an image by observers, see Ernest B. Furgurson, *Westmoreland: The Inevitable General* (Boston: Little, Brown, 1968); Maureen Mylander, *The Generals: Making It, Military-Style* (New York: Dial Press, 1974), p. 6; Halberstam, *The Best and the Brightest,* pp. 670-671. Westmoreland was an Eagle Scout.
67. Morris Janowitz, *The Professional Soldier: A Social and Political Portrait* (New York: Free Press, 1960), esp. chaps. 2-3.
68. Williams interview.
69. Davidson to Westmoreland, 22 June 1960.
70. Brig. Gen. William W. Bessell, "First Year of Evolution," *Assembly* 20 (Fall 1961): 11.
71. Westmoreland to Corps of Cadets, quoted in Halberstam, *The Best and the Brightest,* pp. 679-680.

Chapter Six

1. W. K. Earle, Captain USCG (ret.), "The Academy Today—An Old Grad's View," USCGA *Alumni Bulletin* 37 (Nov.-Dec. 1975): 57.
2. W. K. Earle to author, 1 June 1976.
3. W. K. Earle, "Professional Studies at the Academy—A 20-Year Survey," USCGA *Alumni Bulletin* 37 (March-April 1975): 14-19.
4. Albert A. Lawrence, Captain USCG (ret.), to author, 6 June 1976.
5. Admission Office USCGA.
6. Captain Joseph Pois, a reserve officer, was the management specialist who advanced the argument. A. C. Richmond, Admiral USCG (ret.), to author, 2 July 1976.
7. Richmond to author, 2 July 1976.
8. Richmond to author, 2 July 1976.
9. Richmond to author, 2 July 1976.
10. Earle to author, 1 June 1976. Earle, an instructor in law at the time (and acting Commandant of Cadets during the summer months), was the officer who made the suggestion, and was appointed to head the committee to study the problem.
11. Richmond to author, 2 July 1976; Earle to author, 1 June 1976.
12. Richmond to author, 2 July 1976.
13. Biographical data in this and subsequent paragraphs are from U.S. Coast Guard Headquarters, Public Information Division, "Rear Admiral Frank A. Leamy," mimeographed biographical sketch (Washington, D.C.: USCG, Feb. 1964). See also the epitaph by P. F. Foye, "Rear Admiral Frank A. Leamy, USCG (Ret.)," USCGA *Alumni Association Bulletin* 28 (1966): 254-255.
14. Willard Smith interview.
15. Willard Smith interview.
16. Willard Smith interview. Leamy sent a letter to then-Commander Earle thanking him for his work as chairman of the committee whose report to Mauerman had been suppressed, but which Leamy had seen, and noting that

many of the committee's recommendations, including the establishment of a system of company officers, had been put into effect. Earle to author, 1 June 1976.

17. U.S., Congress, "Report of the Board of Visitors to the U.S. Coast Guard Academy, New London, Conn., May 2, 1958," Senate, *Congressional Record*, 86th Cong., 1st sess., 1959, p. 9302.

18. Willard Smith interview.

19. Otto Graham, Captain USCG, to author, 19 July 1976.

20. Albert A. Lawrence, Capt. USCG (ret.), to author, 6 June 1976.

21. Willard Smith interview.

22. The two faculty members who were given this assignment by Admiral Leamy were Stanley Smith, then head of the Department of Mathematics, and Paul Foye, then an instructor in the Department of Humanities. Each of these officers, in turn, later became Dean of the Academy. Stanley Smith interview.

23. U.S. Armed Service Academies, "Report of the Conference of Superintendents, Academies of the Armed Forces, 1958," mimeographed (New London, Conn.: USCGA, 1958).

24. Earle to author, 1 June 1976. Also Stanley Smith interview.

25. Stanley Smith, Captain USCG (ret.), to author, 29 June 1976.

26. Captain W. R. Richards, "ECPD and the Academy," USCGA *Alumni Association Bulletin* 19 (Jan.-Feb. 1958): 14-18.

27. Richmond to author, 2 July 1976; Lawrence to author, 6 June 1976; Stanley Smith to author, 29 June 1976.

28. Foye and Wells interviews.

29. U.S., Armed Services Academies, "Report of the Conference of Superintendents, Academies of the Armed Forces, 1960," mimeographed (Annapolis, Md.: USNA, 1960), pp. 49-50.

30. Lawrence to author, 6 June 1976. Leamy died in 1966. Virtually everyone associated with the Coast Guard Academy with whom I have communicated has nothing but high praise for Leamy. It seems clear that his reputation as the "father of the modern Coast Guard Academy" is well established.

31. Biographical data on Evans in this and subsequent paragraphs are from U.S. Coast Guard Headquarters, Public Information Division, "Rear Admiral Stephen H. Evans," mimeographed biographical sketch (Washington, D.C.: USCG, June 1960).

32. Stephen Hadley Evans, Captain USCG, *The United States Coast Guard 1790–1915, A Definitive History*, with a Postscript: 1915–1949 (Annapolis, Md.: U.S. Naval Institute, 1949).

33. Richmond to author, 2 July 1976; Evans to author, 20 June 1976.

34. Stephen H. Evans, Rear Admiral USCG, "Remarks at a General Muster of Personnel on the Occasion of His Assuming Command of the 14th Coast Guard District, 5 July 1957," mimeographed copy made available to the author by Admiral Evans. Emphasis in the original text.

35. See S. H. Evans, "Honor in the Corps," USCGA Alumni Association *Bulletin* 38 (Jan.-Feb. 1976): 2-3. Also, Evans to author, 20 June 1976.

36. USCGA, "Minutes of the Academy Advisory Committee, Meeting Number 63, 25-26 April 1960."

37. Willard Smith interview.

38. USCGA, "Minutes of the Academy Advisory Committee, Meeting Number 65, 15-16 March 1961," Appendix B, Dean's Report to the Committee, prepared 8 March 1961.

39. USCGA, Superintendent, "Faculty: Improvement of," typewritten memorandum to Commandant USCG, 22 August 1960. Also, USCGA, "Proposed Plan for an Adequate Faculty for 600 Cadets," mimeographed (New London, Conn.: revised 22 August 1960). The two documents are enclosures to USCGA, "Minutes of the Academy Advisory Committee, Meeting Number 64, 28-29 November 1960," Appendix IX.

40. USCGA, 1961–1962 *Catalogue of Courses* (Washington, D.C.; USGPO, 1961) pp. 69-71.

41. USCGA, Superintendent, "House Committee on Appropriations Interest in Service Academies," letter to Commandant USCG, 5 February 1962.

42. Robert A. Wallace, Special Assistant to the Secretary of the Treasury, to Rear Admiral Stephen H. Evans, Superintendent, USCGA, 3 August 1961; and Wallace to Fred Dutton, Cabinet Secretary, The White House, 7 August 1961. Copies of the letters provided to the author by Admiral Evans. Also, USCGA, "Minutes, Meeting Number 65."

43. This dialogue with the White House has been recounted by Evans to author, 20 June 1976, and by Lawrence to author, 6 June 1976.

44. "This was accomplished by increasing our efforts to disseminate cadet procurement information in those high schools and colleges with heavy black enrollments and by enlisting the interest and cooperation of well-known and respected blacks in the national community." Evans to author, 20 June 1976.

45. Evans to author, 20 June 1976.

46. S. Hadley Evans, Rear Admiral USCG, Superintendent USCGA, "Remarks to the Senior Cadets Regimental Officers, 11 Feb. 1961," mimeographed.

47. USCGA, "Minutes of the Academic Advisory Committee, Meeting Number 64, 28-29 Nov. 1960," Appendix V, Table 12. Data in the table appeared originally in *Life* (3 Oct. 1960).

48. Clarence Cannon to Robert S. McNamara, 29 November 1961. Also, Roswell L. Gilpatrick (Deputy Secretary of Defense) to Clarence Cannon, 12 December 1961. Copies of the correspondence provided to the author by Admiral Evans.

49. USCGA, "Minutes of the Academy Advisory Committee, Meeting Number 67, 5-6 March 1962," Appendices B and I.

50. USCGA, "Minutes, Meeting Number 67," App. I.

51. Roland as quoted by Willard Smith, interview.

52. Willard Smith interview.

53. Willard Smith interview. At cadet initiative, comprehensive surveys were undertaken in January 1964 of the reactions of cadets to the recent reforms of the swab system. Swabs themselves (Class of 1967) reacted favor-

ably. One of them asked, rhetorically, "Are we bound to tradition or to reason?" Members of the classes of 1965 and 1966, however, which had the major responsibility for administering the system, feared that the "slackening" of requirements had dangerously undermined the value of the system. Reports on the surveys supplied to the author by William C. Carr (USCGA '65).

54. Willard Smith interview.

55. Adams to author, 14 Aug. 1976. Details of Adams's career are contained in *Who's Who in America*, 39th ed., 1976-77 (Chicago: Marquis, 1976).

56. See "Dean of Instruction Retires," USCGA *Alumni Association Bulletin* 23 (Nov.-Dec. 1961): 16. Also Stanley Smith interview.

57. Stanley L. Smith, Captain USCG, Dean of Academics USCGA, "U.S. Coast Guard Academy Modified Curriculum," USCGA *Alumni Association Bulletin* 27 (1965): 262. Foye interview. Adams to author, 14 Aug. 1976. Stanley Smith to author, 29 June 1976.

58. Willard Smith interview.

59. Evans to author, 20 June 1976; Willard Smith interview.

60. Earle, USCGA *Alumni Bulletin* 37 (March-April 1975): 14-19. S. L. Smith, USCGA *Alumni Association Bulletin* 27 (1965): 262. William B. Hewitt, Lt. USCG, "Where Are We Going?" USCGA *Alumni Association Bulletin* 28 (1966): 259-265.

61. Stanley Smith interview.

62. Raymond J. Perry, Captain USCG, "An Educational Approach to Computers," USCGA *Alumni Association Bulletin* 27 (1965): 266-279. John H. Hanna III, Ensign USCG, and Harold E. Millan, Jr., Ensign USCG, "The Academy Scholars Program, A Total Educational Experience," USCGA *Alumni Association Bulletin* 28 (1966): 357-360.

63. See Gary Russell, "The Bureaucratic Culture and Personality of the United States Coast Guard" (M.A. thesis, University of Connecticut, 1973).

64. Lawrence to author, 6 June 1976.

65. As Admiral Smith himself has acknowledged, emphasizing the fundamental conservatism in such matters of cadets themselves.

66. Joseph Henry Hughes, Jr., *The Making of a Coast Guard Officer: A Covenant with Honor* (New York: Philosophical Library, 1966), 79. Most of the book consists of letters written to Hughes by his son, Joseph III.

67. Hughes, *The Making*, pp. 296-297.

68. Willard Smith has indicated that initially he had hoped that the company officers could serve as effective counsellors, but eventually came to believe that Captain Malcolm Williams, the staff psychologist, had been accurate in cautioning that the task was incompatible with the disciplinary role. Smith and Williams interviews.

Chapter Seven

1. U.S., Congress, House of Representatives, Committee on Appropriations, *Hearings: Report on Russia by Vice Admiral Hyman G. Rickover,* 86th Cong., 1st sess., 1959, esp. pp. 67-72.

2. These events are described in detail by Clay Blair, Jr., *The Atomic Submarine and Admiral Rickover* (New York: Holt, 1954), pp. 193-255.

3. James Calvert, Commander USN, *Surface at the Pole* (New York: McGraw-Hill, 1960), p. 14.

4. See Elmo R. Zumwalt, Jr., *On Watch* (New York: Quadrangle, New York Times Book Co., 1976), chap. 5: "The Rickover Complication."

5. William E. Simons, *Liberal Education in the Service Academies* (New York: Teachers College, Columbia University, for the Institute of Higher Education, 1965), p. 141n.

6. U.S., *Report of the Board of Visitors to the Naval Academy,* 15 March 1957, p. 5.

7. Melson interview.

8. See Wayne Hughes, Jr., Lt. USN, "New Directions in Naval Academy Education," *U.S. Naval Institute Proceedings* 86 (May 1960): 37-45. Also Joseph H. Nevins, Jr., Rear Admiral USN (ret.), "The United States Naval Academy and its Curriculum: A Chronology of Changes and Some Problems," *Shipmate* 31 (March 1968): 7-10.

9. The number of commissioned officers with Ph.D.'s was approximately 1 in 300 in the Army; 1 in 500 in the Air Force; and 1 in 800 in the Navy. "Questions and Answers at the Alumni Assembly," *Shipmate* 24 (Nov. 1961): 10.

10. Shinn interview.

11. Melson interview.

12. Melson interview.

13. J. F. Davidson interview.

14. *New York Times,* 12 Feb. 1961, p. 1; 13 Feb. 1961, p. 20.

15. U.S., Congress, House, Committee on Appropriations, Subcommittee on Department of Defense, *Hearings: DOD Appropriations 1962,* 87th Cong., 1st sess., Part 6, p. 32.

16. *Hearings: DOD Appropriations 1962,* Part 6, pp. 46-49, 31.

17. U.S., Secretary of the Navy, Directive of 22 May 1962, excerpted in *Army, Navy, Air Force Journal and Register* 94 (2 June 1962): 1.

18. J. F. Davidson interview.

19. *Army, Navy, Air Force Journal and Register* 94 (2 June 1962): 1, 46.

20. USNA, Superintendent, "Report to Board of Trustees," in *Shipmate* 26 (Jan. 1963): 15.

21. See the analysis by Hanson W. Baldwin of factors involved in Korth's resignation; *New York Times,* 15 Oct. 1963, p. 33.

22. Robert W. McNitt, Rear Adm., USN (ret.), "Challenge and Change, The Naval Academy—1959–1968," *Shipmate* 35 (April 1972): 3-6. McNitt served

as Secretary of the Academic Board at the time of the 1964 Naval Academy reorganization.

23. Robert W. McNitt, Captain, USN, "Tecumseh, God of the 'C'," *Shipmate* 26 (Sept.-Oct. 1963): 12-13, 18.

24. C. C. Kirkpatrick, Rear Admiral, USN, Supt. USNA, "A Century of Academic Achievements," *Shipmate* 26 (Sept.-Oct. 1963): 4. Hanson W. Baldwin, "Changes at Annapolis: Two Years Under Admiral Kirkpatrick Revolutionize the System of Teaching," *New York Times*, 21 Dec. 1963, p. 10.

25. Huston interview.

26. David Boroff, "Annapolis: Teaching Young Sea Dogs Old Tricks," *Harper's* 226 (Jan. 1963): 46-52. See also his "West Point: Ancient Incubator for a New Breed," *Harper's* 225 (Dec. 1962): 51-59; and "Air Force Academy: A Slight Gain in Altitude," *Harper's* 226 (Feb. 1963): 86-98.

27. Kirk interview.

28. There is a variation of as much as three percentage points among the various reports of the use of grading quotas at Annapolis. The apparent discrepancies may be explicable in terms of actual variations among departments in any given year, and in terms of slight changes that occurred in the quotas from one year to the next. Cf. Leroy F. Aarons, "Improve Studies, Report Urges Naval Academy," *Washington Post*, 2 April 1966; Ben A. Franklin, "Grades Inflated For Midshipmen," *New York Times*, 10 April 1966, p. 1; Luther J. Carter, "U.S. Naval Academy: Faculty Unrest," *Science* 152 (20 May 1966): 1043-1045.

29. Carter, *Science* 152 (20 May 1966): 1044.

30. Ben A. Franklin, "Annapolis Chided on Grade Quotas," *New York Times*, 1 May 1966, pp. 1, 39.

31. Richard C. Vitzthum, quoted in *New York Times*, 12 April 1966, p. 20.

32. Quoted by Franklin, *New York Times*, 1 May 1966, p. 1.

33. Evaluation Team Report, quoted by Leroy F. Aarons, "Accrediting of Academy Reaffirmed," *Washington Post*, 1 May 1966, p. A1.

34. Evaluation Team Report, quoted by Aarons, *Washington Post*, 1 May 1966, p. A1.

35. Commission on Institutions of Higher Education, Middle States Association of Colleges and Secondary Schools, Official Press Release, 30 April 1966, reprinted in U.S., President, *Report of the Board of Visitors to the United States Naval Academy 1966*, App. I.

36. Ben A. Franklin, "Rift at Annapolis Remains After Mediators Leave," *New York Times*, 20 Nov. 1966, p. 77.

37. *New York Times*, 20 Nov. 1966, p. 77.

38. *New York Times*, 20 Nov. 1966, p. 77.

39. Kauffman, written statement, U.S., Congress, House Committee on Armed Services, Special Subcommittee on Service Academies, *Report and Hearings: Administration of the Service Academies*, 90th Cong., 1st and 2d sess., 1968, p. 10372.

40. Aarons, *Washington Post*, 1 May 1966, p. A1.

41. The "Hawthorne effect" was first observed in managerial experiments

with working conditions at the Hawthorne plant of Western Electric in the 1920s. Management altered environmental conditions such as lighting in an effort to improve morale and production, but discovered that morale and productivity in the control group as well as the experimental group improved. They concluded—perhaps fallaciously—that the increased attention that management devoted to workers in both control and experimental groups led to increased morale and productivity. For a critique of this view, see Charles Perrow, *Complex Organizations* (Glenview, Ill.; Scott Foresman, 1972).

42. Kauffman statement, *Report and Hearings, 1967–1968*, p. 10372.

43. *Report and Hearings, 1967–1968*, pp. 10287-10288.

44. USNA, Professional Training and Education Committee, "The Professional Training and Education of Midshipmen at the U.S. Naval Academy: A Final Report," 20 Nov. 1967, in *Report and Hearings, 1967–1968*, App. B, pp. VII-XLIX. The final report of the PT&E Committee includes the preliminary report, completed in May 1967, and an interim report, completed in August 1967.

45. Kauffman testimony, *Report and Hearings, 1967–1968*, p. 10414.

46. W. F. V. Bennett, "The Professional Training and Education of Midshipmen at the United States Naval Academy Today," *Shipmate* 31 (June 1968): 7-19. Bennett chaired the 1967 PT&E Committee. The article provides a valuable detailed account for the years 1958–1968.

47. U.S., President, *Report of the Board of Visitors to the United States Naval Academy* (Annapolis, Md.: USNA, 30 April 1967), p. 24.

48. Calvert, *Surface at the Pole*, pp. 11-16.

49. Calvert, *Surface at the Pole*.

50. James Calvert, *The Naval Profession* (New York: McGraw-Hill, 1965). When Calvert became Academy Superintendent, a revised edition was published in 1971.

51. Details of Calvert's career are provided in "Rear Admiral James F. Calvert, Superintendent Designate," *Shipmate* 31 (June 1968): 5. See also comments by Zumwalt, *On Watch*, p. 96.

52. Calvert interview.

53. Calvert, *The Naval Profession*, revised ed., pp. 107-108. Also, Calvert, "The Fine Line at the Naval Academy," *U.S. Naval Institute Proceedings* 96 (Oct. 1970): 63-68.

54. Collectively, the four ships were designated the Naval Academy Training Squadron (NATRON). USNA, Superintendent, *Report to the Board of Visitors* (Annapolis, Md.: USNA, 25 April 1969; also, 10 Dec. 1971).

55. USNA, Superintendent, *Report to the Board of Visitors* (Annapolis, Md.: USNA, 25 April 1969), pp. 15-16.

56. Calvert to newly reporting faculty, Orientation Program, 19 Aug. 1971, USNA, Annapolis, Md. The author was in attendance. Also, "'Failure' of Parents Cited by Calvert," Annapolis *Evening Capital*, 19 Jan. 1972. p. 3.

57. Peters interview.

58. Detailed admissions and attrition data for the USNA classes of 1947–1972 are provided by the Assistant Registrar, USNA, in *Shipmate* 31 (Dec.

1968): 13. These data include the number who were qualified "mentally," but not the number who were fully qualified in all respects each year. The former, not the latter, typically are reported in an annual report by the USNA Dean of Admissions. However, the Superintendent revealed the magnitude of the problem regarding fully qualified applicants in an address to the alumni, reported in *Shipmate* 35 (Jan. 1972): 5.

59. Unlike the Coast Guard Academy, which relies entirely upon a system of nationwide presidential appointments, the Naval, Military, and Air Force academies have appointments from a variety of sources, roughly two-thirds of them being from U.S. senators and representatives.

60. "A Serious and Growing Problem: A Message from the Superintendent," *Shipmate* 31 (April 1968): 5-6. Also Kauffman testimony, *Report and Hearings, 1967–1968*, esp. pp. 10377-10389.

61. Presentation by Thomas C. Lynch, Lieutenant Commander USN, Assistant Director, Recruitment and Candidate Guidance, USNA, at orientation for new faculty, USNA, 20 Aug. 1971.

62. "The Alumni Assembly," *Shipmate* 35 (Jan. 1972): 5.

63. Kauffman testimony, *Report and Hearings*, p. 10533.

64. *Shipmate* 32 (Nov. 1969): 14.

65. Presentation by Coppedge to parents of the Class of 1975, Parents Forum, 28 Aug. 1971.

66. Observed by the author during his year as a visiting professor at the Academy, and reported to him by numerous midshipmen.

67. Calvert to newly reporting faculty, Orientation Program, 19 Aug. 1971.

68. A perceptive discussion of the "hairy chest syndrome" and related components of "the operational code of the national security managers" (civilian and military) is provided by Richard J. Barnet, *Roots of War: The Men and Institutions Behind U.S. Foreign Policy* (New York: Atheneum, 1972; Penguin, 1973), esp. chap. 5.

69. Calvert, address to USNA alumni, "Homecoming 1968," *Shipmate* 31 (Dec. 1968): 3.

70. Calvert interview.

71. Megargee interview.

72. Calvert interview. The other three boards that were established were the Facilities Review Board (real estate, buildings and grounds); the Financial Resources Board (budget review); and the Manpower Review Board (enlisted, civil service, faculty, and officer personnel).

73. Details of curriculum developments at the Naval Academy from the 1950s through 1975 are described in USNA, *Report to the Commission on Institutions of Higher Education, The Middle States Association of Colleges and Secondary Schools* (Annapolis, Md.: USNA, Feb. 1976), chap. 2: "USNA Curriculum Development," prepared by John D. Yarbro, Professor Area-Language Studies, USNA. The Yarbro report was prepared initially in the summer of 1966 for the newly created Academic Advisory Board. It was updated in January 1974 for the Clements Committee, and then again in December 1975 for inclusion in the above report. Yarbro to author, 16 Nov. 1976

74. Conord interview. Conord had been one of the nucleus of instructors who participated in the summer 1966 workshop with Quinn. See also USNA, Academic Computing Center, "Tutors for Midshipmen? Modern Technology Brings Back Advantages of Individual Instruction," *Shipmate* 32 (May 1969): 4-8.

75. Bruce Johnson interview.

76. See R. C. Gentz, Commander USN, "The Current Status of the Engineering Curricula at the U.S. Naval Academy," *Shipmate* 35 (March 1972): 17-20.

77. Pitt interview.

78. Reported to the author by his midshipmen students during 1971–1972.

79. Huston interview. Huston cited as one example of the "Neanderthals" a Navy captain whose response to the prospective hiring of a historian on the faculty was, "Couldn't we find someone without a beard?"

80. Calvert interview.

81. Calvert interview.

82. A detailed chronology of the proposals and counterproposals that were generated in the weeks leading up to the reorganization is provided in AAUP, President USNA Chapter, to Superintendent, USNA, 31 Dec. 1969. Correspondence provided to the author by Richard Megargee, who was the AAUP chapter president.

83. AAUP, President USNA Chapter, to Superintendent, USNA, 31 Dec. 1969.

84. Calvert interview.

85. Presentation by A. Stuart Pitt, Professor and Chairman, Department of English, USNA, at orientation for new faculty, 23 Aug. 1971.

86. The expression was used frequently by insiders at Annapolis during the Calvert years.

87. "Admiral Calvert Shares 'Memories of the Future,'" *Anne Arundel* (Maryland) *Times*, 29 June 1972, pp. 13-14. Emphasis is added.

88. *Anne Arundel Times*, 29 June 1972, pp. 13-14.

Chapter Eight

1. In Janowitz's terms, career choices that for the "first wave" had been "adaptive" departures from the norm had come to be among the "prescribed" alternatives. Morris Janowitz, *The Professional Soldier: A Social and Political Portrait* (Glencoe, Ill: Free Press, 1960), pp. 165-171.

2. Eleven of every twelve college presidents before the Civil War were clergymen. Frederick Rudolph, *The American College and University: A History* (New York: Vintage Books, 1962), p. 170.

3. Quote attributed to an unnamed friend of Goodpaster's, by Linda Charlton, "New West Point Superintendent," *New York Times*, 5 April 1977, sec. 1, p. 26. Defense Secretary Harold Brown observed that "General Goodpaster possesses the unique blend of military and educational background and

experience deemed necessary for the superintendency today and reflects [*sic*] the best qualified individual available to the Army at this time." "Goodpaster Will Leave Retirement to be West Point Superintendent," *New York Times,* 5 April 1977, sec. 1, pp. 1, 26.

4. However, there have been important fluctuations over time in the seniority gaps between Superintendents and Commandants. The Air Force Academy provides the most extreme example. The first Superintendent, Hubert Harmon, had graduated from West Point a full twenty years ahead of his Commandant of Cadets, Robert Stillman. Yet the fourth Commandant of Cadets, Robert Strong, Jr., was a West Point classmate of the Superintendent under whom he served beginning in 1963, Robert Warren. The modal pattern at the academies has been roughly midway between these extremes.

5. Previous research suggests that the warrior image of the infantry is perpetuated not only by its mission ("Close with and destroy the enemy"), but also by self-selection into the infantry of officers whose values support the image. West Point cadets whose orientation toward the military profession emphasized "heroic" values were most inclined to select infantry as their branch of service preference (and to a somewhat lesser extent, armor and artillery, other combat arms). In contrast, cadets with a "managerial" orientation were likely to prefer service in the Army Engineers or the Air Force (then an option for West Pointers). John P. Lovell, "The Professional Socialization of the West Point Cadet," in *The New Military,* ed. Morris Janowitz (New York: Russell Sage, 1964), pp. 119-157.

6. Numerous participants and observers have commented on the problem. For example, see William L. Hauser, *America's Army in Crisis* (Baltimore: Johns Hopkins University Press, 1973), esp. chaps. 10-12. Franklin D. Margiotta, "A Military Elite in Transition: Air Force Leaders in the 1980s," *Armed Forces and Society* 2 (Feb. 1976): 155-184. David W. Moore and B. Thomas Trout, "Military Advancement: The Visibility Theory of Promotion," *American Political Science Review* 72 (June 1978): 452-468.

7. See Henry Kissinger, "The Policymaker and the Intellectual," *The Reporter* 20 (5 March 1959): 30-35. Also see Chris Argyris, *Some Causes of Organizational Ineffectiveness within the Department of State,* Center for International Systems Research Occasional Paper No. 2 (Washington: Dept. of State Pub. 8180, Jan. 1967).

8. Janowitz, *The Professional Soldier,* esp. chaps. 6-8.

9. Maureen Mylander, *The Generals: Making It, Military-Style* (New York: Dial Press, 1974).

10. Lt. Gen. Garrison H. Davidson, "Tomorrow's Leaders," *Army* 15 (Oct. 1964): 21-29. Davidson pointed out that the average tenure since World War II of Commandants of the Army War College and of the Command and General Staff College, and of Superintendents of the Military Academy, was 1.8 years, as compared to a 9.7 year average tenure for the presidents of sixteen major civilian educational institutions in the same time period. The Department of the Army had denied Davidson permission to publish the article until after his 1964 retirement. Davidson to author, 18 Dec. 1974.

11. Charles C. Moskos's useful review notes the tendency of scholars in the late 1960s to detect a growing convergence of American military and civilian institutions. By the early 1970s, however, several observers were identifying a halt or reversal of the convergence trend. Moskos, "The Military," *Annual Review of Sociology* 2 (1976): 55-77. See also David R. Segal *et. al.*, "Convergence, Isomorphism and Interdependence at the Civil-Military Interface," *Journal of Political and Military Sociology* 2 (1974): 157-172.

Chapter Nine

1. Stanley Smith, Capt. USCG (ret.), to author, 29 June 1976. Albert A. Lawrence, Capt. USCG (ret.), to author, 6 June 1976. Willard Smith, Melson, Huston, and Pitt interviews.

2. The academies constitute what Caplow describes as an "organization set," consisting of "two or more organizations of the same type, each of which is continuously visible to every other." Theodore Caplow, *Principles of Organization* (New York: Harcourt, Brace & World, 1964), chap. 6. Also William M. Evans, "The Organization-Set: Toward a Theory of Interorganizational Relations," in *Approaches to Organizational Design*, ed. James D. Thompson (Pittsburgh: University of Pittsburgh Press, 1966), pp. 173-191.

3. James Feron, "Cadet Committee at West Point Does Away With 'The Silence,'" *New York Times*, 12 Sept. 1973, p. 1. In November 1975, the NBC television network released a dramatized version of the silencing of Cadet Pelosi, with actor Richard Thomas portraying Pelosi, and Cliff Gorman portraying Stanley Greenberg, whose interviews with Pelosi provided the primary basis for the script. See also, "USMA Comments on NBC's 'The Silence,'" *Assembly* 34 (Dec. 1975): 28.

4. Col. Lee D. Olvey, "Remarks in introducing Lt. Gen. William A. Knowlton," USMA, *Senior Conference 13-15 June 1974, Educating the Professional Soldier: Final Report* (West Point, N.Y.: USMA, 1974), p. 43.

5. This episode is described in amusing detail by Stuart H. Loory, *Defeated: Inside America's Military Machine* (New York: Random House, 1973), chap. 9: "Hair and Harassment in the Garrison-Ghettos."

6. A complete list of the 120 "Z-grams" that were issued is provided as an appendix to Elmo R. Zumwalt, Jr., *On Watch* (New York: Quadrangle, 1976).

7. U.S., Congress, House of Representatives, Committee on Government Operations, Subcommittees on Legislation and National Security, *Hearings: Problems in Administration of the Military Service Academies*, 94th Cong., 2d sess., 17 March 1976, pp. 58-59.

8. Thomas W. Carr, Director of Defense Education, presentation in a workshop on "Future Military Education Systems," conference on "The Changing American Military Profession," Air Command and Staff College, Maxwell Air Force Base, Alabama, 22 Oct. 1976.

The author also has benefitted from discussions with Major James R. Golden of the Military Academy faculty, who prepared several detailed working papers for the committee, and with Air Force LTC David H. Roe and Army Major John A. Montefusco, who served as key members of the committee staff in the Pentagon.

9. U.S., Comptroller General, *Financial Operations of the Five Service Academies* (Washington, D.C.: General Accounting Office, 6 Feb. 1975), p. 19. Academic costs for USMA and USAFA were estimated at $16 million and $17 million annually, as compared to $13 million for USNA.

10. Amendment No. 1690 to HR12438 as reported in U.S., Congress, Senate, *Congressional Record*, 94th Cong., 2d sess., 25 May 1976, p. S7960. Emphasis is added. Roy Werner, Senator Glenn's legislative assistant, has assured the author, in response to an inquiry, that the amendment was introduced without any previous contact or prompting from the Navy or any other service. Telephone conversation 14 April 1977.

11. James Feron, "New Cheating Case Erupts at West Point," *New York Times*, 7 April 1976, sec. 1, p. 1. Also, Feron, "More Than 90 Cadets at West Point Face Charges of Cheating on a Test," *New York Times*, 8 April 1976, sec. 1, p. 39.

12. Albert to Hoffmann, quoted in "Army Chief Asked to Enter Academy Scandal," Louisville *Courier-Journal*, 12 Aug. 1976, p. A4.

13. U.S., Secretary of the Army, Special Commission on the United States Military Academy, *Report*, prepared by Colonel Frank Borman *et al.*, 15 Dec. 1976.

14. James Feron, "Army to Let Ousted Cadets Return But Not Under Borman's Formula," *New York Times*, 6 Jan. 1977, sec. 1, p. 32.

15. Michael T. Rose, *A Prayer for Relief: The Constitutional Infirmities of the Military Academies' Conduct, Honor and Ethics Systems* (New York: New York University School of Law, 1973). A somewhat sensationalized account of the Air Force reaction to the publication of Rose's study is provided by Robert J. Flood, "'Busted' out of the United States Air Force," *True* 55 (July 1974): 33 ff.

16. See Linda Greenhouse, "Five Accused West Point Cadets Contest Academy Panel's Administration of Honor Code as Unconstitutional," *New York Times*, 27 May 1973, Sec. 2, p. 25. Also, *White v. Knowlton*, 361 F. Supp. 445 (1973).

17. LTJG. Fred N. Squires, "Women at the Coast Guard Academy: One Officer's Opinion," USCGA *Alumni Bulletin* 36 (May-June 1974): 54-55.

18. Lt. Paul M. Regan, "Revolutio Copernicanus De Academia," USCGA *Alumni Bulletin* 37 (Sept.-Oct. 1975): 45-51.

19. See for example by Earle, "Where Have All the Sailors Gone?" *Alumni Bulletin* 36 (Jan.-Feb. 1974): 48-56; and "The Academy Today—An Old Grad's View," *Alumni Bulletin* 37 (Nov.-Dec. 1975): 56-60.

20. "Sea Breezes," *Shipmate* 35 (April 1972): 37.

21. "The Naval Academy of the Future," *Shipmate* 35 (April 1972): 11-15. This was a follow-up on "And in Twenty-Five More Years," by Schratz, in *Shipmate* 33 (Sept.-Oct. 1970): 20-23.

22. Both of these speeches, and a description of events in MacArthur's life

from the time of his return home from Korea in 1951, are contained in the final chapter of his *Reminiscences*.

23. (New York: Simon and Schuster, 1973).

24. As I have observed elsehwere, "Unfortunately, [the Galloway and Johnson] account takes us from the virtuous stereotype of the West Pointer provided by academy public relations to an antithetical portrait that also is a stereotype." John P. Lovell, "No Tunes of Glory: America's Military in the Aftermath of Vietnam," *Indiana Law Journal* 49 (Summer 1974): 698-717, 706. The essay reviews the Galloway and Johnson book and six others on the American military which were published in the early 1970s.

25. In the order mentioned, the works are: Arthur Heise, *The Brass Factories* (Washington, D.C.: Public Affairs Press, 1969); Joseph Ellis and Robert Moore, *School for Soldiers: West Point and the Profession of Arms* (New York: Oxford University Press, 1974); Richard F. Rosser, "The Future of the American Military Academies," a paper presented at a conference of the Inter-University Seminar on Armed Forces and Society, Chicago, October 1973; Richard C. U'Ren, *Ivory Fortress: A Psychiatrist Looks at West Point* (Indianapolis: Bobbs-Merrill, 1974); Josiah Bunting, "The Humanities in the Education of the Military Professional," in *The System for Educating Military Officers in the U.S.*, ed. Lawrence J. Korb (Pittsburgh: International Studies Assn., 1976), pp. 155-158.

26. Representatives Aspin and Hébert, respectively, quoted by Congressional Quarterly, *The Power of the Pentagon* (Washington, D.C.: Congressional Quarterly, Inc., 1972), pp. 4-5.

27. U.S., Congress, House, *Congressional Record*, 85th Cong., 1st sess., 1957, p. 13175; *Congressional Record*, 86th Cong., 1st sess., 1959, p. 9301. (The 1958 board report was entered belatedly into the *Record*.)

28. Douglas MacArthur, *Reminiscences* (New York: McGraw-Hill, 1964), pp. 25-26. Historian Stephen Ambrose has discovered from his own investigation of the hearings in which Cadet MacArthur had been called to testify that MacArthur actually did divulge "quite a few names, but they were all of cadets who had either already admitted their guilt or who, for one reason or another, had resigned." Ambrose, *Duty, Honor, Country: A History of West Point* (Baltimore: Johns Hopkins University Press, 1966), p. 231n.

29. U.S., Congress, House, Committee on Armed Services, Special Subcommittee on Service Academies, *Report and Hearings: Administration of the Service Academies*, 90th Cong., 1st and 2d sess., 1967–1968, p. 10677.

30. *Report and Hearings*, 1967–1968, pp. 10793-10795.

31. *Report and Hearings*, 1967–1968, pp. 10224b-10224c.

32. Hébert as quoted by Daniel Rapoport, UPI dispatch, "Fighting Men, Not Rhodes Scholars—Hébert Zeroes in on Service Schools," reprinted in *Report and Hearings, 1967–1968*, p. 10741.

33. *Report and Hearings, 1967–1968*, p. 10224e.

34. *Report and Hearings, 1967–1968*, pp. 10635, 10920.

35. U.S., Comptroller General, *Financial Operations of the Five Service Academies* (Washington, D.C.: GAO, 6 Feb. 1975); *Academic and Military*

Programs of the Five Service Academies (31 Oct. 1975); *Student Attrition at the Five Federal Service Academies* (5 March 1976). The last report had three enclosures: (A) "Technical Report on Survey of Factors Related to Attrition"; (B) "Review of Studies on Academy Attrition and Related Issues"; (C) "Characteristics of Academy Students."

36. Comptroller General, *Student Attrition*, Encl. B.

37. Elmer B. Staats, prepared statement in U.S., Congress, House, Committee on Government Operations, Legislation and National Security Subcommittee, *Hearing: Problems in Administration of the Military Service Academies*, 94th Cong., 2d sess., 17 March 1976, pp. 45-46.

38. Comptroller General, *Student Attrition*, pp. 85, 57.

Chapter Ten

1. The following excerpt by Nathan Hale is contained in the 1940 edition of *Reef Points*, the "plebe bible" of the Naval Academy, but not in the editions in the postwar period.

> —And for your country, boy, and for that flag, never dream a dream but of serving her as she bids you, though the service carry you through a thousand hells. No matter what happens to you, no matter who flatters you or abuses you, never look at another flag, never let a night pass but you pray God to bless that flag. Remember, boy, that behind all these men you have to do with, behind officers, and government, and people even, there is the Country Herself, your country, and that you belong to her as to your own Mother. Stand by Her, boy, as you would stand by your mother.

2. The landmark research undertaken during World War II by Stouffer *et al.* demonstrated that combat motivation seldom could be attributed to patriotism; soldiers were intolerant of idealistic appeals, especially from civilians or those outside the unit with which they identified. Primary-group loyalty (to one's buddies and one's unit) was a major motivating factor. Samuel A. Stouffer *et al.*, *The American Soldier*, 4 vols. (Princeton, N.J.: Princeton University Press, 1949), esp. vol. 2. Studies of the German army showed also that solidarity at the small-unit level typically was more important than ideological fervor in sustaining soldiers under the stresses of combat. Edward A. Shils and Morris Janowitz, "Cohesion and Disintegration in the Wehrmacht in World War II," *Public Opinion Quarterly* 12 (1948): 280-315. During the Korean War, a participant-observer study documented the importance of "buddy relations" to combat morale. Roger W. Little, "Buddy Relations and Combat Role Performance," *The New Military: Changing Patterns of Organization*, ed. Morris Janowitz (New York: Russell Sage, 1965), pp. 194-224. The Vietnam war demonstrated again the relative ineffectiveness of patriotic appeals. However, the lack of unit cohesiveness due to the personnel rotation system contributed to the reduction of primary-group loyalty. Charles C. Moskos, Jr., *The American Enlisted Man* (New York: Russell Sage, 1970).

3. Dwight D. Eisenhower, *At Ease: Stories I Tell to Friends* (New York: Avon Books, 1968), p. 18.

4. Eisenhower, *At Ease*, p. 18.

5. Letter, Baker as Secretary of War to the Chairman of the Committee on Military Affairs, U.S. House of Representatives, 17 May 1920. Quoted in Sidney Forman, *West Point: A History of the United States Military Academy* (New York: Columbia University Press, 1950), p. 155n. The quote also is included in virtually every pamphlet on the honor system produced at West Point in the past thirty years.

6. Similarly, Burton Clark has noted in a study of the evolution of Antioch, Reed, and Swarthmore colleges:

> Indications of organizational legend are pride and exaggeration; the most telling symptom is an intense sense of the unique. Men behave as if they knew a beautiful secret that no one outside the lucky few could ever share. An organizational saga turns an organization into a community, even a cult.

Clark, *The Distinctive College: Antioch, Reed, & Swarthmore* (Chicago: Aldine 1970), p. 235.

7. In 1961 boards of visitors proposed that the Air Force and Military academies be increased to a size equal to that of the Naval Academy. In March 1964, a law was enacted authorizing an increase in the enrollment of each of the three DOD academies to approximately 4,000.

8. James Calvert, Vice Adm. USN, "The Fine Line at the Naval Academy," U.S. Naval Institute *Proceedings* 96 (Oct. 1970): 63-68.

9. For example, a West Point tactical officer, told by the staff psychiatrist that a plebe who was considering resignation had been psychologically unprepared for the harsh environment of "Beast Barracks," replied, "Why don't you put it in plain English, Doc; the kid doesn't have guts." Reported to the author by the staff psychiatrist, autumn 1961. Also at work, of course, is the syllogistic reasoning that the plebe system is designed to "weed out the unfit"; therefore, those who drop out or are dismissed must be "unfit."

10. Col. Hugh T. Reed, *Cadet Life at West Point*, 3d ed. (Richmond, Ind.: Irvin Reed & Son, 1896 and 1911), p. 167. Reed entered the Academy in 1869, graduating in 1873. The book provides a useful picture, through narrative, photographs, and illustrations, of cadet life in the late nineteenth and early twentieth centuries.

11. John McA. Palmer, "How Football Came to West Point," *The Pointer* 29 (7 Sept. 1951): 18; reprinted from *Assembly* (Jan. 1943).

12. Captain W. D. Puleston (USN), *Annapolis: Gangway to the Quarterdeck* (New York: Appleton-Century, 1942), p. 165.

13. LTG Robert Eichelberger to Earl Blaik, 5 July 1942, quoted in Earl "Red" Blaik, *The Red Blaik Story* (New Rochelle, N.Y.: Arlington House, 1974), p. 181. *The Red Blaik Story* incorporates and supplements *You Have to Pay the Price*, which Blaik wrote with Tim Cohane, with a foreword by Douglas MacArthur (New York: Holt, Rinehart and Winston, 1960).

14. Blaik, *The Red Blaik Story*, p. 281.

15. Maxwell D. Taylor, *Swords and Plowshares* (New York: Norton, 1972), pp. 115-116.

16. The *New York Times* gave front-page coverage to the incident, including the views of Academy officials and those of critics for a week beginning 4 August 1951.

17. Incidentally, Red Blaik was retained as football coach, and charged with the responsibility of rebuilding the football team virtually from scratch. He had considered resignation, but had been persuaded by supporters such as Generals Eichelberger and MacArthur to remain rather than leave under circumstances that would seem to confirm his acceptance, at least in part, of blame for the breakdown in the observance of the honor code. His departure during the Davidson years was described in chapter 5.

18. U.S., Congress, House of Representatives, Committee on Armed Services, Special Subcommittee on Service Academies, *Report and Hearings: Administration of the Service Academies*, 90th Cong., 1st and 2d sess., testimony of Brigadier General Robert McDermott, Academic Dean USAFA, 5 March 1968, pp. 19098-10909.

19. U.S., Secretary of the Army, Special Commission on the U.S. Military Academy, *Report*, prepared by Frank Borman, Col. USAF (ret.) *et al.*, 15 Dec. 1976, pp. 65-69. Referred to hereafter as *Borman Commission Report*.

20. USMA, "Report of Superintendent's Special Study Group on Honor at West Point," Colonel Harry A. Buckley, Jr. and Cadet William J. Reid, co-chairmen, mimeographed (West Point, N.Y.: USMA, 23 May 1975), p. 9.

21. *Borman Commission Report*.

22. *Borman Commission Report*, pp. 38, 40.

23. *Borman Commission Report*, p. 6. Emphasis is added.

24. Earl "Red" Blaik, "A Cadet Under MacArthur," *Assembly* 23 (Spring 1964): 9.

25. Brigadier General Fred W. Sladen, who had graduated from West Point in 1890, thirteen years earlier than MacArthur.

26. Erving Goffman, "Characteristics of Total Institutions," in *Symposium on Preventive and Social Psychiatry 15-17 April 1957* (Washington, D.C.: Walter Reed Army Institute of Research, 1957), pp. 43-84. An expanded version of the essay, from which the quote is taken, appears in Goffman's collection of essays, *Asylums* (Garden City, N.Y.: Doubleday, Anchor, 1961), pp. 14-15.

27. As staff psychiatrists at West Point noted:

> A recognition of his "proper place" in the group includes the ascription to the plebe of gross incompetence, stupidity, low rank and low prestige. This type of position is largely a game; otherwise, the appropriate role behaviors become very difficult to carry out. Thus, it is quite clear that the qualities of incompetence ascribed to the plebe would, if really true, disqualify him as an Army officer. Valid role enactments in this area, then, require an *as-if* attitude on the part of both upperclassmen and plebes: both groups, for example, must act as-if the plebe were indeed incompetent, but neither can believe it if the group is going to accept the aspirant as a member or if the aspirant is going to feel qualified to become a member.

USMA, *Adaptation to West Point: A Study of Some Psychological Factors Associated with Adjustment at the United States Military Academy* (Interim

Report, Medical Research Project, U.S. Army Hospital, West Point: USMA, June 1959).

28. U.S., Comptroller General, *Student Attrition at the Five Federal Service Academies* (Washington, D.C.: GAO, 5 March 1976), p. 55.

29. For example, the book of regulations at the Naval Academy had grown from a 66-page document at the turn of the century to one of 161 pages at the eve of World War II. A passion for prescribing details of conduct was far more evident in the latter book than in the former. By 1951, there were 322 pages of regulations, and by 1955, 576 pages. Books of regulations are on file in the archives of the academies.

30. Knowlton, letter to alumni, *Assembly* 32 (Dec. 1973); inside front cover. Emphasis in original text. Also, "Regulations USCC," *Assembly* 32 (Dec. 1973); 13, 34.

31. U.S., Comptroller General, *Student Attrition at the Five Federal Service Academies* (Washington, D.C.: GAO, 5 March 1976), pp. 73-74.

32. Survey cited by James P. Sterba, "Dropouts Plague the Air Academy," *New York Times*, 3 June 1973, sec. 1, p. 1.

Chapter Eleven

1. Calvert interview.

2. Edward Bernard Glick, *Soldiers, Scholars and Society: The Social Impact of the American Military* (Pacific Palisades, Cal.: Goodyear, 1971), chap. 8. Also, Glick, "Do We Really Need Our Military Academies," *The Chronicle of Higher Education* 13 (22 Feb. 1977): 34.

3. John W. Masland and Laurence I. Radway, *Soldiers and Scholars: Military Education and National Policy* (Princeton, N.J.: Princeton University Press, 1957), p. 114.

4. Laurence I. Radway, "Cadet Education in a Liberal Society," in *Centennial Symposium on Military Education, Military Academies: Problems and Prospects* (Kingston, Ontario: The Royal Military College of Canada, 14-17 June 1976), pp. 77-93.

5. See Masland and Radway, *Soldiers and Scholars*. Also Laurence I. Radway, "Recent Trends at American Service Academies," in *Public Opinion and the Military Establishment*, ed. Charles C. Moskos, Jr. (Beverly Hills, Cal.: Sage, 1971), pp. 3-35.

6. Radway, *Centennial Symposium*, p. 90.

7. Richard F. Rosser, "The Future of the American Military Academies," paper presented at the annual meeting of the Inter-University Seminar on Armed Forces and Society, Chicago, Oct. 1973.

8. Masland and Radway, *Soldiers and Scholars*, pp. 114-115.

9. Melvin J. Maas, "Minority Report," *Report of the Board of Visitors to the Naval Academy*, 2 May 1940.

10. U.S., Secretary of Defense, *A Report and Recommendation to the Secretary of Defense by the Service Academy Board*, prepared by Robert L. Stearns, Dwight D. Eisenhower, *et al.* (Washington, D.C.: Jan. 1950), pp. 2-8.

11. James D. Thompson, *Organizations in Action* (New York: McGraw-Hill, 1967), p. 91. See also Raymond E. Callahan, *Education and the Cult of Efficiency* (Chicago: University of Chicago Press, 1962).

12. Paul R. Schratz, "And in Twenty-Five More Years," *Shipmate* 33 (Sept.-Oct. 1970): 20-23. Frederick C. Thayer, "AFROTC and USAFA: Time for a Change," *Air Force Magazine* (July 1972), reprinted in USAFA *Association of Graduates Magazine* 2 (Sept. 1972): 11-13. Fred Harburg, "Should USAFA Become a Graduate School?" *The Talon* 17 (April 1972): 28. Paul M. Regan "Revolutio Copernicanus De Academia," USCGA *Alumni Bulletin* 37 (Sept.-Oct. 1975): 45-51. Roger A. Beaumont and William P. Snyder, "A Fusion Strategy for Pre-Commissioning Training," *Journal of Political and Military Sociology* 5 (Autumn 1977): 259-277.

13. Regan added, "Although the Academy has an excellent permanent staff and an extremely dedicated rotating staff, no one would claim that the education provided is on a par with that of MIT or CalTech." Regan, *Alumni Bulletin* 37 (Sept.-Oct. 1975): 46.

14. Edmund Burke, "Reflections on the French Revolution," in William Elliott and Neil McDonald, eds., *Western Political Heritage* (New York: Prentice-Hall, 1955), p. 684.

15. For a critical review of the growth of the influence of the human relations movement on managerial ideologies in America, see Charles Perrow, *Complex Organizations: A Critical Essay* (Glenview, Ill.: Scott, Foresman, 1972), esp. chaps. 2 and 3.

16. As noted in chapter one, the Coast Guard preempted congressional action by opening its academy to women. At the other three academies, the admission of women followed legislative action requiring it.

17. The classic explanation of the human tendency in decision making toward incrementalism is that provided by Charles E. Lindblom, "The Science of 'Muddling Through,'" *Public Administration Review* 20 (Spring 1959): 79-88. See also David Braybrooke and Charles E. Lindblom, *A Strategy of Decision: Policy Evaluation as a Social Process* (New York: Free Press, 1963).

18. Morris Janowitz, "U.S. Forces and the Zero Draft," *Adelphi Paper No.* 94 (London: International Institute for Strategic Studies, 1973). Janowitz, banquet address, text contained in USMA, *Final Report: Senior Conference, 13-15 June 1974, Educating the Professional Soldier* (West Point, N.Y.: USMA, 1974), pp. 31-38.

INDEX